Kalman Filtering

Kalman Filtering: Theory and Practice

Using MATLAB

Second Edition

MOHINDER S. GREWAL
California State University at Fullerton

ANGUS P. ANDREWS
Rockwell Science Center

A Wiley-Interscience Publication
John Wiley & Sons, Inc.
NEW YORK • CHICHESTER • WEINHEIM • BRISBANE • SINGAPORE • TORONTO

This book is printed on acid-free paper. 

Published simultaneously in Canada.

For ordering and customer service, call 1-800-CALL-WILEY.

Library of Congress Cataloging-in-Publication Data:

Grewal, Mohinder S.
Kalman filtering : theory and practice using MATLAB, second edition / Mohinder S. Grewal, Angus P. Andrews.
p. cm.
Includes bibliographical references and index.
ISBN 0-471-39254-5 (cloth : alk. paper)
1. Kalman filtering. 2. MATLAB. I. Andrews, Angus P. II. Title.

QA402.3 .G697 2001
003′.5–dc 21 00-043689

Printed in the United States of America.

10 9 8 7 6 5

Contents

PREFACE ix
ACKNOWLEDGMENTS xiii

1 General Information 1

1.1 On Kalman Filtering 1
1.2 On Estimation Methods 5
1.3 On the Notation Used in This Book 20
1.4 Summary 22
Problems 23

2 Linear Dynamic Systems 25

2.1 Chapter Focus 25
2.2 Dynamic Systems 26
2.3 Continuous Linear Systems and Their Solutions 30
2.4 Discrete Linear Systems and Their Solutions 41
2.5 Observability of Linear Dynamic System Models 42
2.6 Procedures for Computing Matrix Exponentials 48
2.7 Summary 50
Problems 53

3 Random Processes and Stochastic Systems 56

3.1 Chapter Focus 56
3.2 Probability and Random Variables 58
3.3 Statistical Properties of Random Variables 66

3.4 Statistical Properties of Random Processes 68
3.5 Linear System Models of Random Processes and Sequences 76
3.6 Shaping Filters and State Augmentation 84
3.7 Covariance Propagation Equations 88
3.8 Orthogonality Principle 97
3.9 Summary 102
Problems 104

4 Linear Optimal Filters and Predictors 114

4.1 Chapter Focus 114
4.2 Kalman Filter 116
4.3 Kalman–Bucy Filter 126
4.4 Optimal Linear Predictors 128
4.5 Correlated Noise Sources 129
4.6 Relationships between Kalman and Wiener Filters 130
4.7 Quadratic Loss Functions 131
4.8 Matrix Riccati Differential Equation 133
4.9 Matrix Riccati Equation in Discrete Time 148
4.10 Relationships between Continuous and Discrete Riccati Equations 153
4.11 Model Equations for Transformed State Variables 154
4.12 Application of Kalman Filters 155
4.13 Smoothers 160
4.14 Summary 164
Problems 165

5 Nonlinear Applications 169

5.1 Chapter Focus 169
5.2 Problem Statement 170
5.3 Linearization Methods 171
5.4 Linearization about a Nominal Trajectory 171
5.5 Linearization about the Estimated Trajectory 175
5.6 Discrete Linearized and Extended Filtering 176
5.7 Discrete Extended Kalman Filter 178
5.8 Continuous Linearized and Extended Filters 181
5.9 Biased Errors in Quadratic Measurements 182
5.10 Application of Nonlinear Filters 184
5.11 Summary 198
Problems 200

6 Implementation Methods 202

6.1 Chapter Focus 202
6.2 Computer Roundoff 204
6.3 Effects of Roundoff Errors on Kalman Filters 209
6.4 Factorization Methods for Kalman Filtering 216

6.5 Square-Root and UD Filters 238
6.6 Other Alternative Implementation Methods 252
6.7 Summary 265
Problems 266

7 Practical Considerations 270

7.1 Chapter Focus 270
7.2 Detecting and Correcting Anomalous Behavior 271
7.3 Prefiltering and Data Rejection Methods 294
7.4 Stability of Kalman Filters 298
7.5 Suboptimal and Reduced-Order Filters 299
7.6 Schmidt–Kalman Filtering 309
7.7 Memory, Throughput, and Wordlength Requirements 316
7.8 Ways to Reduce Computational Requirements 326
7.9 Error Budgets and Sensitivity Analysis 332
7.10 Optimizing Measurement Selection Policies 336
7.11 Application to Aided Inertial Navigation 342
7.12 Summary 346
Problems 347

Appendix A MATLAB Software 350

A.1 Notice 350
A.2 General System Requirements 350
A.3 Diskette Directory Structure 351
A.4 MATLAB Software for Chapter 2 351
A.5 MATLAB Software for Chapter 4 351
A.6 MATLAB Software for Chapter 5 352
A.7 MATLAB Software for Chapter 6 352
A.8 MATLAB Software for Chapter 7 353
A.9 Other Sources of Software 353

Appendix B A Matrix Refresher 355

B.1 Matrix Forms 355
B.2 Matrix Operations 359
B.3 Block Matrix Formulas 363
B.4 Functions of Square Matrices 366
B.5 Norms 370
B.6 Cholesky Decomposition 373
B.7 Orthogonal Decompositions of Matrices 375
B.8 Quadratic Forms 377
B.9 Derivatives of Matrices 379

REFERENCES 381

INDEX 395

Preface

The first edition of this book was published by Prentice-Hall in 1993. With this second edition, as with the first, our primary objective is to provide our readers a working familiarity with both the *theoretical* and *practical* aspects of Kalman filtering by including "real-world" problems in practice as illustrative examples. We are pleased to have this opportunity to incorporate the many helpful corrections and suggestions from our colleagues and students over the last several years for the overall improvement of the textbook. The book covers the historical background of Kalman filtering and the more practical aspects of implementation: how to represent the problem in a mathematical model, analyze the performance of the estimator as a function of model parameters, implement the mechanization equations in numerically stable algorithms, assess its computational requirements, test the validity of results, and monitor the filter performance in operation. These are important attributes of the subject that are often overlooked in theoretical treatments but are necessary for application of the theory to real-world problems.

We have converted all algorithm listings and all software to MATLAB®[1], so that users can take advantage of its excellent graphing capabilities and a programming interface that is very close to the mathematical equations used for defining Kalman filtering and its applications. See Appendix A, Section A.2, for more information on MATLAB.

The inclusion of the software is practically a matter of necessity, because Kalman filtering would not be very useful without computers to implement it. It is a better learning experience for the student to discover how the Kalman filter works by observing it in action.

The implementation of Kalman filtering on computers also illuminates some of the practical considerations of finite-wordlength arithmetic and the need for alter-

[1] MATLAB is a registered trademark of The Mathworks, Inc.

native algorithms to preserve the accuracy of the results. If the student wishes to apply what she or he learns, then it is essential that she or he experience its workings and failings—and learn to recognize the difference.

The book is organized for use as a text for an introductory course in stochastic processes at the senior level and as a first-year graduate-level course in Kalman filtering theory and application. It could also be used for self-instruction or for purposes of review by practicing engineers and scientists who are not intimately familiar with the subject. The organization of the material is illustrated by the following chapter-level dependency graph, which shows how the subject of each chapter depends upon material in other chapters. The arrows in the figure indicate the recommended order of study. Boxes above another box and connected by arrows indicate that the material represented by the upper boxes is background material for the subject in the lower box.

Chapter 1 provides an informal introduction to the general subject matter by way of its history of development and application. Chapters 2 and 3 and Appendix B cover the essential background material on linear systems, probability, stochastic processes, and modeling. These chapters could be covered in a senior-level course in electrical, computer, and systems engineering.

Chapter 4 covers linear optimal filters and predictors, with detailed examples of applications. Chapter 5 is devoted to nonlinear estimation by "extended" Kalman

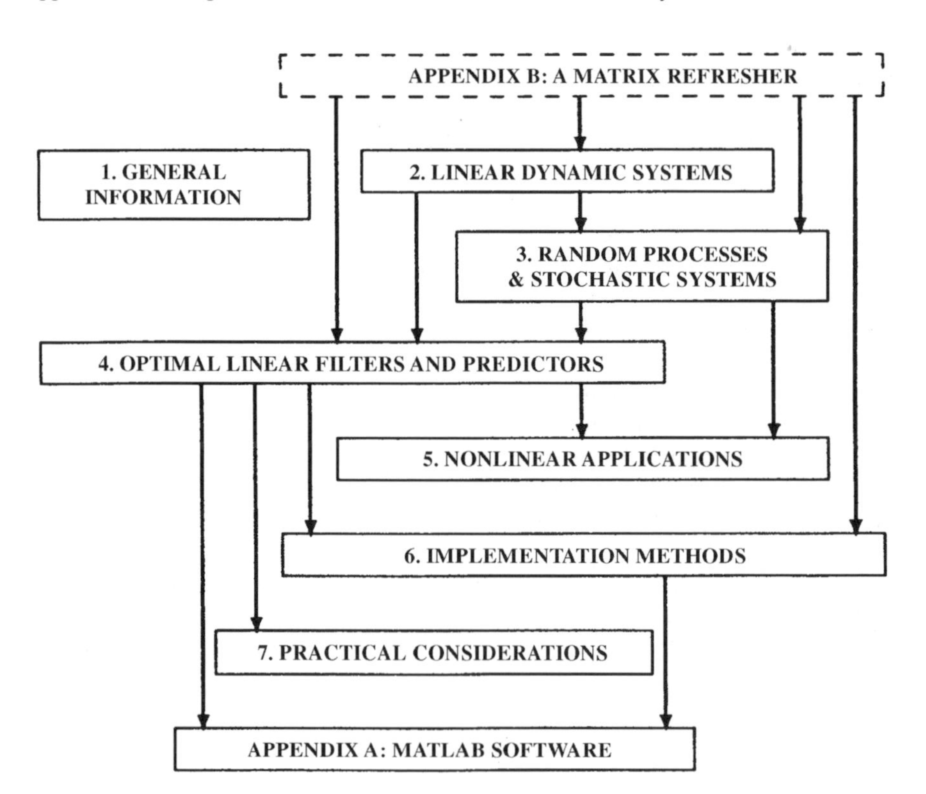

filters. Applications of these techniques to the identification of unknown parameters of systems are given as examples. Chapter 6 covers the more modern implementation techniques, with algorithms provided for computer implementation.

Chapter 7 deals with more practical matters of implementation and use beyond the numerical methods of Chapter 6. These matters include memory and throughput requirements (and methods to reduce them), divergence problems (and effective remedies), and practical approaches to suboptimal filtering and measurement selection.

Chapters 4–7 cover the essential material for a first-year graduate class in Kalman filtering theory and application or as a basic course in digital estimation theory and application. A solutions manual for each chapter's problems is available.

PROF. MOHINDER S. GREWAL, PHD, PE

California State University at Fullerton

ANGUS P. ANDREWS, PHD

Rockwell Science Center, Thousand Oaks, California

Acknowledgments

The authors express their appreciation to the following individuals for their contributions during the preparation of the first edition: Robert W. Bass, E. Richard Cohen, Thomas W. De Vries, Reverend Joseph Gaffney, Thomas L. Gunckel II, Dwayne Heckman, Robert A. Hubbs, Thomas Kailath, Rudolf E. Kalman, Alan J. Laub, Robert F. Nease, John C. Pinson, John M. Richardson, Jorma Rissanen, Gerald E. Runyon, Joseph Smith and Donald F. Wiberg. We also express our appreciation to Donald Knuth and Leslie Lamport for TEX and LATEX, respectively.

In addition, the following individuals deserve special recognition for their careful review, corrections, and suggestions for improving the second edition: Dean Dang and Gordon Inverarity.

Most of all, for their dedication, support, and understanding through both editions, we dedicate this book to Sonja Grewal and Jeri Andrews.

M. S. G., A. P. A.

1

General Information

. . . the things of this world cannot be made known without mathematics.
—Roger Bacon (1220–1292), Opus Majus, transl. R. Burke, 1928

1.1 ON KALMAN FILTERING

1.1.1 First of All: What Is a Kalman Filter?

Theoretically the Kalman Filter is an estimator for what is called the *linear-quadratic problem*, which is the problem of estimating the instantaneous "state" (a concept that will be made more precise in the next chapter) of a linear dynamic system perturbed by white noise—by using measurements linearly related to the state but corrupted by white noise. The resulting estimator is statistically optimal with respect to any quadratic function of estimation error.

Practically, it is certainly one of the greater discoveries in the history of statistical estimation theory and possibly the greatest discovery in the twentieth century. It has enabled humankind to do many things that could not have been done without it, and it has become as indispensable as silicon in the makeup of many electronic systems. Its most immediate applications have been for the control of complex dynamic systems such as continuous manufacturing processes, aircraft, ships, or spacecraft. To control a dynamic system, you must first know what it is doing. For these applications, it is not always possible or desirable to measure every variable that you want to control, and the Kalman filter provides a means for inferring the missing information from indirect (and noisy) measurements. The Kalman filter is also used for predicting the likely future courses of dynamic systems that people are not likely to control, such as the flow of rivers during flood, the trajectories of celestial bodies, or the prices of traded commodities.

From a practical standpoint, these are the perspectives that this book will present:

- *It is only a tool.* It does not solve any problem all by itself, although it can make it easier for you to do it. It is not a *physical* tool, but a *mathematical* one. It is made from mathematical models, which are essentially tools for the mind. They make mental work more efficient, just as mechanical tools make physical work more efficient. As with any tool, it is important to understand its use and function before you can apply it effectively. The purpose of this book is to make you sufficiently familiar with and proficient in the use of the Kalman filter that you can apply it correctly and efficiently.
- *It is a computer program.* It has been called "ideally suited to digital computer implementation" [21], in part because it uses a *finite representation* of the estimation problem—by a *finite* number of variables. It does, however, assume that these variables are *real numbers*—with *infinite* precision. Some of the problems encountered in its use arise from the distinction between finite dimension and finite information, and the distinction between "finite" and "manageable" problem sizes. These are all issues on the practical side of Kalman filtering that must be considered along with the theory.
- *It is a complete statistical characterization of an estimation problem.* It is much more than an *estimator,* because it propagates the entire *probability distribution* of the variables it is tasked to estimate. This is a complete characterization of the current *state of knowledge* of the dynamic system, including the influence of all past measurements. These probability distributions are also useful for statistical analysis and the predictive design of sensor systems.
- *In a limited context, it is a learning method.* It uses a model of the estimation problem that distinguishes between *phenomena* (what one is able to observe), *noumena* (what is really going on), and the state of knowledge about the noumena that one can deduce from the phenomena. That state of knowledge is represented by probability distributions. To the extent that those probability distributions represent *knowledge* of the real world and the cumulative processing of knowledge is *learning,* this is a learning process. It is a fairly simple one, but quite effective in many applications.

If these answers provide the level of understanding that you were seeking, then there is no need for you to read the rest of the book. If you need to understand Kalman filters well enough to use them, then read on!

1.1.2 How It Came to Be Called a Filter

It might seem strange that the term "filter" would apply to an estimator. More commonly, a filter is a physical device for removing unwanted fractions of mixtures. (The word *felt* comes from the same medieval Latin stem, for the material was used as a filter for liquids.) Originally, a filter solved the problem of separating unwanted components of gas–liquid–solid mixtures. In the era of crystal radios and vacuum tubes, the term was applied to analog circuits that "filter" electronic signals. These

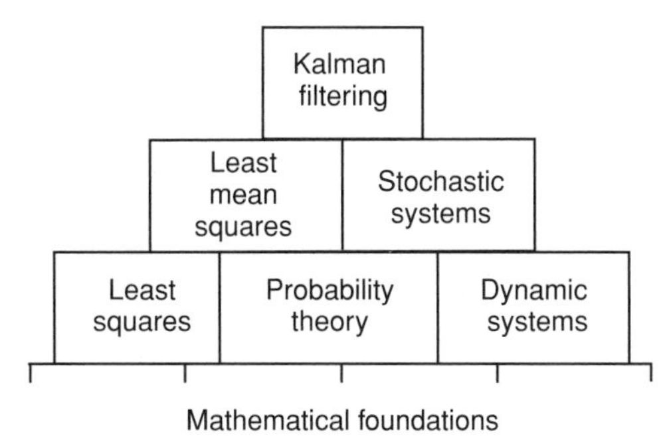

Fig. 1.1 *Foundational concepts in Kalman filtering.*

signals are mixtures of different frequency components, and these physical devices preferentially attenuate unwanted frequencies.

This concept was extended in the 1930s and 1940s to the separation of "signals" from "noise," both of which were characterized by their power spectral densities. Kolmogorov and Wiener used this statistical characterization of their probability distributions in forming an optimal estimate of the signal, given the sum of the signal and noise.

With Kalman filtering the term assumed a meaning that is well beyond the original idea of *separation* of the components of a mixture. It has also come to include the solution of an *inversion problem,* in which one knows how to represent the measurable variables as functions of the variables of principal interest. In essence, it inverts this functional relationship and estimates the independent variables as inverted functions of the dependent (measurable) variables. These variables of interest are also allowed to be dynamic, with dynamics that are only partially predictable.

1.1.3 Its Mathematical Foundations

Figure 1.1 depicts the essential subjects forming the foundations for Kalman filtering theory. Although this shows Kalman filtering as the apex of a pyramid, it is itself but part of the foundations of another discipline—"modern" control theory—and a proper subset of statistical decision theory.

We will examine only the top three layers of the pyramid in this book, and a little of the underlying mathematics[1] (matrix theory) in Appendix B.

1.1.4 What It Is Used For

The applications of Kalman filtering encompass many fields, but its use as a tool is almost exclusively for two purposes: *estimation* and *performance analysis* of estimators.

[1]It is best that one not examine the bottommost layers of these mathematical foundations too carefully, anyway. They eventually rest on human intellect, the foundations of which are not as well understood.

Role 1: Estimating the State of Dynamic Systems What is a dynamic system? Almost everything, if you are picky about it. Except for a few fundamental physical constants, there is hardly anything in the universe that is truly *constant.* The orbital parameters of the asteroid Ceres are not constant, and even the "fixed" stars and continents are moving. Nearly all physical systems are dynamic to some degree. If one wants very precise estimates of their characteristics over time, then one has to take their dynamics into consideration.

The problem is that one does not always know their dynamics very precisely either. Given this state of partial ignorance, the best one can do is express our ignorance more precisely—using probabilities. The Kalman filter allows us to estimate the state of dynamic systems with certain types of random behavior by using such statistical information. A few examples of such systems are listed in the second column of Table 1.1.

Role 2: The Analysis of Estimation Systems. The third column of Table 1.1 lists some possible sensor types that might be used in estimating the state of the corresponding dynamic systems. The objective of design analysis is to determine how best to use these sensor types for a given set of design criteria. These criteria are typically related to estimation accuracy and system cost.

The Kalman filter uses a complete description of the probability distribution of its estimation errors in determining the optimal filtering gains, and this probability distribution may be used in assessing its performance as a function of the "design parameters" of an estimation system, such as

- the types of sensors to be used,
- the locations and orientations of the various sensor types with respect to the system to be estimated,

TABLE 1.1 Examples of Estimation Problems

Application	Dynamic System	Sensor Types
Process control	Chemical plant	Pressure
		Temperature
		Flow rate
		Gas analyzer
Flood prediction	River system	Water level
		Rain gauge
		Weather radar
Tracking	Spacecraft	Radar
		Imaging system
Navigation	Ship	Sextant
		Log
		Gyroscope
		Accelerometer
		Global Positioning System (GPS) receiver

- the allowable noise characteristics of the sensors,
- the prefiltering methods for smoothing sensor noise,
- the data sampling rates for the various sensor types, and
- the level of model simplification to reduce implementation requirements.

The analytical capability of the Kalman filter formalism also allows a system designer to assign an "error budget" to subsystems of an estimation system and to trade off the budget allocations to optimize cost or other measures of performance while achieving a required level of estimation accuracy.

1.2 ON ESTIMATION METHODS

We consider here just a few of the sources of intellectual material presented in the remaining chapters and principally those contributors[2] whose lifelines are shown in Figure 1.2. These cover only 500 years, and the study and development of mathematical concepts goes back beyond history. Readers interested in more detailed histories of the subject are referred to the survey articles by Kailath [25, 176], Lainiotis [192], Mendel and Geiseking [203], and Sorenson [47, 224] and the personal accounts of Battin [135] and Schmidt [216].

1.2.1 Beginnings of Estimation Theory

The first method for forming an *optimal* estimate from noisy data is the *method of least squares*. Its discovery is generally attributed to Carl Friedrich Gauss (1777–1855) in 1795. The inevitability of measurement errors had been recognized since the time of Galileo Galilei (1564–1642) , but this was the first formal method for dealing with them. Although it is more commonly used for linear estimation problems, Gauss first used it for a nonlinear estimation problem in mathematical astronomy, which was part of a dramatic moment in the history of astronomy. The following narrative was gleaned from many sources, with the majority of the material from the account by Baker and Makemson [97]:

> On January 1, 1801, the first day of the nineteenth century, the Italian astronomer Giuseppe Piazzi was checking an entry in a star catalog. Unbeknown to Piazzi, the entry had been added erroneously by the printer. While searching for the "missing" star, Piazzi discovered, instead, a new planet. It was *Ceres*—the largest of the minor planets and the first to be discovered—but Piazzi did not know that yet. He was able to track and measure its apparent motion against the "fixed" star background during 41 nights of viewing from Palermo before his work was interrupted. When he returned to his work, however, he was unable to find Ceres again.

[2]The only contributor after R. E. Kalman on this list is Gerald J. Bierman, an early and persistent advocate of *numerically stable* estimation methods. Other recent contributors are acknowledged in Chapter 6.

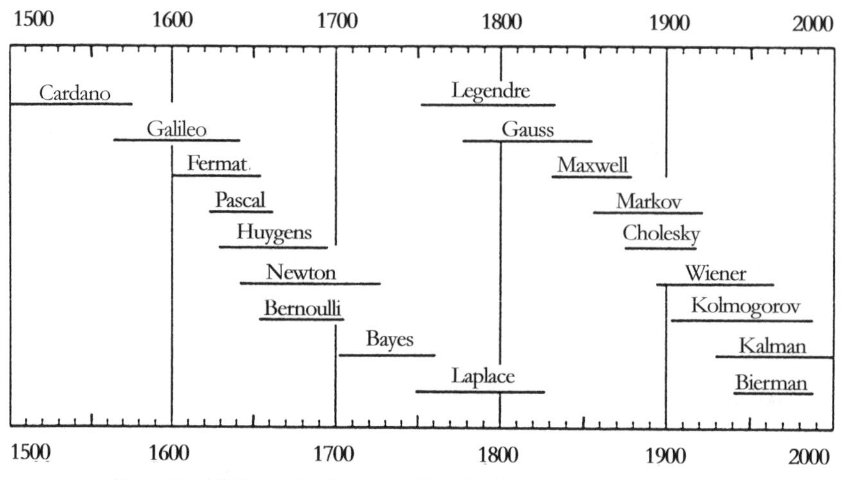

Fig. 1.2 *Lifelines of referenced historical figures and R. E. Kalman.*

On January 24, Piazzi had written of his discovery to Johann Bode. Bode is best known for *Bode's law*, which states that the distances of the planets from the sun, in astronomical units, are given by the sequence

$$d_n = \tfrac{1}{10}(4 + 3 \times 2^n) \quad \text{for } n = -\infty, 0, 1, 2, ?, 4, 5, \ldots . \tag{1.1}$$

Actually, it was not Bode, but Johann Tietz who first proposed this formula, in 1772. At that time there were only six known planets. In 1781, Friedrich Herschel discovered Uranus, which fit nicely into this formula for $n = 6$. No planet had been discovered for $n = 3$. Spurred on by Bode, an association of European astronomers had been searching for the "missing" eighth planet for nearly 30 years. Piazzi was not part of this association, but he did inform Bode of his unintended discovery.

Piazzi's letter did not reach Bode until March 20. (Electronic mail was discovered much later.) Bode suspected that Piazzi's discovery might be the missing planet, but there was insufficient data for determining its orbital elements by the methods then available. It is a problem in nonlinear equations that Newton, himself, had declared as being among the most difficult in mathematical astronomy. Nobody had solved it and, as a result, Ceres was lost in space again.

Piazzi's discoveries were not published until the autumn of 1801. The possible discovery—and subsequent loss—of a new planet, coinciding with the beginning of a new century, was exciting news. It contradicted a philosophical justification for there being only seven planets—the number known before Ceres and a number defended by the respected philosopher Georg Hegel, among others. Hegel had recently published a book in which he chastised the astronomers for wasting their time in searching for an eighth planet when there was a sound philosophical justification for there being only seven. The new planet became a subject of conversation in intellectual circles nearly everywhere. Fortunately, the problem caught the attention of a 24-year-old mathematician at Göttingen named Carl Friedrich Gauss.

> Gauss had toyed with the orbit determination problem a few weeks earlier but had set it aside for other interests. He now devoted most of his time to the problem, produced an estimate of the orbit of Ceres in December, and sent his results to Piazzi. The new planet, which had been sighted on the first day of the year, was found again—by its discoverer—on the last day of the year.
>
> Gauss did not publish his orbit determination methods until 1809.[3] In this publication, he also described the method of least squares that he had discovered in 1795, at the age of 18, and had used it in refining his estimates of the orbit of Ceres.

Although Ceres played a significant role in the history of discovery and it still reappears regularly in the nighttime sky, it has faded into obscurity as an object of intellectual interest. The method of least squares, on the other hand, has been an object of continuing interest and benefit to generations of scientists and technologists ever since its introduction. It has had a profound effect on the history of science. It was the first optimal estimation method, and it provided an important connection between the experimental and theoretical sciences: It gave experimentalists a practical method for estimating the unknown parameters of theoretical models.

1.2.2 Method of Least Squares

The following example of a least-squares problem is the one most often seen, although the *method* of least squares may be applied to a much greater range of problems.

EXAMPLE 1.1: Least-Squares Solution for Overdetermined Linear Systems Gauss discovered that if he wrote a system of equations in matrix form, as

$$\begin{bmatrix} h_{11} & h_{12} & h_{13} & \cdots & h_{1n} \\ h_{21} & h_{22} & h_{23} & \cdots & h_{2n} \\ h_{31} & h_{32} & h_{33} & \cdots & h_{3n} \\ \vdots & \vdots & \vdots & \ddots & \vdots \\ h_{l1} & h_{l2} & h_{l3} & \cdots & h_{ln} \end{bmatrix} \begin{bmatrix} x_1 \\ x_2 \\ x_3 \\ \vdots \\ x_n \end{bmatrix} = \begin{bmatrix} z_1 \\ z_2 \\ z_3 \\ \vdots \\ z_l \end{bmatrix} \tag{1.2}$$

or

$$Hx = z, \tag{1.3}$$

[3]In the meantime, the method of least squares had been discovered independently and published by Andrien-Marie Legendre (1752–1833) in France and Robert Adrian (1775–1855) in the United States [176]. [It had also been discovered and used before Gauss was born by the German-Swiss physicist Johann Heinrich Lambert (1728–1777).] Such Jungian synchronicity (i.e., the phenomenon of multiple, near-simultaneous discovery) was to be repeated for other breakthroughs in estimation theory, as well—for the Wiener filter and the Kalman filter.

then he could consider the problem of solving for that value of an *estimate* $\hat{x}$ (pronounced "x-hat") that minimizes the "estimated measurement error" $H\hat{x} - z$. He could characterize that estimation error in terms of its Euclidean vector norm $|H\hat{x} - z|$, or, equivalently, its square:

$$\varepsilon^2(\hat{x}) = |H\hat{x} - z|^2 \tag{1.4}$$

$$= \sum_{i=1}^{m}\left[\sum_{j=1}^{n} h_{ij}\hat{x}_j - z_i\right]^2, \tag{1.5}$$

which is a continuously differentiable function of the n unknowns $\hat{x}_1, \hat{x}_2, \hat{x}_3, \ldots, \hat{x}_n$. This function $\varepsilon^2(\hat{x}) \to \infty$ as any component $\hat{x}_k \to \pm\infty$. Consequently, it will achieve its minimum value where all its derivatives with respect to the $\hat{x}_k$ are zero. There are n such equations of the form

$$0 = \frac{\partial \varepsilon^2}{\partial \hat{x}_k} \tag{1.6}$$

$$= 2\sum_{i=1}^{m} h_{ik}\left[\sum_{j=1}^{n} h_{ij}\hat{x}_j - z_i\right] \tag{1.7}$$

for $k = 1, 2, 3, \ldots, n$. Note that in this last equation the expression

$$\sum_{j=1}^{n} h_{ij}\hat{x}_j - z_i = \{H\hat{x} - z\}_i, \tag{1.8}$$

the ith row of $H\hat{x} - z$, and the outermost summation is equivalent to the dot product of the kth column of H with $H\hat{x} - z$. Therefore Equation 1.7 can be written as

$$0 = 2H^{\mathrm{T}}[H\hat{x} - z] \tag{1.9}$$

$$= 2H^{\mathrm{T}}H\hat{x} - 2H^{\mathrm{T}}z \tag{1.10}$$

or

$$H^{\mathrm{T}}H\hat{x} = H^{\mathrm{T}}z,$$

where the matrix transpose H^{T} is defined as

$$H^{\mathrm{T}} = \begin{bmatrix} h_{11} & h_{21} & h_{31} & \cdots & h_{m1} \\ h_{12} & h_{22} & h_{32} & \cdots & h_{m2} \\ h_{13} & h_{23} & h_{33} & \cdots & h_{m3} \\ \vdots & \vdots & \vdots & \ddots & \vdots \\ h_{1n} & h_{2n} & h_{3n} & \cdots & h_{mn} \end{bmatrix} \tag{1.11}$$

The normal equation of the linear least squares problem. The equation

$$H^{\mathrm{T}} H \hat{x} = H^{\mathrm{T}} z \tag{1.12}$$

is called the *normal equation* or the *normal form* of the equation for the linear least-squares problem. It has precisely as many equivalent scalar equations as unknowns.

The Gramian of the linear least squares problem. The normal equation has the solution

$$\hat{x} = (H^{\mathrm{T}} H)^{-1} H^{\mathrm{T}} z,$$

provided that the matrix

$$\mathcal{G} = H^{\mathrm{T}} H \tag{1.13}$$

is *nonsingular* (i.e., invertible). The matrix product $\mathcal{G} = H^{\mathrm{T}} H$ in this equation is called the *Gramian matrix*.[4] The determinant of the Gramian matrix characterizes whether or not the column vectors of H are linearly independent. If its determinant is zero, the column vectors of H are linearly dependent, and $\hat{x}$ cannot be determined uniquely. If its determinant is nonzero, then the solution $\hat{x}$ is uniquely determined.

Least-squares solution. In the case that the Gramian matrix *is* invertible (i.e., nonsingular), the solution $\hat{x}$ is called the least-squares solution of the overdetermined linear inversion problem. It is an estimate that makes no assumptions about the nature of the unknown measurement errors, although Gauss alluded to that possibility in his description of the method. The formal treatment of uncertainty in estimation would come later.

This form of the Gramian matrix will be used in Chapter 2 to define the observability matrix of a linear dynamic system model in discrete time.

Least Squares in Continuous Time. The following example illustrates how the principle of least squares can be applied to fitting a vector-valued parametric model to data in continuous time. It also illustrates how the issue of *determinacy* (i.e., whether there is a *unique* solution to the problem) is characterized by the Gramian matrix in this context.

[4]Named for the Danish mathematician Jorgen Pedersen Gram (1850–1916). This matrix is also related to what is called the *unscaled Fisher information matrix*, named after the English statistician Ronald Aylmer Fisher (1890–1962). Although information matrices and Gramian matrices have different definitions and uses, they can amount to almost the same thing in this particular instance. The formal statistical definition of the term *information matrix* represents the information obtained from a sample of values from a known probability distribution. It corresponds to a scaled version of the Gramian matrix when the measurement errors in z have a joint Gaussian distribution, with the scaling related to the uncertainty of the measured data. The information matrix is a *quantitative* statistical characterization of the "information" (in some sense) that is in the data z used for estimating x. The Gramian, on the other hand, is used as an *qualitative* algebraic characterization of the uniqueness of the solution.

EXAMPLE 1.2: Least-Squares Fitting of Vector-Valued Data in Continuous Time Suppose that, for each value of time t on an interval $t_0 \leq t \leq t_f$, $z(t)$ is an ℓ-dimensional signal vector that is modeled as a function of an unknown n-vector x by the equation

$$z(t) = H(t)x,$$

where $H(t)$ is a known $\ell \times n$ matrix. The squared error in this relation at each time t will be

$$\begin{aligned}\varepsilon^2(t) &= |z(t) - H(t)x|^2 \\ &= x^{\mathrm{T}}[H^{\mathrm{T}}(t)H(t)]x - 2x^{\mathrm{T}}H^{\mathrm{T}}(t)z(t) + |z(t)|^2.\end{aligned}$$

The squared integrated error over the interval will then be the integral

$$\begin{aligned}\|\varepsilon\|^2 &= \int_{t_0}^{t_f} \varepsilon^2(t)\,dt \\ &= x^{\mathrm{T}}\left[\int_{t_0}^{t_f} H^{\mathrm{T}}(t)H(t)\,dt\right]x - 2x^{\mathrm{T}}\left[\int_{t_0}^{t_f} H^{\mathrm{T}}(t)z(t)\,dt\right] + \int_{t_0}^{t_f} |z(t)|^2\,dt,\end{aligned}$$

which has exactly the same array structure with respect to x as the algebraic least-squares problem. The least-squares solution for x can be found, as before, by taking the derivatives of $\|\varepsilon\|^2$ with respect to the components of x and equating them to zero. The resulting equations have the solution

$$\hat{x} = \left[\int_{t_0}^{t_f} H^{\mathrm{T}}(t)H(t)\,dt\right]^{-1}\left[\int_{t_0}^{t_f} H^{\mathrm{T}}(t)z(t)\,dt\right],$$

provided that the corresponding Gramian matrix

$$\mathcal{G} = \int_{t_0}^{t_f} H^{\mathrm{T}}(t)H(t)\,dt$$

is nonsingular.

This form of the Gramian matrix will be used in Chapter 2 to define the *observability matrix* of a linear dynamic system model in continuous time.

1.2.3 Gramian Matrix and Observability

For the examples considered above, observability does not depend upon the measurable data (z). It depends only on the nonsingularity of the Gramian matrix ($\mathcal{G}$), which depends only on the linear constraint matrix (H) between the unknowns and knowns.

Observability of a set of unknown variables is the issue of whether or not their values are *uniquely determinable* from a given set of *constraints,* expressed as equations involving functions of the unknown variables. The unknown variables are said to be *observable* if their values are uniquely determinable from the given constraints, and they are said to be *unobservable* if they are not uniquely determinable from the given constraints.

The condition of *nonsingularity* (or "*full rank*") of the Gramian matrix is an *algebraic* characterization of observability when the constraining equations are *linear* in the unknown variables. It also applies to the case that the constraining equations are not exact, due to errors in the values of the allegedly known parameters of the equations.

The Gramian matrix will be used in Chapter 2 to define observability of the states of dynamic systems in continuous time and discrete time.

1.2.4 Introduction of Probability Theory

Beginnings of Probability Theory. Probabilities represent the state of knowledge about physical phenomena by providing something more useful than "I don't know" to questions involving uncertainty. One of the mysteries in the history of science is why it took so long for mathematicians to formalize a subject of such practical importance. The Romans were selling insurance and annuities long before expectancy and risk were concepts of serious mathematical interest. Much later, the Italians were issuing insurance policies against business risks in the early Renaissance, and the first known attempts at a theory of probabilities—for games of chance—occurred in that period. The Italian Girolamo Cardano[5] (1501–1576) performed an accurate analysis of probabilities for games involving dice. He assumed that successive tosses of the dice were statistically independent events. He and the contemporary Indian writer Brahmagupta stated without proof that the accuracies of empirical statistics tend to improve with the number of trials. This would later be formalized as a *law of large numbers.*

More general treatments of probabilities were developed by Blaise Pascal (1623–1662), Pierre de Fermat (1601–1655), and Christiaan Huygens (1629–1695). Fermat's work on combinations was taken up by Jakob (or James) Bernoulli (1654–1705), who is considered by some historians to be the founder of probability theory. He gave the first rigorous proof of the law of large numbers for repeated independent trials (now called *Bernoulli trials*). Thomas Bayes (1702–1761) derived his famous rule for statistical inference sometime after Bernoulli. Abraham de Moivre (1667–1754), Pierre Simon Marquis de Laplace (1749–1827), Adrien Marie Legendre (1752–1833), and Carl Friedrich Gauss (1777–1855) continued this development into the nineteenth century.

[5]Cardano was a practicing physician in Milan who also wrote books on mathematics. His book *De Ludo Hleae,* on the mathematical analysis of games of chance (principally dice games), was published nearly a century after his death. Cardano was also the inventor of the most common type of universal joint found in automobiles, sometimes called the *Cardan joint* or *Cardan shaft.*

Between the early nineteenth century and the mid-twentieth century, the probabilities themselves began to take on more meaning as physically significant attributes. The idea that the laws of nature embrace random phenomena, and that these are treatable by probabilistic models began to emerge in the nineteenth century. The development and application of probabilistic models for the physical world expanded rapidly in that period. It even became an important part of sociology. The work of James Clerk Maxwell (1831–1879) in statistical mechanics established the probabilistic treatment of natural phenomena as a scientific (and successful) discipline.

An important figure in probability theory and the theory of random processes in the twentieth century was the Russian academician Andrei Nikolaeovich Kolmogorov (1903–1987). Starting around 1925, working with H. Ya. Khinchin and others, he reestablished the foundations of probability theory on measurement theory, which became the accepted mathematical basis of probability and random processes. Along with Norbert Wiener (1894–1964), he is credited with founding much of the theory of prediction, smoothing and filtering of Markov processes, and the general theory of ergodic processes. His was the first formal theory of optimal estimation for systems involving random processes.

1.2.5 Wiener Filter

Norbert Wiener (1894–1964) is one of the more famous prodigies of the early twentieth century. He was taught by his father until the age of 9, when he entered high school. He finished high school at the age of 11 and completed his undergraduate degree in mathematics in three years at Tufts University. He then entered graduate school at Harvard University at the age of 14 and completed his doctorate degree in the philosophy of mathematics when he was 18. He studied abroad and tried his hand at several jobs for six more years. Then, in 1919, he obtained a teaching appointment at the Massachusetts Institute of Technology (MIT). He remained on the faculty at MIT for the rest of his life.

In the popular scientific press, Wiener is probably more famous for naming and promoting *cybernetics* than for developing the Wiener filter. Some of his greatest mathematical achievements were in generalized harmonic analysis, in which he extended the Fourier transform to functions of finite *power*. Previous results were restricted to functions of finite *energy,* which is an unreasonable constraint for signals on the real line. Another of his many achievements involving the generalized Fourier transform was proving that the transform of white noise is also white noise.[6]

Wiener Filter Development. In the early years of the World War II, Wiener was involved in a military project to design an automatic controller for directing antiaircraft fire with radar information. Because the speed of the airplane is a

[6]He is also credited with the discovery that the power spectral density of a signal equals the Fourier transform of its autocorrelation function, although it was later discovered that Einstein had known it before him.

nonnegligible fraction of the speed of bullets, this system was required to "shoot into the future." That is, the controller had to predict the future course of its target using noisy radar tracking data.

Wiener derived the solution for the least-mean-squared prediction error in terms of the autocorrelation functions of the signal and the noise. The solution is in the form of an integral operator that can be synthesized with analog circuits, given certain constraints on the regularity of the autocorrelation functions or, equivalently, their Fourier transforms. His approach represents the probabilistic nature of random phenomena in terms of power spectral densities.

An analogous derivation of the optimal linear predictor for discrete-time systems was published by A. N. Kolmogorov in 1941, when Wiener was just completing his work on the continuous-time predictor.

Wiener's work was not declassified until the late 1940s, in a report titled "Extrapolation, interpolation, and smoothing of stationary time series." The title was subsequently shortened to "Time series." An early edition of the report had a yellow cover, and it came to be called "the yellow peril." It was loaded with mathematical details beyond the grasp of most engineering undergraduates, but it was absorbed and used by a generation of dedicated graduate students in electrical engineering.

1.2.6 Kalman Filter

Rudolf Emil Kalman was born on May 19, 1930, in Budapest, the son of Otto and Ursula Kalman. The family emigrated from Hungary to the United States during World War II. In 1943, when the war in the Mediterranean was essentially over, they traveled through Turkey and Africa on an exodus that eventually brought them to Youngstown, Ohio, in 1944. Rudolf attended Youngstown College there for three years before entering MIT.

Kalman received his bachelor's and master's degrees in electrical engineering at MIT in 1953 and 1954, respectively. His graduate advisor was Ernst Adolph Guillemin, and his thesis topic was the behavior of solutions of second-order difference equations [114]. When he undertook the investigation, it was suspected that second-order difference equations might be modeled by something analogous to the describing functions used for second-order differential equations. Kalman discovered that their solutions were not at all like the solutions of differential equations. In fact, they were found to exhibit chaotic behavior.

In the fall of 1955, after a year building a large analog control system for the E. I. DuPont Company, Kalman obtained an appointment as lecturer and graduate student at Columbia University. At that time, Columbia was well known for the work in control theory by John R. Ragazzini, Lotfi A. Zadeh,[7] and others. Kalman taught at Columbia until he completed the Doctor of Science degree there in 1957.

For the next year, Kalman worked at the research laboratory of the International Business Machines Corporation in Poughkeepsie and for six years after that at the

[7]Zadeh is perhaps more famous as the "father" of fuzzy systems theory and interpolative reasoning.

research center of the Glenn L. Martin company in Baltimore, the Research Institute for Advanced Studies (RIAS).

Early Research Interests. The algebraic nature of systems theory first became of interest to Kalman in 1953, when he read a paper by Ragazzini published the previous year. It was on the subject of sampled-data systems, for which the time variable is discrete valued. When Kalman realized that linear discrete-time systems could be solved by transform methods, just like linear continuous-time systems, the idea occurred to him that there is no fundamental difference between continuous and discrete linear systems. The two must be equivalent in some sense, even though the solutions of linear differential equations cannot go to zero (and stay there) in finite time and those of discrete-time systems can. That started his interest in the connections between systems theory and algebra.

In 1954 Kalman began studying the issue of *controllability*, which is the question of whether there exists an input (control) function to a dynamic system that will drive the state of that system to zero. He was encouraged and aided by the work of Robert W. Bass during this period. The issue of eventual interest to Kalman was whether there is an *algebraic* condition for controllability. That condition was eventually found as the rank of a matrix.[8] This implied a connection between algebra and systems theory.

Discovery of the Kalman Filter. In late November of 1958, not long after coming to RIAS, Kalman was returning by train to Baltimore from a visit to Princeton. At around 11 PM, the train was halted for about an hour just outside Baltimore. It was late, he was tired, and he had a headache. While he was trapped there on the train for that hour, an idea occurred to him: *Why not apply the notion of state variables*[9] *to the Wiener filtering problem?* He was too tired to think much more about it that evening, but it marked the beginning of a great exercise to do just that. He read through Loève's book on probability theory [68] and equated expectation with projection. That proved to be pivotal in the derivation of the Kalman filter. With the additional assumption of finite dimensionality, he was able to derive the Wiener filter as what we now call the Kalman filter. With the change to state-space form, the mathematical background needed for the derivation became much simpler, and the proofs were within the mathematical reach of many undergraduates.

Introduction of the Kalman Filter. Kalman presented his new results in talks at several universities and research laboratories before it appeared in print.[10] His ideas were met with some skepticism among his peers, and he chose a mechanical

[8]The *controllability matrix,* a concept defined in Chapter 2.

[9]Although function-space methods were then the preferred approach to the filtering problem, the use of state-space models for time-varying systems had already been introduced (e.g., by Laning and Battin [67] in 1956).

[10]In the meantime, some of the seminal ideas in the Kalman filter had been published by Swerling [227] in 1959 and Stratonovich [25, 226] in 1960.

engineering journal (rather than an electrical engineering journal) for publication, because "When you fear stepping on hallowed ground with entrenched interests, it is best to go sideways." [11] His second paper, on the continuous-time case, was once rejected because—as one referee put it—one step in the proof "cannot possibly be true." (It was true.) He persisted in presenting his filter, and there was more immediate acceptance elsewhere. It soon became the basis for research topics at many universities and the subject of dozens of doctoral theses in electrical engineering over the next several years.

Early Applications. Kalman found a receptive audience for his filter in the fall of 1960 in a visit to Stanley F. Schmidt at the Ames Research Center of NASA in Mountain View, California [118]. Kalman described his recent result and Schmidt recognized its potential applicability to a problem then being studied at Ames—the trajectory estimation and control problem for the Apollo project, a planned manned mission to the moon and back. Schmidt began work immediately on what was probably the first full implementation of the Kalman filter. He soon discovered what is now called "extended Kalman filtering," which has been used ever since for most real-time nonlinear applications of Kalman filtering. Enthused over his own success with the Kalman filter, he set about proselytizing others involved in similar work. In the early part of 1961, Schmidt described his results to Richard H. Battin from the MIT Instrumentation Laboratory (later renamed the Charles Stark Draper Laboratory). Battin was already using state space methods for the design and implementation of astronautical guidance systems, and he made the Kalman filter part of the Apollo onboard guidance,[12] which was designed and developed at the Instrumentation Laboratory. In the mid-1960s, through the influence of Schmidt, the Kalman filter became part of the Northrup-built navigation system for the C5A air transport, then being designed by Lockheed Aircraft Company. The Kalman filter solved the *data fusion problem* associated with combining radar data with inertial sensor data to arrive at an overall estimate of the aircraft trajectory and the *data rejection problem* associated with detecting exogenous errors in measurement data. It has been an integral part of nearly every onboard trajectory estimation and control system designed since that time.

Other Research Interests. Around 1960, Kalman showed that the related notion of observability for dynamic systems had an algebraic dual relationship with controllability. That is, by the proper exchange of system parameters, one problem could be transformed into the other, and vice versa.

Richard S. Bucy was also at RIAS in that period, and it was he who suggested to Kalman that the Wiener–Hopf equation is equivalent to the matrix Riccati equa-

[11] The two quoted segments in this paragraph are from a talk on *System Theory: Past and Present* given by Kalman at the University of California at Los Angeles (UCLA) on April 17, 1991, in a symposium organized and hosted by A. V. Balakrishnan at UCLA and sponsored jointly by UCLA and the National Aeronautics and Space Administration (NASA) Dryden Laboratory.

[12] Another fundamental improvement in Kalman filter implementation methods was made soon after by James E. Potter at the MIT Instrumentation Laboratory. This will be discussed in the next subsection.

tion—if one assumes a finite-dimensional state-space model. The general nature of this relationship between integral equations and differential equations first became apparent around that time. One of the more remarkable achievements of Kalman and Bucy in that period was proving that the Riccati equation can have a stable (steady-state) solution even if the dynamic system is unstable—provided that the system is observable and controllable.

Kalman also played a leading role in the development of realization theory, which also began to take shape around 1962. This theory addresses the problem of finding a system model to explain the observed input–output behavior of a system. This line of investigation led to a *uniqueness principle* for the mapping of exact (i.e., noiseless) data to linear system models.

In 1985, Kalman was awarded the Kyoto Prize, considered by some to be the Japanese equivalent of the Nobel Prize. On his visit to Japan to accept the Kyoto Prize, he related to the press an epigram that he had first seen in a pub in Colorado Springs in 1962, and it had made an impression on him. It said:

Little people discuss other people.
Average people discuss events.
Big people discuss ideas.

His own work, he felt, had been concerned with ideas.

In 1990, on the occasion of Kalman's sixtieth birthday, a special international symposium was convened for the purpose of honoring his pioneering achievements in what has come to be called *mathematical system theory,* and a *Festschrift* with that title was published soon after [3].

Impact of Kalman Filtering on Technology. From the standpoint of those involved in estimation and control problems, at least, this has to be considered the greatest achievement in estimation theory of the twentieth century. Many of the achievements since its introduction would not have been possible without it. It was one of the enabling technologies for the Space Age, in particular. The precise and efficient navigation of spacecraft through the solar system could not have been done without it.

The principal uses of Kalman filtering have been in "modern" control systems, in the tracking and navigation of all sorts of vehicles, and in predictive design of estimation and control systems. These technical activities were made possible by the introduction of the Kalman filter. (If you need a demonstration of its impact on technology, enter the keyword "Kalman filter" in a technical literature search. You will be overwhelmed by the sheer number of references it will generate.)

Relative Advantages of Kalman and Wiener Filtering

1. The Wiener filter implementation in analog electronics can operate at much higher effective throughput than the (digital) Kalman filter.
2. The Kalman filter is implementable in the form of an algorithm for a digital computer, which was replacing analog circuitry for estimation and control at

the time that the Kalman filter was introduced. This implementation may be slower, but it is capable of much greater accuracy than had been achievable with analog filters.

3. The Wiener filter does not require finite-dimensional stochastic process models for the signal and noise.
4. The Kalman filter does not require that the deterministic dynamics or the random processes have stationary properties, and many applications of importance include nonstationary stochastic processes.
5. The Kalman filter is compatible with the state-space formulation of optimal controllers for dynamic systems, and Kalman was able to prove useful dual properties of estimation and control for these systems.
6. For the modern controls engineering student, the Kalman filter requires less additional mathematical preparation to learn and use than the Wiener filter. As a result, the Kalman filter can be taught at the undergraduate level in engineering curricula.
7. The Kalman filter provides the necessary information for mathematically sound, statistically-based decision methods for detecting and rejecting anomalous measurements.

1.2.7 Square-Root Methods and All That

Numerical Stability Problems. The great success of Kalman filtering was not without its problems, not the least of which was marginal stability of the numerical solution of the associated Riccati equation. In some applications, small roundoff errors tended to accumulate and eventually degrade the performance of the filter. In the decades immediately following the introduction of the Kalman filter, there appeared several better numerical implementations of the original formulas. Many of these were adaptations of methods previously derived for the least squares problem.

Early ad hoc Fixes. It was discovered early on[13] that forcing symmetry on the solution of the matrix Riccati equation improved its apparent numerical stability—a phenomenon that was later given a more theoretical basis by Verhaegen and Van Dooren [232]. It was also found that the influence of roundoff errors could be ameliorated by artificially increasing the covariance of process noise in the Riccati equation. A symmetrized form of the discrete-time Riccati equation was developed by Joseph [15] and used by R. C. K. Lee at Honeywell in 1964. This "structural" reformulation of the Kalman filter equations improved robustness against roundoff errors in some applications, although later methods have performed better on some problems [125].

[13]These fixes were apparently discovered independently by several people. Schmidt [118] and his colleagues at NASA had discovered the use of forced symmetry and "pseudonoise" to counter roundoff effects and cite R. C. K. Lee at Honeywell with the independent discovery of the symmetry effect.

Square-Root Filtering. These methods can also be considered as "structural" reformulations of the Riccati equation, and they predate the Bucy–Joseph form. The first of these was the "square-root" implementation by Potter and Stern [208], first published in 1963 and successfully implemented for space navigation on the Apollo manned lunar exploration program. Potter and Stern introduced the idea of factoring the covariance matrix into *Cholesky factors*,[14] in the format

$$P = CC^{\mathrm{T}}, \tag{1.14}$$

and expressing the observational update equations in terms of the Cholesky factor C, rather than P. The result was better numerical stability of the filter implementation at the expense of added computational complexity. A generalization of the Potter and Stern method to handle vector-valued measurements was published by one of the authors [130] in 1968, but a more efficient implementation—in terms of *triangular* Cholesky factors—was published by Bennet in 1967 [138].

Square-Root and UD Filters. There was a rather rapid development of faster algorithmic methods for square-root filtering in the 1970s, following the work at NASA/JPL (then called the Jet Propulsion Laboratory, at the California Institute of Technology) in the late 1960s by Dyer and McReynolds [156] on temporal update methods for Cholesky factors. Extensions of square-root covariance and information filters were introduced in Kaminski's 1971 thesis [115] at Stanford University. The first of the triangular factoring algorithms for the observational update was due to Agee and Turner [106], in a 1972 report of rather limited circulation. These algorithms have roughly the same computational complexity as the conventional Kalman filter, but with better numerical stability. The "fast triangular" algorithm of Carlson was published in 1973 [149], followed by the "square-root-free" algorithm of Bierman in 1974 [7] and the associated temporal update method introduced by Thornton [124]. The computational complexity of the square-root filter for time-invariant systems was greatly simplified by Morf and Kailath [204] soon after that. Specialized parallel processing architectures for fast solution of the square-root filter equations were developed by Jover and Kailath [175] and others over the next decade, and much simpler derivations of these and earlier square-root implementations were discovered by Kailath [26].

Factorization Methods. The square-root methods make use of matrix decomposition[15] methods that were originally derived for the least-squares problem. These

[14]A square root S of a matrix P satisfies the equation $P = SS$ (i.e., without the transpose on the second factor). Potter and Stern's derivation used a special type of symmetric matrix called an *elementary matrix*. They factored an elementary matrix as a square of another elementary matrix. In this case, the factors were truly square roots of the factored matrix. This square-root appellation has stuck with extensions of Potter and Stern's approach, even though the factors involved are Cholesky factors, not matrix square roots.

[15]The term "decomposition" refers to the representation of a matrix (in this case, a covariance matrix) as a product of matrices having more useful computational properties, such as sparseness (for triangular factors) or good numerical stability (for orthogonal factors). The term "factorization" was used by Bierman [7] for such representations.

include the so-called QR decomposition of a matrix as the product of an orthogonal matrix (Q) and a "triangular"[16] matrix (R). The matrix R results from the application of orthogonal transformations of the original matrix. These orthogonal transformations tend to be well conditioned numerically. The operation of applying these transformations is called the "triangularization" of the original matrix, and triangularization methods derived by Givens [164], Householder [172], and Gentleman [163] are used to make Kalman filtering more robust against roundoff errors.

1.2.8 Beyond Kalman Filtering

Extended Kalman Filtering and the Kalman–Schmidt Filter. Although it was originally derived for a linear problem, the Kalman filter is habitually applied with impunity—and considerable success—to many nonlinear problems. These extensions generally use partial derivatives as linear approximations of nonlinear relations. Schmidt [118] introduced the idea of evaluating these partial derivatives at the *estimated* value of the state variables. This approach is generally called the *extended Kalman filter*, but it was called the *Kalman–Schmidt filter* in some early publications. This and other methods for approximate linear solutions to nonlinear problems are discussed in Chapter 5, where it is noted that these will not be adequate for all nonlinear problems. Mentioned here are some investigations that have addressed estimation problems from a more general perspective, although they are not covered in the rest of the book.

Nonlinear Filtering Using Higher Order Approximations. Approaches using higher order expansions of the filter equations (i.e., beyond the linear terms) have been derived by Stratonovich [78], Kushner [191], Bucy [147], Bass et al. [134], and others for quadratic nonlinearities and by Wiberg and Campbell [235] for terms through third order.

Nonlinear Stochastic Differential Equations. Problems involving nonlinear and random dynamic systems have been studied for some time in statistical mechanics. The propagation over time of the *probability distribution* of the state of a nonlinear dynamic system is described by a nonlinear partial differential equation called the *Fokker–Planck equation.* It has been studied by Einstein [157], Fokker [160], Planck [207], Kolmogorov [187], Stratonovich [78], Baras and Mirelli [52], and others. Stratonovich modeled the effect on the probability distribution of information obtained through noisy measurements of the dynamic system, an effect called *conditioning.* The partial differential equation that includes these effects is called the *conditioned Fokker–Planck equation.* It has also been studied by Kushner [191], Bucy [147], and others using the *stochastic calculus* of Kiyosi Itô—also called the "Itô calculus." It is a non-Riemannian calculus developed specifically for stochastic differential systems with noise of infinite bandwidth. This general approach results in a stochastic partial differential equation describing

[16]See Chapter 6 and Appendix B for discussions of triangular forms.

the evolution over time of the probability distribution over a "state space" of the dynamic system under study. The resulting model does *not* enjoy the finite representational characteristics of the Kalman filter, however. The computational complexity of obtaining a solution far exceeds the already considerable burden of the conventional Kalman filter. These methods are of significant interest and utility but are beyond the scope of this book.

Point Processes and the Detection Problem. A *point process* is a type of random process for modeling events or objects that are distributed over time or space, such as the arrivals of messages at a communications switching center or the locations of stars in the sky. It is also a model for the initial states of systems in many estimation problems, such as the locations of aircraft or spacecraft under surveillance by a radar installation or the locations of submarines in the ocean. The *detection problem* for these surveillance applications must usually be solved before the *estimation problem* (i.e., tracking of the objects with a Kalman filter) can begin. The Kalman filter requires an initial state for each object, and that initial state estimate must be obtained by detecting it. Those initial states are distributed according to some point process, but there are no technically mature methods (comparable to the Kalman filter) for estimating the state of a point process. A unified approach combining detection and tracking into one optimal estimation method was derived by Richardson [214] and specialized to several applications. The detection and tracking problem for a *single object* is represented by the conditioned Fokker–Planck equation. Richardson derived from this one-object model an infinite hierarchy of partial differential equations representing *object densities* and truncated this hierarchy with a simple closure assumption about the relationships between orders of densities. The result is a single partial differential equation approximating the evolution of the density of objects. It can be solved numerically. It provides a solution to the difficult problem of detecting dynamic objects whose initial states are represented by a point process.

1.3 ON THE NOTATION USED IN THIS BOOK

1.3.1 Symbolic Notation

The fundamental problem of symbolic notation, in almost any context, is that there are never enough symbols to go around. There are not enough letters in the Roman alphabet to represent the sounds of standard English, let alone all the variables in Kalman filtering and its applications. As a result, some symbols must play multiple roles. In such cases, their roles will be defined as they are introduced. It is sometimes confusing, but unavoidable.

"Dot" Notation for Derivatives. Newton's notation using $\dot{f}(t)$, $\ddot{f}(t)$ for the first two derivatives of f with respect to t is used where convenient to save ink.

TABLE 1.2 Standard Symbols of Kalman Filtering

Symbols			Symbol Definition
I[a]	II[b]	III[c]	
F	F	A	Dynamic coefficient matrix of continuous linear differential equation defining dynamic system
G	I	B	Coupling matrix between random process noise and state of linear dynamic system
H	M	C	Measurement sensitivity matrix, defining linear relationship between state of the dynamic system and measurements that can be made
$\overline{K}$	Δ	K	Kalman gain matrix
P	P		Covariance matrix of state estimation uncertainty
Q	Q		Covariance matrix of process noise in the system state dynamics
R	0		Covariance matrix of observational (measurement) uncertainty
x	x		State vector of a linear dynamic system
z	y		Vector (or scalar) of measured values
Φ	Φ		State transition matrix of a discrete linear dynamic system

[a] This book [1, 13, 16, 21]. [b] Kalman [23, 179]. [c] Other sources [4, 10, 18, 65].

Standard Symbols for Kalman Filter Variables. There appear to be two "standard" conventions in technical publications for the symbols used in Kalman filtering. The one used in this book is similar to the original notation of Kalman [179]. The other standard notation is sometimes associated with applications of Kalman filtering in control theory. It uses the first few letters of the alphabet in place of the Kalman notation. Both sets of symbol usages are presented in Table 1.2, along with the original (Kalman) notation.

State Vector Notation for Kalman Filtering. The state vector x has been adorned with all sorts of other appendages in the usage of Kalman filtering. Table 1.3 lists the notation used in this book (left column) along with notations found in some other sources (second column). The state vector wears a "hat" as the estimated value, $\hat{x}$, and subscripting to denote the sequence of values that the estimate assumes over time. The problem is that it has two values at the same time: the *a priori*[17] value (before the measurement at the current time has been used in refining the estimate) and the *a posteriori* value (after the current measurement has been used in refining the estimate). These distinctions are indicated by the signum. The negative sign $(-)$ indicates the a priori value, and the positive sign $(+)$ indicates the *a posteriori* value.

[17]This use of the full Latin phrases as adjectives for the prior and posterior statistics is an unfortunate choice of standard notation, because there is no easy way to shorten it. (Even their initial abbreviations are the same.) If those who initiated this notation had known how commonplace it would become, they might have named them otherwise.

TABLE 1.3 Special State-Space Notation

This book	Other sources	Definition of Notational Usage
x	$\underline{x}$ $\vec{x}$ $\mathbf{x}$	Vector
x_k		The kth component of the vector x
x_k	$x[k]$	The kth element of the sequence $\ldots, x_{k-1}, x_k, x_{k+1}, \ldots$ of vectors
$\hat{x}$	$E\langle x \rangle$ $\bar{x}$	An estimate of the value of x
$\hat{x}_k(-)$	$\hat{x}_{k\mid k-1}$ $\hat{x}_{k-}$	A priori estimate of x_k, conditioned on all prior measurements except the one at time t_k
$\hat{x}_k(+)$	$\hat{x}_{k\mid k}$ $\hat{x}_{k+}$	A posteriori estimate of x, conditioned on all available measurements at time t_k
$\dot{x}$	x_t dx/dt	Derivative of x with respect to t (time)

TABLE 1.4 Common Notation for Array Dimensions

Symbol	Vector Name	Dimensions	Symbol	Matrix Name	Dimensions Row	Dimensions Column
x	System state	n	Φ	State transition	n	n
w	Process noise	r	G	Process noise coupling	n	r
u	Control input	s	Q	Process noise covariance	r	r
z	Measurement	ℓ	H	Measurement sensitivity	ℓ	n
v	Measurement noise	ℓ	R	Measurement noise covariance	ℓ	ℓ

Common Notation for Array Dimensions. Symbols used for the *dimensions* of the "standard" arrays in Kalman filtering will also be standardized, using the notation of Gelb et al. [21] shown in Table 1.4. These symbols are not used exclusively for these purposes. (Otherwise, one would soon run out of alphabet.) However, whenever one of these arrays is used in the discussion, these symbols will be used for their dimensions.

1.4 SUMMARY

The *Kalman filter* is an estimator used to estimate the *state* of a *linear dynamic system* perturbed by *Gaussian white noise* using measurements that are *linear* functions of the system state but corrupted by *additive* Gaussian white noise. The mathematical model used in the derivation of the Kalman filter is a reasonable representation for many problems of practical interest, including *control problems* as well as *estimation problems.* The Kalman filter model is also used for the *analysis* of *measurement and estimation problems.*

The *method of least squares* was the first "optimal" estimation method. It was discovered by Gauss (and others) around the end of the eighteenth century, and it is still much in use today. If the associated *Gramian matrix* is *nonsingular*, the method of least squares determines the *unique* values of a set of unknown variables such that the *squared deviation* from a set of constraining equations is minimized.

Observability of a set of unknown variables is the issue of whether or not they are *uniquely determinable* from a given set of *constraining equations.* If the constraints are *linear* functions of the unknown variables, then those variables are *observable if and only if* the associated Gramian matrix is nonsingular. If the Gramian matrix is *singular,* then the unknown variables are *unobservable.*

The *Wiener–Kolmogorov filter* was derived in the 1940s by Norbert Wiener (using a model in continuous time) and Andrei Kolmogorov (using a model in discrete time) working independently. It is a *statistical estimation method.* It estimates the state of a dynamic process so as to minimize the *mean-squared estimation error.* It can take advantage of *statistical* knowledge about random processes in terms of their power spectral densities in the *frequency domain.*

The *"state-space" model* of a dynamic process uses differential equations (or difference equations) to represent both deterministic and random phenomena. The *state variables* of this model are the variables of interest and their derivatives of interest. Random processes are characterized in terms of their statistical properties in the *time domain,* rather than the frequency domain. The Kalman filter was derived as the solution to the Wiener filtering problem using the state-space model for dynamic and random processes. The result is easier to derive (and to use) than the Wiener–Kolmogorov filter.

Square-root filtering is a reformulation of the Kalman filter for better numerical stability in finite-precision arithmetic. It is based on the same mathematical model, but it uses an equivalent statistical parameter that is less sensitive to roundoff errors in the computation of optimal filter gains. It incorporates many of the more numerically stable computation methods that were originally derived for solving the least-squares problem.

PROBLEMS

1.1 Jean Baptiste Fourier (1768–1830) was studying the problem of approximating a function $f(\theta)$ on the circle $0 \leq \theta < 2\pi$ by a linear combination of trigonometric functions:

$$f(\theta) \approx a_0 + \sum_{j=1}^{n} [a_j \cos(j\theta) + b_j \sin(j\theta)]. \tag{1.15}$$

See if you can help him on this problem. Use the method of least squares to demonstrate that the values

$$\hat{a}_0 = \frac{1}{2\pi}\int_0^{2\pi} f(\theta)\, d\theta,$$

$$\hat{a}_j = \frac{1}{\pi}\int_0^{2\pi} f(\theta)\cos(j\theta)\, d\theta,$$

$$\hat{b}_j = \frac{1}{\pi}\int_0^{2\pi} f(\theta)\sin(j\theta)\, d\theta$$

of the coefficients a_j and b_j for $1 \le j \le n$ give the least integrated squared approximation error

$$\begin{aligned}
\varepsilon^2(a, b) &= \| f - \hat{f}(a, b)\|^2_{\mathscr{L}_2} \\
&= \int_0^{2\pi} \left[\hat{f}(\theta) - f(\theta)\right]^2 d\theta \\
&= \int_0^{2\pi} \left\{ a_0 + \sum_{j=1}^{n} [a_j \cos(j\theta) + b_j \sin(j\theta)] \right\}^2 d\theta \\
&\quad - 2\int_0^{2\pi} \left\{ a_0 + \sum_{j=1}^{n} [a_j \cos(j\theta) + b_j \sin(j\theta)] \right\} f(\theta)\, d\theta \\
&\quad + \int_0^{2\pi} f^2(\theta)\, d\theta.
\end{aligned}$$

You may assume the equalities

$$\int_0^{2\pi} d\theta = 2\pi$$

$$\int_0^{2\pi} \cos(j\theta)\cos(k\theta)\, d\theta = \begin{cases} 0, & j \ne k \\ \pi, & j = k, \end{cases}$$

$$\int_0^{2\pi} \sin(j\theta)\sin(k\theta)\, d\theta = \begin{cases} 0, & j \ne k \\ \pi, & j = k \end{cases}$$

$$\int_0^{2\pi} \cos(j\theta)\sin(k\theta)\, d\theta = 0, \qquad 0 \le j \le n, \qquad 1 \le k \le n$$

as given.

2

Linear Dynamic Systems

What we experience of nature is in models, and all of nature's models are so beautiful.[1]
R. Buckminster Fuller (1895–1983)

2.1 CHAPTER FOCUS

Models for Dynamic Systems. Since their introduction by Isaac Newton in the seventeenth century, differential equations have provided concise mathematical models for many dynamic systems of importance to humans. By this device, Newton was able to model the motions of the planets in our solar system with a small number of variables and parameters. Given a finite number of initial conditions (the initial positions and velocities of the sun and planets will do) and these equations, one can uniquely determine the positions and velocities of the planets for all time. The finite-dimensional representation of a problem (in this example, the problem of predicting the future course of the planets) is the basis for the so-called *state-space approach* to the representation of differential equations and their solutions, which is the focus of this chapter. The dependent variables of the differential equations become *state variables* of the dynamic system. They explicitly represent all the important characteristics of the dynamic system at any time.

The whole of dynamic system theory is a subject of considerably more scope than one needs for the present undertaking (Kalman filtering). This chapter will stick to just those concepts that are essential for that purpose, which is the development of the state-space representation for dynamic systems described by systems of linear differential equations. These are given a somewhat heuristic treatment, without the mathematical rigor often accorded the subject, omitting the development and use of the transform methods of functional analysis for solving differential equations when they serve no purpose in the derivation of the Kalman filter. The interested reader will find a more formal and thorough presentation in most upper-level and graduate-level textbooks on

[1]From an interview quoted by Calvin Tomkins in "From in the outlaw area," *The New Yorker*, January 8, 1966.

ordinary differential equations. The objective of the more engineering-oriented treatments of dynamic systems is usually to solve the *controls problem,* which is the problem of defining the *inputs* (i.e., control settings) that will bring the state of the dynamic system to a desirable condition. That is not the objective here, however.

2.1.1 Main Points to Be Covered

The objective in this chapter is to characterize the measurable *outputs* of dynamic systems as functions of the internal *states* and *inputs* of the system. (The italicized terms will be defined more precisely further along.) The treatment here is deterministic, in order to define functional relationships between inputs and outputs. In the next chapter, the inputs are allowed to be nondeterministic (i.e., random), and the objective of the following chapter will be to estimate the states of the dynamic system in this context.

Dynamic Systems and Differential Equations. In the context of Kalman filtering, a *dynamic system* has come to be synonymous with a system of ordinary differential equations describing the evolution over time of the state of a physical system. This mathematical model is used to derive its solution, which specifies the functional dependence of the state variables on their initial values and the system inputs. This solution defines the functional dependence of the measurable outputs on the inputs and the coefficients of the model.

Mathematical Models for Continuous and Discrete Time. The principal dynamic system models are summarized in Table 2.1.[2] For implementation in digital computers, the problem representation is transformed from an *analog* model (functions of continuous time) to a *digital* model (functions defined at discrete times).

Observability characterizes the *feasibility* of uniquely determining the state of a given dynamic system if its outputs are known. This characteristic of a dynamic system is determinable from the parameters of its mathematical model.

2.2 DYNAMIC SYSTEMS

2.2.1 Dynamic Systems Represented by Differential Equations

A *system* is an assemblage of interrelated entities that can be considered as a whole. If the attributes of interest of a system are changing with time, then it is called a *dynamic system*. A *process* is the evolution over time of a dynamic system.

Our *solar system,* consisting of the sun and its planets, is a physical example of a dynamic system. The motions of these bodies are governed by laws of motion that depend only upon their current relative positions and velocities. Sir Isaac Newton (1642–1727) discovered these laws and expressed them as a system of differential equations—another of his discoveries. From the time of Newton, engineers and scientists have learned to define dynamic systems in terms of the differential equations that govern their behavior. They have also learned how to solve many of these differential equations to obtain formulas for predicting the future behavior of dynamic systems.

[2]These include nonlinear models, which are discussed in Chapter 5. The primary interest in this chapter will be in linear models.

TABLE 2.1 Mathematical Models of Dynamic Systems

	Continuous	Discrete
Time invariant		
Linear	$\dot{x}(t) = Fx(t) + Cu(t)$	$x_k = \Phi x_{k-1} + \Gamma u_{k-1}$
General	$\dot{x}(t) = f(x(t), u(t))$	$x_k = f(x_{k-1}, u_{k-1})$
Time varying		
Linear	$\dot{x}(t) = F(t)x(t) + C(t)u(t)$	$x_k = \Phi_{k-1}x_{k-1} + \Gamma_{k-1}u_{k-1}$
General	$\dot{x}(t) = f(t, x(t), u(t))$	$x_k = f(k, x_{k-1}, u_{k-1})$

EXAMPLE 2.1 (below, left): Newton's Model for a Dynamic System of n Massive Bodies For a planetary system with n *bodies* (idealized as point masses), the acceleration of the ith body in any *inertial* (i.e., non-rotating and non-accelerating) Cartesian coordinate system is given by Newton's third law as the second-order differential equation

$$\frac{d^2 r_i}{dt^2} = C_g \sum_{\substack{j=1 \\ j \neq i}}^{n} \frac{m_j [r_j - r_i]}{|r_j - r_i|^3}, \ 1 \leq i \leq n,$$

where r_j is the position coordinate vector of the jth body, m_j is the mass of the jth body, and C_g is the gravitational constant. This set of n differential equations, plus the associated initial conditions of the bodies (i.e., their initial positions and velocities) theoretically determines the future history of the planetary system.

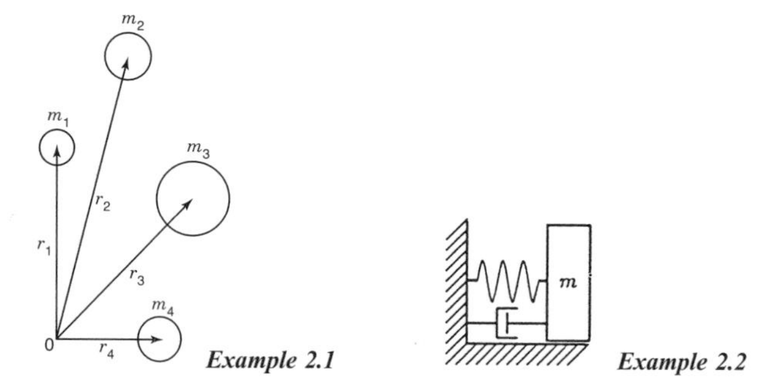

Example 2.1 *Example 2.2*

EXAMPLE 2.2 (above, right): The Harmonic Resonator with Linear Damping Consider the accompanying diagram of an idealized apparatus with a mass m attached through a spring to an immovable base and its frictional contact to its support base represented by a dashpot. Let δ be the displacement of the mass from its position at rest, $d\delta/dt$ be the velocity of the mass, and $a(t) = d^2\delta/dt^2$ its acceleration. The force F acting on the mass can be represented by Newton's second law as

$$\begin{aligned} F(t) &= ma(t) \\ &= m\left[\frac{d^2\delta}{dt^2}(t)\right] \\ &= -k_s\delta(t) - k_d\,\frac{d\delta}{dt}(t), \end{aligned}$$

where k_s is the spring constant and k_d is the drag coefficient of the dashpot. This relationship can be written as a differential equation

$$m\frac{d^2\delta}{dt^2} = -k_s\delta - k_d\frac{d\delta}{dt}$$

in which time (t) is the differential variable and displacement (δ) is the dependent variable. This equation constrains the dynamical behavior of the damped harmonic resonator. The *order* of a differential equation is the order of the highest derivative, which is *2* in this example. This one is called a *linear* differential equation, because both sides of the equation are linear combinations of δ and its derivatives. (That of Example 2.1 is a *nonlinear* differential equation.)

Not All Dynamic Systems Can Be Modeled by Differential Equations. There are other types of dynamic systems, such as those modeled by Petri nets or inference nets. However, the only types of dynamic systems considered in this book will be modeled by differential equations or by discrete-time linear state dynamic equations derived from linear differential or difference equations.

2.2.2 State Variables and State Equations

The second-order differential equation of the previous example can be transformed to a system of two first-order differential equations in the two dependent variables $x_1 = \delta$ and $x_2 = d\delta/dt$. In this way, one can reduce the form of any system of higher order differential equations to an equivalent system of first-order differential equations. These systems are generally classified into the types shown in Table 2.1, with the most general type being a *time-varying* differential equation for representing a dynamic system with time-varying dynamic characteristics. This is represented in vector form as

$$\dot{x}(t) = f(t, x(t), u(t)), \tag{2.1}$$

where Newton's "dot" notation is used as a shorthand for the derivative with respect to time, and a vector-valued function f to represent a system of n equations

$$\begin{aligned}
\dot{x}_1 &= f_1(t, x_1, x_2, x_3, \ldots, x_n, u_1, u_2, u_3, \ldots, u_r, t), \\
\dot{x}_2 &= f_2(t, x_1, x_2, x_3, \ldots, x_n, u_1, u_2, u_3, \ldots, u_r, t), \\
\dot{x}_3 &= f_3(t, x_1, x_2, x_3, \ldots, x_n, u_1, u_2, u_3, \ldots, u_r, t), \\
&\vdots \\
\dot{x}_n &= f_n(t, x_1, x_2, x_3, \ldots, x_n, u_1, u_2, u_3, \ldots, u_r, t)
\end{aligned} \tag{2.2}$$

in the independent variable t (time), n dependent variables $\{x_i | 1 \leq i \leq n\}$, and r known inputs $\{u_i | 1 \leq i \leq r\}$. These are called the *state equations* of the dynamic system.

State Variables Represent the Degrees of Freedom of Dynamic Systems. The variables $x_1, \ldots, x_n$ are called the *state variables* of the dynamic system defined by Equation 2.2. They are collected into a single n-vector

$$x(t) = [x_1(t) \quad x_2(t) \quad x_3(t) \quad \cdots \quad x_n(t)]^{\mathrm{T}} \tag{2.3}$$

called the *state vector* of the dynamic system. The n-dimensional domain of the state vector is called the *state space* of the dynamic system. Subject to certain continuity conditions on the functions f_i and u_i, the values $x_i(t_0)$ at some initial time t_0 will uniquely determine the values of the solutions $x_i(t)$ on some closed time interval $t \in [t_0, t_f]$ with initial time t_0 and final time t_f [57]. In that sense, the initial value of each state variable represents an independent degree of freedom of the dynamic system. The n values $x_1(t_0), x_2(t_0), x_3(t_0), \ldots, x_n(t_0)$ can be varied independently, and they uniquely determine the state of the dynamic system over the time interval $t \in [t_0, t_f]$.

EXAMPLE 2.3: State Space Model of the Harmonic Resonator For the second-order differential equation introduced in Example 2.2, let the state variables $x_1 = \delta$ and $x_2 = \dot{\delta}$. The first state variable represents the displacement of the mass from static equilibrium, and the second state variable represents the instantaneous velocity of the mass. The system of first-order differential equations for this dynamic system can be expressed in matrix form as

$$\frac{d}{dt}\begin{bmatrix} x_1(t) \\ x_2(t) \end{bmatrix} = F_c \begin{bmatrix} x_1(t) \\ x_2(t) \end{bmatrix},$$

$$F_c = \begin{bmatrix} 0 & 1 \\ -\dfrac{k_s}{m} & -\dfrac{k_d}{m} \end{bmatrix},$$

where F_c is called the *coefficient matrix* of the system of first-order linear differential equations. This is an example of what is called the *companion form* for higher order linear differential equations expressed as a system of first-order differential equations.

2.2.3 Continuous Time and Discrete Time

The dynamic system defined by Equation 2.2 is an example of a *continuous* system, so called because it is defined with respect to an independent variable t that varies continuously over some real interval $t \in [t_0, t_f]$. For many practical problems, however, one is only interested in knowing the state of a system at a discrete set of times $t \in \{t_1, t_2, t_3, \ldots\}$. These discrete times may, for example, correspond to the times at which the outputs of a system are sampled (such as the times at which Piazzi recorded the direction to Ceres). For problems of this type, it is convenient to order the times t_k according to their integer subscripts:

$$t_0 < t_1 < t_2 < \cdots t_{k-1} < t_k < t_{k+1} < \cdots.$$

That is, the time sequence is ordered according to the subscripts, and the subscripts take on all successive values in some range of integers. For problems of this type, it suffices to define the state of the dynamic system as a recursive relation,

$$x(t_{k+1}) = f(x(t_k), t_k, t_{k+1}), \tag{2.4}$$

by means of which the state is represented as a function of its previous state. This is a definition of a *discrete dynamic system*. For systems with *uniform time intervals* Δt

$$t_k = k\Delta t.$$

Shorthand Notation for Discrete-Time Systems. It uses up a lot of ink if one writes $x(t_k)$ when all one cares about is the sequence of values of the state variable x. It is more efficient to shorten this to x_k, so long as it is understood that it stands for $x(t_k)$, and not the kth component of x. If one must talk about a particular component at a particular time, one can always resort to writing $x_i(t_k)$ to remove any ambiguity. Otherwise, let us drop t as a symbol whenever it is clear from the context that we are talking about discrete-time systems.

2.2.4 Time-Varying Systems and Time-Invariant Systems

The term "physical plant" or "plant" is sometimes used in place of "dynamic system," especially for applications in manufacturing. In many such applications, the dynamic system under consideration is literally a physical plant—a fixed facility used in the manufacture of materials. Although the input $u(t)$ may be a function of time, the functional dependence of the state dynamics on u and x does not depend upon time. Such systems are called *time invariant* or *autonomous*. Their solutions are generally easier to obtain than those of time-varying systems.

2.3 CONTINUOUS LINEAR SYSTEMS AND THEIR SOLUTIONS

2.3.1 Input–Output Models of Linear Dynamic Systems

The block diagram in Figure 2.1 represents a linear continuous system with three types of variables:

- Inputs, which are under our control, and therefore known to us, or at least measurable by us. (In the next chapter, however, they will be assumed to be known only statistically. That is, individual samples of u are random but with known statistical properties.)
- State variables, which were described in the previous section. In most applications, these are "hidden variables," in the sense that they cannot generally be measured directly but must be somehow inferred from what can be measured.
- Outputs, which are those things that can be known through measurements.

These concepts are discussed in greater detail in the following subsections.

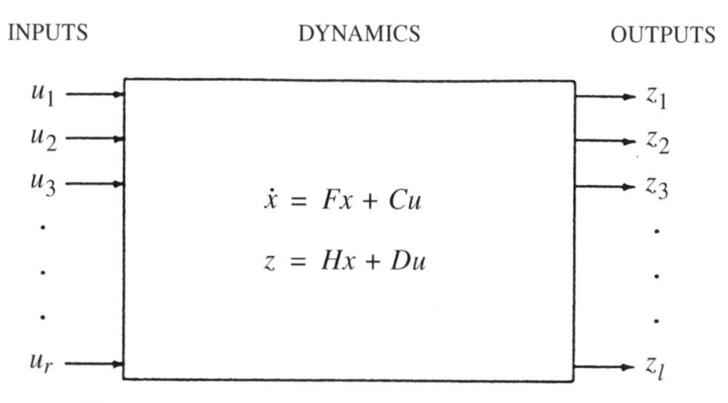

Fig. 2.1 Block diagram of a linear dynamic system.

2.3.2 Dynamic Coefficient Matrices and Input Coupling Matrices

The dynamics of linear systems are represented by a set of n first-order linear differential equations expressible in vector form as

$$\begin{aligned}\dot{x}(t) &= \frac{d}{dt}x(t) \\ &= F(t)x(t) + C(t)u(t),\end{aligned} \tag{2.5}$$

where the elements and components of the matrices and vectors can be functions of time:

$$F(t) = \begin{bmatrix} f_{11}(t) & f_{12}(t) & f_{13}(t) & \cdots & f_{1n}(t) \\ f_{21}(t) & f_{22}(t) & f_{23}(t) & \cdots & f_{2n}(t) \\ f_{31}(t) & f_{32}(t) & f_{33}(t) & \cdots & f_{3n}(t) \\ \vdots & \vdots & \vdots & \ddots & \vdots \\ f_{n1}(t) & f_{n2}(t) & f_{n3}(t) & \cdots & f_{nn}(t) \end{bmatrix},$$

$$C(t) = \begin{bmatrix} c_{11}(t) & c_{12}(t) & c_{13}(t) & \cdots & c_{1r}(t) \\ c_{21}(t) & c_{22}(t) & c_{23}(t) & \cdots & c_{2r}(t) \\ c_{31}(t) & c_{32}(t) & c_{33}(t) & \cdots & c_{3r}(t) \\ \vdots & \vdots & \vdots & \ddots & \vdots \\ c_{n1}(t) & c_{n2}(t) & c_{n3}(t) & \cdots & c_{nr}(t) \end{bmatrix},$$

$$u(t) = [u_1(t) \quad u_2(t) \quad u_3(t) \quad \cdots \quad u_r(t)]^{\mathrm{T}}.$$

The matrix $F(t)$ is called the *dynamic coefficient matrix*, or simply the *dynamic matrix*. Its elements are called the *dynamic coefficients*. The matrix $C(t)$ is called the *input coupling matrix*, and its elements are called *input coupling coefficients*. The r-vector u is called the *input vector*.

EXAMPLE 2.4: Dynamic Equation for a Heating/Cooling System Consider the temperature T in a heated enclosed room or building as the state variable of a dynamic system. A simplified plant model for this dynamic system is the linear equation

$$\dot{T}(t) = -k_c[T(t) - T_o(t)] + k_h u(t),$$

where the constant "cooling coefficient" k_c depends on the quality of thermal insulation from the outside, T_o is the temperature outside, k_h is the heating/cooling rate coefficient of the heater or cooler, and u is an input function that is either $u = 0$ (off) or $u = 1$ (on) and can be defined as a function of any measurable quantities. The outside temperature T_o, on the other hand, is an example of an input function which may be directly measurable at any time but is not predictable in the future. It is effectively a random process.

2.3.3 Companion Form for Higher Order Derivatives

In general, the nth-order linear differential equation

$$\frac{d^n y(t)}{dt^n} + f_1(t)\frac{d^{n-1}y(t)}{dt^{n-1}} + \cdots + f_{n-1}(t)\frac{dy(t)}{dt} + f_n(t)y(t) = u(t) \tag{2.6}$$

can be rewritten as a system of n first-order differential equations. Although the state variable representation as a first-order system is not unique [56], there is a unique way of representing it called the *companion form.*

Companion Form of the State Vector. For the nth-order linear dynamic system shown above, the companion form of the state vector is

$$x(t) = \left[y(t),\quad \frac{d}{dt}y(t),\quad \frac{d^2}{dt^2}y(t),\quad \ldots,\quad \frac{d^{n-1}}{dt^{n-1}}y(t) \right]^{\mathrm{T}}. \tag{2.7}$$

Companion Form of the Differential Equation. The nth-order linear differential equation can be rewritten in terms of the above state vector $x(t)$ as the vector differential equation

$$\frac{d}{dt}\begin{bmatrix} x_1(t) \\ x_2(t) \\ \vdots \\ x_{n-1}(t) \\ x_n(t) \end{bmatrix} = \begin{bmatrix} 0 & 1 & 0 & \cdots & 0 \\ 0 & 0 & 1 & \cdots & 0 \\ \vdots & \vdots & \vdots & \ddots & \vdots \\ 0 & 0 & 0 & \cdots & 1 \\ -f_n(t) & -f_{n-1}(t) & -f_{n-2}(t) & \cdots & -f_1(t) \end{bmatrix} \begin{bmatrix} x_1(t) \\ x_2(t) \\ x_3(t) \\ \vdots \\ x_n(t) \end{bmatrix} + \begin{bmatrix} 0 \\ 0 \\ \vdots \\ 0 \\ 1 \end{bmatrix} u(t). \tag{2.8}$$

When Equation 2.8 is compared with Equation 2.5, the matrices $F(t)$ and $C(t)$ are easily identified.

The Companion Form is Ill-conditioned. Although it simplifies the relationship between higher order linear differential equations and first-order systems of differential equations, the companion matrix is not recommended for implementation. Studies by Kenney and Liepnik [185] have shown that it is poorly conditioned for solving differential equations.

2.3.4 Outputs and Measurement Sensitivity Matrices

Measurable Outputs and Measurement Sensitivities. Only the inputs and outputs of the system can be measured, and it is usual practice to consider the variables z_i as the measured values. For linear problems, they are related to the state variables and the inputs by a system of linear equations that can be represented in vector form as

$$z(t) = H(t)x(t) + D(t)u(t), \tag{2.9}$$

where

$$z(t) = [z_1(t) \quad z_2(t) \quad z_3(t) \quad \cdots \quad z_\ell(t)]^\mathrm{T},$$

$$H(t) = \begin{bmatrix} h_{11}(t) & h_{12}(t) & h_{13}(t) & \cdots & h_{1n}(t) \\ h_{21}(t) & h_{22}(t) & h_{23}(t) & \cdots & h_{2n}(t) \\ h_{31}(t) & h_{32}(t) & h_{33}(t) & \cdots & h_{3n}(t) \\ \vdots & \vdots & \vdots & \ddots & \vdots \\ h_{\ell 1}(t) & h_{\ell 2}(t) & h_{\ell 3}(t) & \cdots & h_{\ell n}(t) \end{bmatrix},$$

$$D(t) = \begin{bmatrix} d_{11}(t) & d_{12}(t) & d_{13}(t) & \cdots & d_{1r}(t) \\ d_{21}(t) & d_{22}(t) & d_{23}(t) & \cdots & d_{2r}(t) \\ d_{31}(t) & d_{32}(t) & d_{33}(t) & \cdots & d_{3r}(t) \\ \vdots & \vdots & \vdots & \ddots & \vdots \\ d_{\ell 1}(t) & d_{\ell 2}(t) & d_{\ell 3}(t) & \cdots & d_{\ell r}(t) \end{bmatrix}.$$

The ℓ-vector $z(t)$ is called the *measurement vector*, or the *output vector* of the system. The coefficient $h_{ij}(t)$ represents the *sensitivity* (measurement sensor scale factor) of the ith measured output to the jth internal state. The matrix $H(t)$ of these values is called the *measurement sensitivity matrix*, and $D(t)$ is called the *input–output coupling matrix*. The *measurement sensitivities* $h_{ij}(t)$ and input/output coupling coefficients $d_{ij}(t)$, $1 \le i \le \ell$, $1 \le j \le r$, are known functions of time. The state equation 2.5 and the output equation 2.9 together form the dynamic equations of the system shown in Figure 2.1.

2.3.5 Difference Equations and State Transition Matrices (STMs)

Difference equations are the discrete-time versions of differential equations. They are usually written in terms of *forward differences* $x(t_{k+1}) - x(t_k)$ of the state variable (the dependent variable), expressed as a function ψ of all independent variables or of the forward value $x(t_{k+1})$ as a function ϕ of all independent variables (including the previous value as an independent variable):

$$x(t_{k+1}) - x(t_k) = \psi(t_k, x(t_k), u(t_k)),$$

or

$$\begin{aligned} x(t_{k+1}) &= \phi(t_k, x(t_k), u(t_k)), \\ \phi(t_k, x(t_k), u(t_k)) &= x(t_k) + \psi(t_k, x(t_k), u(t_k)). \end{aligned} \tag{2.10}$$

The second of these (Equation 2.10) has the same general form of the recursive relation shown in Equation 2.4, which is the one that is usually implemented for discrete-time systems.

For linear dynamic systems, the functional dependence of $x(t_{k+1})$ on $x(t_k)$ and $u(t_k)$ can be represented by matrices:

$$\begin{aligned} x(t_{k+1}) - x(t_k) &= \Psi(t_k)x(t_k) + C(t_k)u(t_k), \\ x_{k+1} &= \Phi_k x_k + C_k u_k, \\ \Phi_k &= I + \Psi(t_k), \end{aligned} \tag{2.11}$$

where the matrices Ψ and Φ replace the functions ψ and ϕ, respectively. The matrix Φ is called the *state transition matrix (STM)*. The matrix C is called the *discrete-time input coupling matrix*, or simply the *input coupling matrix*—if the discrete-time context is already established.

2.3.6 Solving Differential Equations for STMs

A state transition matrix is a solution of what is called the "homogeneous"[3] matrix equation associated with a given linear dynamic system. Let us define first what homogeneous equations are, and then show how their solutions are related to the solutions of a given linear dynamic system.

Homogeneous Systems. The equation $\dot{x}(t) = F(t)x(t)$ is called the *homogeneous part* of the linear differential equation $\dot{x}(t) = F(t)x(t) + C(t)u(t)$. The solution of the homogeneous part can be obtained more easily than that of the full equation, and its solution is used to define the solution to the general (nonhomogeneous) linear equation.

[3]This terminology comes from the notion that every term in the expression so labeled contains the dependent variable. That is, the expression is *homogeneous* with respect to the dependent variable.

Fundamental Solutions of Homogeneous Equations. An $n \times n$ matrix-valued function $\Phi(t)$ is called a *fundamental solution* of the homogeneous equation $\dot{x}(t) = F(t)x(t)$ on the interval $t \in [0, T]$ if $\dot{\Phi}(t) = F(t)\Phi(t)$ and $\Phi(0) = I_n$, the $n \times n$ identity matrix. Note that, for any possible initial vector $x(0)$, the vector $x(t) = \Phi(t)x(0)$ satisfies the equation

$$\begin{aligned}
\dot{x}(t) &= \frac{d}{dt}[\Phi(t)x(0)] && (2.12)\\
&= \left[\frac{d}{dt}\Phi(t)\right]x(0) && (2.13)\\
&= [F(t)\Phi(t)]x(0) && (2.14)\\
&= F(t)[\Phi(t)x(0)] && (2.15)\\
&= F(t)x(t). && (2.16)
\end{aligned}$$

That is, $x(t) = \Phi(t)x(0)$ is the solution of the homogeneous equation $\dot{x} = Fx$ with initial value $x(0)$.

EXAMPLE 2.5 The unit upper triangular Toeplitz matrix

$$\Phi(t) = \begin{bmatrix}
1 & t & \frac{1}{2}t^2 & \frac{1}{1\cdot 2\cdot 3}t^3 & \cdots & \frac{1}{(n-1)!}t^{n-1} \\
0 & 1 & t & \frac{1}{2}t^2 & \cdots & \frac{1}{(n-2)!}t^{n-2} \\
0 & 0 & 1 & t & \cdots & \frac{1}{(n-3)!}t^{n-3} \\
0 & 0 & 0 & 1 & \cdots & \frac{1}{(n-4)!}t^{n-4} \\
\vdots & \vdots & \vdots & \vdots & \ddots & \vdots \\
0 & 0 & 0 & 0 & \cdots & 1
\end{bmatrix}$$

is the fundamental solution of $\dot{x} = Fx$ for the strictly upper triangular Toeplitz dynamic coefficient matrix

$$F = \begin{bmatrix}
0 & 1 & 0 & \cdots & 0 \\
0 & 0 & 1 & \cdots & 0 \\
\vdots & \vdots & \vdots & \ddots & \vdots \\
0 & 0 & 0 & \cdots & 1 \\
0 & 0 & 0 & \cdots & 0
\end{bmatrix},$$

which can be verified by showing that $\Phi(0) = I$ and $\dot{\Phi} = F\Phi$. This dynamic coefficient matrix, in turn, is the companion matrix for the nth-order linear homogeneous differential equation $(d/dt)^n y(t) = 0$.

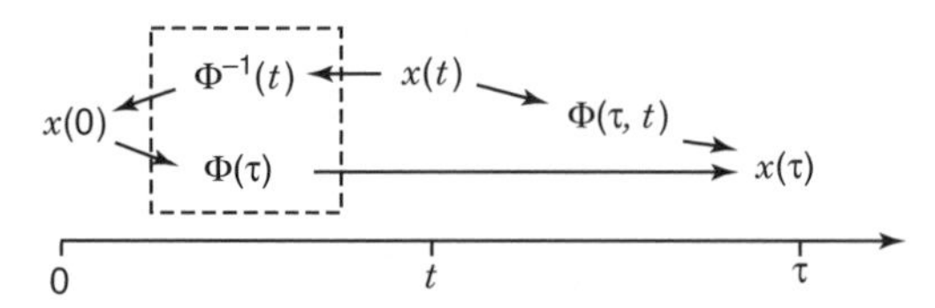

Fig. 2.2 The STM as a composition of fundamental solution matrices.

Existence and Nonsingularity of Fundamental Solutions. If the elements of the matrix $F(t)$ are continuous functions on some interval $0 \leq t \leq T$, then the fundamental solution matrix $\Phi(t)$ is guaranteed to exist and to be nonsingular on an interval $0 \leq t \leqslant \tau$ for some $\tau > 0$. These conditions also guarantee that $\Phi(t)$ will be nonsingular on some interval of nonzero length, as a consequence of the continuous dependence of the solution $\Phi(t)$ of the matrix equation on its (nonsingular) initial conditions $[\Phi(0) = I]$ [57].

State Transition Matrices. Note that the fundamental solution matrix $\Phi(t)$ transforms any initial state $x(0)$ of the dynamic system to the corresponding state $x(t)$ at time t. If $\Phi(t)$ is nonsingular, then the products $\Phi^{-1}(t)x(t) = x(0)$ and $\Phi(\tau)\Phi^{-1}(t)x(t) = x(\tau)$. That is, the matrix product

$$\Phi(\tau, t) = \Phi(\tau)\Phi^{-1}(t) \tag{2.17}$$

transforms a solution from time t to the corresponding solution at time τ, as diagrammed in Figure 2.2. Such a matrix is called the state transition matrix[4] for the associated linear homogeneous differential equation. The state transition matrix $\Phi(\tau, t)$ represents the transition to the state at time τ from the state at time t.

Properties of STMs and Fundamental Solution Matrices. The same symbol (Φ) has been used for *fundamental solution matrices* and for state transition matrices, the distinction being made by the number of arguments. By convention, then,

$$\Phi(\tau, 0) = \Phi(\tau).$$

Other useful properties of Φ include the following:

1. $\Phi(\tau, \tau) = \Phi(0) = I$,
2. $\Phi^{-1}(\tau, t) = \Phi(t, \tau)$,
3. $\Phi(\tau, \sigma)\Phi(\sigma, t) = \Phi(\tau, t)$,
4. $(\partial/\partial\tau)\Phi(\tau, t) = F(\tau)\Phi(\tau, t)$,

[4]Formally, an operator $\Phi(t, t_0, x(t_0))$ such that $x(t) = \Phi(t, t_0, x(t_0))$ is called an *evolution operator* for a dynamic system with state x. A state transition matrix is a linear evolution operator.

and

5. $(\partial/\partial t)\Phi(\tau, t) = -\Phi(\tau, t)F(t)$.

EXAMPLE 2.6: Fundamental Solution Matrix for the Underdamped Harmonic Resonator *The general solution of the differential equation.* In Examples 2.2 and 2.3, the displacement δ of the damped harmonic resonator was modeled by the state equation

$$x = \begin{bmatrix} \delta \\ \dot{\delta} \end{bmatrix},$$

$$\dot{x} = Fx,$$

$$F = \begin{bmatrix} 0 & 1 \\ -\dfrac{k_s}{m} & -\dfrac{k_d}{m} \end{bmatrix}.$$

The characteristic values of the dynamic coefficient matrix F are the roots of its characteristic polynomial

$$\det(\lambda I - F) = \lambda^2 + \frac{k_d}{m}\lambda + \frac{k_s}{m},$$

which is a quadratic polynomial with roots

$$\lambda_1 = \frac{1}{2}\left(-\frac{k_d}{m} + \sqrt{\frac{k_d^2}{m^2} - \frac{4k_s}{m}}\right),$$

$$\lambda_2 = \frac{1}{2}\left(-\frac{k_d}{m} - \sqrt{\frac{k_d^2}{m^2} - \frac{4k_s}{m}}\right).$$

The general solution for the displacement δ can then be written in the form

$$\delta(t) = \alpha e^{\lambda_1 t} + \beta e^{\lambda_2 t},$$

where α and β are (possibly complex) free variables.

The underdamped solution. The resonator is considered *underdamped* if the discriminant

$$\frac{k_d^2}{m^2} - \frac{4k_s}{m} < 0,$$

in which case the roots are a conjugate pair of nonreal complex numbers and the general solution can be rewritten in "real form" as

$$\delta(t) = ae^{-t/\tau}\cos(\omega t) + be^{-t/\tau}\sin(\omega t),$$
$$\tau = \frac{2m}{k_d},$$
$$\omega = \sqrt{\frac{k_s}{m} - \frac{k_d^2}{4m^2}},$$

where a and b are now real variables, τ is the decay time constant, and ω is the resonator resonant frequency. This solution can be expressed in state-space form in terms of the real variables a and b:

$$\begin{bmatrix} \delta(t) \\ \dot{\delta}(t) \end{bmatrix} = e^{-t/\tau} \begin{bmatrix} \cos(\omega t) & \sin(\omega t) \\ -\dfrac{\cos(\omega t)}{\tau} - \omega\sin(\omega t) & \omega\cos(\omega t) - \dfrac{\sin(\omega t)}{\tau} \end{bmatrix} \begin{bmatrix} a \\ b \end{bmatrix}.$$

Initial value constraints. The initial values

$$\delta(0) = a, \qquad \dot{\delta}(0) = -\frac{a}{\tau} + \omega b$$

can be solved for a and b as

$$\begin{bmatrix} a \\ b \end{bmatrix} = \begin{bmatrix} 1 & 0 \\ \dfrac{1}{\omega\tau} & \dfrac{1}{\omega} \end{bmatrix} \begin{bmatrix} \delta(0) \\ \dot{\delta}(0) \end{bmatrix}.$$

This can then be combined with the solution for $x(t)$ in terms of a and b to yield the fundamental solution

$$x(t) = \Phi(t)x(0),$$
$$\Phi(t) = \frac{e^{-t/\tau}}{\omega\tau^2} \begin{bmatrix} \tau[\omega\tau\cos(\omega t) + \sin(\omega t)] & \tau^2\sin(\omega t) \\ -(1+\omega^2\tau^2)\sin(\omega\tau) & [\omega\tau^2\cos(\omega t) - \tau\sin(\omega t)] \end{bmatrix}$$

in terms of the damping time constant and the resonant frequency.

2.3.7 Solution of Nonhomogeneous Equations

The solution of the nonhomogeneous state equation 2.5 is given by

$$x(t) = \Phi(t, t_0)x(t_0) + \int_{t_0}^{t} \Phi(t, \tau)C(\tau)u(\tau)\,d\tau \tag{2.18}$$

$$= \Phi(t)\Phi^{-1}(t_0)x(t_0) + \Phi(t)\int_{t_0}^{t} \Phi^{-1}(\tau)C(\tau)u(\tau)\,d\tau, \tag{2.19}$$

where $x(t_0)$ is the initial value and $\Phi(t, t_0)$ is the state transition matrix of the dynamic system defined by $F(t)$. (This can be verified by taking derivatives and using the properties of STMs given above.)

2.3.8 Closed-Form Solutions of Time-Invariant Systems

In this case, the coefficient matrix F is a constant function of time. The solution will still be a function of time, but the associated state transition matrices $\Phi(t, \tau)$ will only depend on the differences $t - \tau$. In fact, one can show that

$$\Phi(t, \tau) = e^{F(t-\tau)} \tag{2.20}$$

$$= \sum_{i=0}^{\infty} \frac{(t-\tau)^i}{i!} F^i, \tag{2.21}$$

where $F^0 = I$, by definition. The solution of the nonhomogeneous equation in this case will be

$$x(t) = e^{F(t-\tau)}x(\tau) + \int_{\tau}^{t} e^{F(t-\sigma)}Cu(\sigma)\,d\sigma \tag{2.22}$$

$$= e^{F(t-\tau)}x(\tau) + e^{Ft}\int_{\tau}^{t} e^{-F\sigma}Cu(\sigma)\,d\sigma. \tag{2.23}$$

The following methods have been used for computing matrix exponentials:

1. The approximation of e^{Ft} by a truncated power series expansion is *not* a recommended general-purpose method, but it is useful if the characteristic values of Ft are well inside the unit circle in the complex plane.
2. $\Phi(t) = e^{Ft} = \mathscr{L}^{-1}(sI - F)^{-1}, t \geq 0$, where I is an $n \times n$ identity matrix, $\mathscr{L}^{-1}$ is the inverse Laplacian operator, and s is the Laplace transform variable.
3. The "scaling and squaring" method combined with a Padé approximation is the recommended general-purpose method. This method is discussed in greater detail in Section 2.6.

4. Numerical integration of the homogeneous part of the differential equation,

$$\frac{d}{dt}\Phi(t) = F\Phi(t), \tag{2.24}$$

with initial value $\Phi(0) = I$. (This method also works for time-varying systems.)

There are many other methods,[5] but these are the most important.

EXAMPLE 2.7: Solution of the Damped Harmonic Resonator Problem with Constant Driving Function Consider again the damped resonator model of Examples 2.2, 2.3, and 2.6. The model can be written in the form of a second-order differential equation

$$\ddot{\delta}(t) + 2\zeta w_n \dot{\delta}(t) + w_n^2 \delta(t) = u(t),$$

where

$$\dot{\delta}(t) = \frac{d\delta}{dt}, \qquad \ddot{\delta}(t) = \frac{d^2\delta}{dt^2}, \qquad \zeta = \frac{k_d}{2\sqrt{mk_s}}, \qquad \omega_n = \sqrt{\frac{k_s}{m}}.$$

The parameter ζ is a unitless damping coefficient and w_n the "natural" (i.e., undamped) frequency of the resonator.

This second-order linear differential equation can be rewritten in a state-space form, with states $x_1 = \delta$ and $x_2 = \dot{\delta} = \dot{x}_1$ and parameters ζ and ω_n, as

$$\frac{d}{dt}\begin{bmatrix} x_1(t) \\ x_2(t) \end{bmatrix} = \begin{bmatrix} 0 & 1 \\ -w_n^2 & -2\zeta w_n \end{bmatrix} \begin{bmatrix} x_1(t) \\ x_2(t) \end{bmatrix} + \begin{bmatrix} 0 \\ 1 \end{bmatrix} u(t)$$

with initial conditions

$$\begin{bmatrix} x_1(t_0) \\ x_2(t_0) \end{bmatrix}.$$

As a numerical example, let

$$u(t) \equiv 1, \qquad w_n = 1, \qquad \zeta = 0.5,$$

so that the coefficient matrix

$$F = \begin{bmatrix} 0 & 1 \\ -1 & -1 \end{bmatrix}.$$

[5] See, for example, Brockett [56], DeRusso et al. [59], or Kreindler and Sarachik [189].

Therefore,

$$(sI - F) = \begin{bmatrix} s & -1 \\ 1 & s+1 \end{bmatrix},$$

$$(sI - F)^{-1} = \frac{1}{s^2+s+1}\begin{bmatrix} s+1 & 1 \\ -1 & s \end{bmatrix}$$

$$\begin{aligned}
\Phi(t) &= e^{Ft} \\
&= \mathscr{L}^{-1}(sI - F)^{-1} \\
&= \mathscr{L}^{-1}\begin{bmatrix} \dfrac{s+1}{s^2+s+1} & \dfrac{1}{s^2+s+1} \\ \dfrac{-1}{s^2+s+1} & \dfrac{s}{s^2+s+1} \end{bmatrix}
\end{aligned}$$

$$= \frac{2e^{-t/2}}{\sqrt{3}}\begin{bmatrix} \frac{1}{2}\sqrt{3}\cos\left(\frac{1}{2}\sqrt{3}t\right) + \frac{1}{2}\sin\left(\frac{1}{2}\sqrt{3}t\right) & \sin\left(\frac{1}{2}\sqrt{3}t\right) \\ -\sin\left(\frac{1}{2}\sqrt{3}t\right) & \frac{1}{2}\sqrt{3}\cos\left(\frac{1}{2}\sqrt{3}t\right) - \frac{1}{2}\sin\left(\frac{1}{2}\sqrt{3}t\right) \end{bmatrix}.$$

2.3.9 Time-Varying Systems

If $F(t)$ is not constant, the dynamic system is called time-varying. If $F(t)$ is a piecewise smooth function of t, the $n \times n$ homogeneous matrix differential equation 2.24 can be solved numerically by the fourth-order Runge–Kutta method.[6]

2.4 DISCRETE LINEAR SYSTEMS AND THEIR SOLUTIONS

2.4.1 Discretized Linear Systems

If one is only interested in the system state at discrete times, then one can use the formula

$$x(t_k) = \Phi(t_k, t_{k-1})x(t_{k-1}) + \int_{t_{k-1}}^{t_k} \Phi(t_k, \sigma)C(\sigma)u(\sigma)\, d\sigma \tag{2.25}$$

to propagate the state vector between the times of interest.

[6]Named after the German mathematicians Karl David Tolme Runge (1856–1927) and Wilhelm Martin Kutta (1867–1944).

Simplification for Constant u. If u is constant over the interval $[t_{k-1}, t_k]$, then the above integral can be simplified to the form

$$x(t_k) = \Phi(t_k, t_{k-1})x(t_{k-1}) + \Gamma(t_{k-1})u(t_{k-1}) \tag{2.26}$$

$$\Gamma(t_{k-1}) = \int_{t_{k-1}}^{t_k} \Phi(t_k, \sigma)C(\sigma)\, d\sigma. \tag{2.27}$$

Shorthand Discrete-Time Notation. For discrete-time systems, the indices k in the time sequence $\{t_k\}$ characterize the times of interest. One can save some ink by using the shorthand notation:

$$x_k \stackrel{\text{def}}{=} x(t_k), \qquad z_k \stackrel{\text{def}}{=} z(t_k), \qquad u_k \stackrel{\text{def}}{=} u(t_k), \qquad H_k \stackrel{\text{def}}{=} H(t_k),$$

$$D_k \stackrel{\text{def}}{=} D(t_k), \qquad \Phi_{k-1} \stackrel{\text{def}}{=} \Phi(t_k, t_{k-1}), \qquad \Gamma_k \stackrel{\text{def}}{=} \Gamma(t_k)$$

for discrete-time systems, eliminating t entirely. Using this notation, one can represent the discrete-time state equations in the more compact form

$$x_k = \Phi_{k-1}x_{k-1} + \Gamma_{k-1}u_{k-1}, \tag{2.28}$$

$$z_k = H_k x_k + D_k u_k \tag{2.29}$$

2.4.2 Time-Invariant Systems

For continuous time-invariant systems that have been discretized using fixed time intervals, the matrices Φ, Γ, H, and D are independent of the discrete-time index as well. In that case, the solution can be written in closed form as

$$x_k = \Phi^k x_0 + \sum_{i=0}^{k-1} \Phi^{k-i-1}\Gamma u_i, \tag{2.30}$$

where Φ^k is the kth power of Φ. The matrix Φ^k can also be computed as

$$\Phi^k = \mathscr{Z}^{-1}[(\mathbf{z}I - \Phi)^{-1}\mathbf{z}], \tag{2.31}$$

where $\mathbf{z}$ is the z-transform variable and $\mathscr{Z}^{-1}$ is the inverse z-transform.

2.5 OBSERVABILITY OF LINEAR DYNAMIC SYSTEM MODELS

Observability is the issue of whether the state of a dynamic system is uniquely determinable from its inputs and outputs, given a model for the dynamic system. It is essentially a property of the given *system model.* A given linear dynamic system model with a given linear input/output model is considered *observable* if and only if its state is *uniquely* determinable from the model definition, its inputs, and its outputs. If the system state is *not* uniquely determinable from the system inputs and outputs, then the system model is considered *unobservable*.

How to Determine Whether a Given Dynamic System Model Is Observable. If the measurement sensitivity matrix is invertible at any (continuous or discrete) time, then the system state can be uniquely determined (by inverting it) as $x = H^{-1}z$. In this case, the system model is considered to be *completely observable* at that time. However, the system can still be *observable over a time interval* even if H is not invertible at *any* time. In the latter case, the unique solution for the system state can be defined by using the least-squares methods of Chapter 1, including those of Sections 1.2.2 and 1.2.3. These use the so-called *Gramian matrix* to characterize whether or not a vector variable is determinable from a given linear model. When applied to the problem of the determinacy of the state of a linear dynamic system, the Gramian matrix is called the *observability matrix* of the given system model.

The observability matrix for dynamic system models in continuous time has the form

$$\mathcal{O}(H, F, t_0, t_f) = \int_{t_0}^{t_f} \Phi^{\mathrm{T}}(t) H^{\mathrm{T}}(t) H(t) \Phi(t)\, dt \tag{2.32}$$

for a linear dynamic system with fundamental solution matrix $\Phi(t)$ and measurement sensitivity matrix $H(t)$, defined over the continuous-time interval $t_0 \le t \le t_f$. Note that this depends on the interval over which the inputs and outputs are observed but not on the inputs and outputs per se. In fact, the observability matrix of a dynamic system model *does not* depend on the inputs u, the input coupling matrix C, or the input–output coupling matrix D—even though the outputs and the state vector depend on them. Because the fundamental solution matrix Φ depends only on the dynamic coefficient matrix F, the observability matrix depends *only* on H and F.

The observability matrix of a linear dynamic system model over a discrete-time interval $t_0 \le t \le t_{k_f}$ has the general form

$$\mathcal{O}(H_k, \Phi_k, 1 \le k \le k_f) = \left\{ \sum_{k=1}^{k_f} \left[\prod_{i=0}^{k-1} \Phi_{k-i} \right]^{\mathrm{T}} H_k^{\mathrm{T}} H_k \left[\prod_{i=0}^{k-1} \Phi_{k-i} \right] \right\}, \tag{2.33}$$

where H_k is the observability matrix at time t_k and Φ_k is the state transition matrix from time t_k to time t_{k+1} for $0 \le k \le k_f$. Therefore, the observability of discrete-time system models depends only on the values of H_k and Φ_k over this interval. As in the continuous-time case, observability does not depend on the system inputs.

The derivations of these formulas are left as exercises for the reader.

2.5.1 Observability of Time-Invariant Systems

The formulas defining observability are simpler when the dynamic coefficient matrices or state transition matrices of the dynamic system model are time invariant. In that case, observability can be characterized by the rank of the matrices

$$M = [H^{\mathrm{T}} \quad \Phi^{\mathrm{T}}H^{\mathrm{T}} \quad (\Phi^{\mathrm{T}})^2H^{\mathrm{T}} \quad \cdots \quad (\Phi^{\mathrm{T}})^{n-1}H^{\mathrm{T}}] \tag{2.34}$$

for discrete-time systems and

$$M = [H^{\mathrm{T}} \quad F^{\mathrm{T}}H^{\mathrm{T}} \quad (F^{\mathrm{T}})^2H^{\mathrm{T}} \quad \cdots \quad (F^{\mathrm{T}})^{n-1}H^{\mathrm{T}}] \tag{2.35}$$

for continuous-time systems. The systems are observable if these have rank n, the dimension of the system state vector. The first of these matrices can be obtained by representing the *initial state* of the linear dynamic system as a function of the system inputs and outputs. The initial state can then be shown to be uniquely determinable if and only if the rank condition is met. The derivation of the latter matrix is not as straightforward. Ogata [38] presents a derivation obtained by using properties of the characteristic polynomial of F.

Practicality of the Formal Definition of Observability. Singularity of the observability matrix is a concise mathematical characterization of observability. This can be too fine a distinction for practical application—especially in finite-precision arithmetic—because arbitrarily small changes in the elements of a singular matrix can render it nonsingular. The following practical considerations should be kept in mind when applying the formal definition of observability:

- It is important to remember that the model is only an approximation to a real system, and we are primarily interested in the properties of the real system, not the model. Differences between the real system and the model are called *model truncation errors.* The art of system modeling depends on knowing where to truncate, but there will almost surely be some truncation error in any model.
- Computation of the observability matrix is subject to model truncation errors and *roundoff errors,* which could make the difference between singularity and nonsingularity of the result. Even if the computed observability matrix is *close* to being singular, it is cause for concern. One should consider a system as *poorly observable* if its observability matrix is close to being singular. For that purpose, one can use the *singular-value decomposition* or the *condition number* of the observability matrix to define a more *quantitative* measure of unobservability. The reciprocal of its condition number measures how close the system is to being unobservable.
- Real systems tend to have some amount of unpredictability in their behavior, due to unknown or neglected exogenous inputs. Although such effects cannot be modeled deterministically, they are not always negligible. Furthermore, the process of measuring the outputs with physical sensors introduces some

amount of sensor noise, which will cause errors in the estimated state. It would be better to have a quantitative characterization of observability that takes these types of uncertainties into account. An approach to these issues (pursued in Chapter 4) uses a *statistical* characterization of observability, based on a statistical model of the uncertainties in the measured system outputs and the system dynamics. The degree of uncertainty in the estimated values of the system states can be characterized by an *information matrix*, which is a statistical generalization of the observability matrix.

EXAMPLE 2.8 Consider the following continuous system:

$$\dot{x}(t) = \begin{bmatrix} 0 & 1 \\ 0 & 0 \end{bmatrix} x(t) + \begin{bmatrix} 0 \\ 1 \end{bmatrix} u(t),$$
$$z(t) = [1 \quad 0] x(t).$$

The observability matrix, using Equation 2.35, is

$$M = \begin{bmatrix} 1 & 0 \\ 0 & 1 \end{bmatrix}, \qquad \text{rank of } M = 2.$$

Here, M has rank equal to the dimension of $x(t)$. Therefore, the system is observable.

EXAMPLE 2.9 Consider the following continuous system:

$$\dot{x}(t) = \begin{bmatrix} 0 & 1 \\ 0 & 0 \end{bmatrix} x(t) + \begin{bmatrix} 0 \\ 1 \end{bmatrix} u(t),$$
$$z(t) = [0 \quad 1] x(t).$$

The observability matrix, using Equation 2.35, is

$$M = \begin{bmatrix} 0 & 0 \\ 1 & 1 \end{bmatrix}, \qquad \text{rank of } M = 1.$$

Here, M has rank less than the dimension of $x(t)$. Therefore, the system is not observable.

EXAMPLE 2.10 Consider the following discrete system:

$$x_k = \begin{bmatrix} 0 & 0 & 0 \\ 0 & 0 & 0 \\ 1 & 1 & 0 \end{bmatrix} x_{k-1} + \begin{bmatrix} 1 \\ 1 \\ 0 \end{bmatrix} u_{k-1},$$

$$z_k = [0 \quad 0 \quad 1] x_k.$$

The observability matrix, using Equation 2.34, is

$$M = \begin{bmatrix} 0 & 1 & 0 \\ 0 & 1 & 0 \\ 1 & 0 & 0 \end{bmatrix}, \qquad \text{rank of } M = 2.$$

The rank is less than the dimension of x_k. Therefore, the system is not observable.

EXAMPLE 2.11 Consider the following discrete system:

$$x_k = \begin{bmatrix} 1 & -1 \\ 1 & 1 \end{bmatrix} x_{k-1} + \begin{bmatrix} 2 \\ 1 \end{bmatrix} u_{k-1},$$

$$z_k = \begin{bmatrix} 1 & 0 \\ -1 & 1 \end{bmatrix} x_k.$$

The observability matrix, using Equation 2.34, is

$$M = \begin{bmatrix} 1 & -1 \\ 0 & 1 \end{bmatrix}, \qquad \text{rank of } M = 2$$

The system is observable.

2.5.2 Controllability of Time-Invariant Linear Systems

Controllability in Continuous Time. The concept of observability in estimation theory has algebraic relationships to the concept of *controllability* in control theory. These concepts and their relationships were discovered by R. E. Kalman as what he called the *duality* and *separability* of the estimation and control problems for linear dynamic systems. Kalman's[7] dual concepts are presented here and in the next subsection, although they are not issues for the estimation problem.

[7]The dual relationships between estimation and control given here are those originally defined by Kalman. These concepts have been refined and extended by later investigators to include concepts of *reachability* and *reconstructibility* as well. The interested reader is referred to the more recent textbooks on "modern" control theory for further exposition of these other "-ilities."

A dynamic system defined on the finite interval $t_0 \leq t \leq t_f$ by the linear model

$$\dot{x}(t) = Fx(t) + Cu(t), \qquad z(t) = Hx(t) + Du(t) \tag{2.36}$$

and with initial state vector $x(t_0)$ is said to be *controllable* at time $t = t_0$ if, for any desired final state $x(t_f)$, there exists a piecewise continuous input function $u(t)$ that drives to state $x(t_f)$. If every initial state of the system is controllable in some finite time interval, then the *system* is said to be controllable.

The system given in Equation 2.36 is controllable if and only if matrix S has n linearly independent columns,

$$S = [C \quad FC \quad F^2C \quad \cdots \quad F^{n-1}C]. \tag{2.37}$$

Controllability in Discrete Time. Consider the time-invariant system model given by the equations

$$x_k = \Phi x_{k-1} + \Gamma u_{k-1}, \tag{2.38}$$

$$z_k = Hx_k + Du_k. \tag{2.39}$$

This system model is considered controllable[8] if there exists a set of control signals u_k defined over the discrete interval $0 \leq k \leq N$ that bring the system from an initial state x_0 to a given final state x_N in N sampling instants, where N is a finite positive integer. This condition can be shown to be equivalent to the matrix

$$S = [\Gamma \quad \Phi\Gamma \quad \Phi^2\Gamma \quad \cdots \quad \Phi^{N-1}\Gamma] \tag{2.40}$$

having rank n.

EXAMPLE 2.12 Determine the controllability of Example 2.8. The controllability matrix, using Equation 2.37, is

$$S = \begin{bmatrix} 0 & 1 \\ 1 & 0 \end{bmatrix}, \qquad \text{rank of } S = 2.$$

Here, S has rank equal to the dimension of $x(t)$. Therefore, the system is controllable.

EXAMPLE 2.13 Determine the controllability of Example 2.10. The controllability matrix, using Equation 2.40, is

$$S = \begin{bmatrix} 1 & 0 & 0 \\ 1 & 0 & 0 \\ 0 & 2 & 0 \end{bmatrix}, \qquad \text{rank of } S = 2.$$

The system is not controllable.

[8]This condition is also called reachability, with controllability restricted to $x_N = 0$.

2.6 PROCEDURES FOR COMPUTING MATRIX EXPONENTIALS

In a 1978 journal article titled "Nineteen dubious ways to compute the exponential of a matrix" [205], Moler and Van Loan reported their evaluations of methods for computing matrix exponentials. Many of the methods tested had serious shortcomings, and no method was considered universally superior. The one presented here was recommended as being more reliable than most. It combines several ideas due to Ward [233], including setting the algorithm parameters to meet a prespecified error bound. It combines Padé approximation with a technique called "scaling and squaring" to maintain approximation errors within prespecified bounds.

2.6.1 Padé Approximation of the Matrix Exponential

Padé approximations. These approximations of functions by rational functions (ratios of polynomials) date from a 1892 publication [206] by H. Padé.[9] They have been used in deriving solutions of differential equations, including Riccati equations[10] [69]. They can also be applied to functions of matrices, including the matrix exponential. In the matrix case, the power series is approximated as a "matrix fraction" of the form $\mathscr{D}^{-1}\mathscr{N}$, with the numerator matrix ($\mathscr{N}$) and denominator matrix ($\mathscr{D}$) represented as polynomials with matrix arguments. The "order" of the Padé approximation is two dimensional. It depends on the orders of the polynomials in the numerator and denominator of the rational function. The Taylor series is the special case in which the order of the denominator polynomial of the Padé approximation is zero. Like the Taylor series approximation, the Padé approximation tends to work best for small values of its argument. For matrix arguments, it will be some matrix norm of the argument that will be required to be small.

Padé approximation of exponential function. The exponential function with argument z has the power series expansion

$$e^z = \sum_{k=0}^{\infty} \frac{1}{k!} z^k.$$

The polynomials $\mathscr{N}_p(z)$ and $\mathscr{D}_q(z)$ such that

$$\mathscr{N}_p(z) = \sum_{k=0}^{p} a_k z^k,$$

$$\mathscr{D}_q(z) = \sum_{k=0}^{q} b_k z^k,$$

$$e^z \mathscr{D}_q(z) - \mathscr{N}_p(z) = \sum_{k=p+q+1}^{\infty} c_k z^k$$

[9]Pronounced *pah-DAY.*.

[10]The order of the numerator and denominator of the matrix fraction are reversed here from the order used in linearizing the Riccati equation in Chapter 4.

are the numerator and denominator polynomials, respectively, of the Padé approximation of e^z. The key feature of the last equation is that there are no terms of order $\leq p+q$ on the right-hand side. This constraint is sufficient to determine the coefficients a_k and b_k of the polynomial approximants, except for a common constant factor. The solution (within a common constant factor) will be [69]

$$a_k = \frac{p!(p+q-k)!}{k!(p-k)!}, \qquad b_k = \frac{(-1)^k q!(p+q-k)!}{k!(q-k)!}.$$

Application to Matrix Exponential. The above formulas may be applied to polynomials with scalar coefficients and square matrix arguments. For any $n \times n$ matrix X,

$$\begin{aligned} f_{pq}(X) &= \left(q! \sum_{i=0}^{q} \frac{(p+q-i)!}{i!(q-i)!} (-X)^i \right)^{-1} \left(p! \sum_{i=0}^{p} \frac{(p+q-i)!}{i!(p-i)!} X^i \right) \\ &\approx e^X \end{aligned}$$

is the Padé approximation of e^X of order (p, q).

Bounding Relative Approximation Error. The bound given here is from Moler and Van Loan [205]. It uses the ∞-norm of a matrix, which can be computed[11] as

$$\|X\|_\infty = \max_{1 \leq i \leq n} \left(\sum_{j=1}^{n} |x_{ij}| \right)$$

for any $n \times n$ matrix X with elements x_{ij}. The *relative* approximation error is defined as the ratio of the matrix ∞-norm of the approximation error to the matrix ∞-norm of the right answer. The relative Padé approximation error is derived as an analytical function of X in Moler and Van Loan [205]. It is shown in Golub and Van Loan [89] that it satisfies the inequality bound

$$\begin{aligned} \frac{\| f_{pq}(X) - e^X \|_\infty}{\|e^X\|_\infty} &\leq \varepsilon(p, q, X) e^{\varepsilon(p,q,X)}, \\ \varepsilon(p, q, X) &= \frac{p! q! 2^{3-p-q}}{(p+q)!(p+q+1)!} \|X\|_\infty. \end{aligned}$$

Note that this bound depends only on the *sum* $p+q$. In that case, the computational complexity of the Padé approximation for a given error tolerance is minimized when $p = q$, that is, if the numerator and denominator polynomials have the same order.

[11]This formula is not the definition of the ∞-norm of a matrix, which is defined in Appendix B. However, it is a consequence of the definition, and it can be used for computing it.

Bounding the Argument. The problem with the Padé approximation is that the error bound grows exponentially with the norm $\|X\|_\infty$. Ward [233] combined scaling (to reduce $\|X\|_\infty$ and the Padé approximation error) with squaring (to rescale the answer) to obtain an approximation with a predetermined error bound. In essence, one chooses the polynomial order to achieve the given bound.

2.6.2 Scaling and Squaring

Note that, for any nonnegative integer N,

$$\begin{aligned} e^X &= (e^{2^{-N}X})^{2^N} \\ &= \underbrace{\{[(\cdots e^{2^{-N}X}\cdots)^2]^2\}^2}_{N \text{ squarings}}. \end{aligned}$$

Consequently, X can be "downscaled" by 2^{-N} to obtain a good Padé approximation of $e^{2^{-N}X}$, then "upscaled" again (by N squarings) to obtain a good approximation to e^X.

2.6.3 MATLAB Implementations

The built-in MATLAB function expm(M) is essentially the one recommended by Moler and Van Loan [205], as implemented by Golub and Van Loan [89, Algorithm 11.3.1, page 558]. It combines scaling and squaring with a Padé approximation for the exponential of the scaled matrix, and it is designed to achieve a specified tolerance of the approximation error. The MATLAB m-file expm1.m (Section A.4) is a script implementation of expm.

MATLAB also includes the functions expm2 (Taylor series approximation) and expm3 (alternative implementation using eigenvalue–eigenvector decompositions), which can be used to test the relative accuracies and speeds relative to expm of these alternative implementations of the matrix exponential function.

2.7 SUMMARY

Systems and Processes. A *system* is a collection of interrelated objects treated as a whole for the purpose of modeling its behavior. It is called *dynamic* if attributes of interest are changing with time. A *process* is the evolution over time of a system.

Continuous and Discrete Time. Although it is sometimes convenient to model time as a continuum, it is often more practical to consider it as taking on discrete values. (Most clocks, for example, advance in discrete time steps.)

State Variables and Vectors. The *state* of a dynamic system at a given instant of time is characterized by the instantaneous values of its attributes of interest. For the problems of interest in this book, the attributes of interest can be characterized by real numbers, such as the electric potentials, temperatures, or positions of its component parts—in appropriate units. A *state variable* of a system is the associated real number. The *state vector* of a system has state variables as its component elements. The system is considered *closed* if the future state of the system for all time is uniquely determined by its current state. For example, neglecting the gravity fields from other massive bodies in the universe, the solar system could be considered as a closed system. If a dynamic system is not closed, then the exogenous causes are called "inputs" to the system. This state vector of a system must be *complete* in the sense that the future state of the system is uniquely determined by its current state *and its future inputs*.[12] In order to obtain a complete state vector for a system, one can extend the state variable components to include derivatives of other state variables. This allows one to use velocity (the derivative of position) or acceleration (the derivative of velocity) as state variables, for example.

State-Space Models for Dynamic Systems. In order that the future state of a system may be determinable from its current state and future inputs, the dynamical behavior of each state variable of the system must be a known function of the instantaneous values of other state variables and the system inputs. In the canonical example of our solar system, for instance, the acceleration of each body is a known function of the relative positions of the other bodies. The *state-space model* for a dynamic system represents these functional dependencies in terms of first-order *differential equations* (in continuous time) or *difference equations* (in discrete time). The differential or difference equations representing the behavior of a dynamic system are called its *state equations*. If these can be represented by *linear* functions, then it is called a *linear dynamic system*.

Linear Dynamic System Models. The model for a linear dynamic system in continuous time can be expressed in general form as a first-order vector differential equation

$$\frac{d}{dt}x(t) = F(t)x(t) + C(t)u(t),$$

where $x(t)$ is the n-dimensional *system state vector* at time t, $F(t)$ is its $n \times n$ *dynamic coefficient matrix*, $u(t)$ is the r-dimensional *system input vector*, and $C(t)$ is the $n \times r$ *input coupling matrix*. The corresponding model for a linear dynamic system in discrete time can be expressed in the general form

$$x_k = \Phi_{k-1}x_{k-1} + \Gamma_{k-1}u_{k-1},$$

[12]This concept in the state-space approach will be generalized in the next chapter to the "state of knowledge" about a system, characterized by the *probability distribution* of its state variables. That is, the *future* probability distribution of the system state variables will be uniquely determined by their *present* probability distribution and the probability distributions of *future inputs*.

where x_{k-1} is the n-dimensional system state vector at time t_{k-1}, x_k is its value a time $t_k > t_{k-1}$, Φ_{k-1} is the $n \times n$ *state transition matrix* for the system at time t_k, u_k is the input vector to the system a time t_k, and Γ_k is the corresponding input coupling matrix.

Time-Varying and Time-Invariant Dynamic Systems. If F and C (or Φ and C) do not depend upon t (or k), then the continuous (or discrete) model is called *time invariant.* Otherwise, the model is *time-varying.*

Homogeneous Systems and Fundamental Solution Matrices. The equation

$$\frac{d}{dt}x(t) = F(t)x(t)$$

is called the *homogeneous part* of the model equation

$$\frac{d}{dt}x(t) = F(t)x(t) + C(t)u(t).$$

A solution $\Phi(t)$ to the corresponding $n \times n$ *matrix equation*

$$\frac{d}{dt}\Phi(t) = F(t)\Phi(t)$$

on an interval starting at time $t = t_0$ and with initial condition

$$\Phi(t_0) = I \qquad \text{(the identity matrix)}$$

is called a *fundamental solution matrix* to the homogeneous equation on that interval. It has the property that, if the elements of $F(t)$ are bounded, then $\Phi(t)$ cannot become singular on a finite interval. Furthermore, for any initial value $x(t_0)$,

$$x(t) = \Phi(t)x(t_0)$$

is the solution to the corresponding homogeneous equation.

Fundamental Solution Matrices and State Transition Matrices. For a *homogenous* system, the state transition matrix Φ_{k-1} from time t_{k-1} to time t_k can be expressed in terms of the fundamental solution $\Phi(t)$ as

$$\Phi_{k-1} = \Phi(t_k)\Phi^{-1}(t_{k-1})$$

for times $t_k > t_{k-1} > t_0$.

Transforming Continuous-Time Models to Discrete Time. The model for a dynamic system in continuous time can be transformed into a model in discrete time using the above formula for the state transition matrix and the following formula for the equivalent discrete-time inputs:

$$u_{k-1} = \phi(t_k) \int_{t_{k-1}}^{t_k} \Phi^{-1}(\tau) C(\tau) u(\tau)\, d\tau.$$

Linear System Output Models and Observability. An *output* of a dynamic system is something we can measure directly, such as directions of the lines of sight to the planets (viewing conditions permitting) or the temperature at thermocouple. A dynamic system model is said to be *observable* from a given set of outputs if it is feasible to determine the state of the system from those outputs. If the dependence of an output z on the system state x is linear, it can be expressed in the form

$$z = Hx,$$

where H is called the *measurement sensitivity matrix*. It can be a function of continuous time $[H(t)]$ or discrete time (H_k). Observability can be characterized by the rank of an *observability matrix* associated with a given system model. The observability matrix is defined as

$$\mathcal{O} = \begin{cases} \displaystyle\int_{t_0}^{t} \Phi^{\mathrm{T}}(\tau) H^{\mathrm{T}}(\tau) H(\tau) \Phi(\tau)\, d\tau & \text{for continuous-time models,} \\ \displaystyle\sum_{i=0}^{m} \left[\left(\prod_{k=0}^{i-1} \Phi_k^{\mathrm{T}} \right) H_i^{\mathrm{T}} H_i \left(\prod_{k=0}^{i-1} \Phi_k^{\mathrm{T}} \right)^{\mathrm{T}} \right] & \text{for discrete-time models.} \end{cases}$$

The system is observable if and only if its observability matrix has full rank (n) for some integer $m \geq 0$ or time $t > t_0$. (The test for observability can be simplified for time-invariant systems.) Note that the determination of observability depends on the (continuous or discrete) interval over which the observability matrix is determined.

Reliable Numerical Approximation of Matrix Exponential. The closed-form solution of a system of first-order differential equations with constant coefficients can be expressed symbolically in terms of the exponential function of a matrix, but the problem of numerical approximation of the exponential function of a matrix is notoriously ill-conditioned.

PROBLEMS

2.1 What is a state vector model for the linear dynamic system $\dfrac{dy(t)}{dt} = u(t)$, expressed in terms of y? (Assume the companion form of the dynamic coefficient matrix.)

2.2 What is the companion matrix for the nth-order differential equation $(d/dt)^n y(t) = 0$? What are its dimensions?

2.3 What is the companion matrix of the above problem when $n = 1$? For $n = 2$?

2.4 What is the fundamental solution matrix of Exercise 2.2 when $n = 1$? When $n = 2$?

2.5 What is the state transition matrix of the above problem when $n = 1$? For $n = 2$?

2.6 Find the fundamental solution matrix $\Phi(t)$ for the system

$$\frac{d}{dt}\begin{bmatrix} x_1(t) \\ x_2(t) \end{bmatrix} = \begin{bmatrix} 0 & 0 \\ -1 & -2 \end{bmatrix}\begin{bmatrix} x_1(t) \\ x_2(t) \end{bmatrix} + \begin{bmatrix} 1 \\ 1 \end{bmatrix}$$

and also the solution $x(t)$ for the initial conditions

$$x_1(0) = 1 \quad \text{and} \quad x_2(0) = 2.$$

2.7 Find the total solution and state transition matrix for the system

$$\frac{d}{dt}\begin{bmatrix} x_1(t) \\ x_2(t) \end{bmatrix} = \begin{bmatrix} -1 & 0 \\ 0 & -1 \end{bmatrix}\begin{bmatrix} x_1(t) \\ x_2(t) \end{bmatrix} + \begin{bmatrix} 5 \\ 1 \end{bmatrix}$$

with initial conditions $x_1(0) = 1$ and $x_2(0) = 2$.

2.8 *The reverse problem: from a discrete-time model to a continuous-time model.* For the discrete-time dynamic system model

$$x_k = \begin{bmatrix} 0 & 1 \\ -1 & 2 \end{bmatrix} x_{k-1} + \begin{bmatrix} 0 \\ 1 \end{bmatrix},$$

find the state transition matrix for continuous time and the solution for the continuous-time system with initial conditions

$$x(0) = \begin{bmatrix} 1 \\ 2 \end{bmatrix}.$$

2.9 Find conditions on c_1, c_2, h_1, h_2 such that the following system is completely observable and controllable:

$$\frac{d}{dt}\begin{bmatrix} x_1(t) \\ x_2(t) \end{bmatrix} = \begin{bmatrix} 1 & 1 \\ 0 & 1 \end{bmatrix}\begin{bmatrix} x_1(t) \\ x_2(t) \end{bmatrix} + \begin{bmatrix} c_1 \\ c_2 \end{bmatrix} u(t),$$

$$z(t) = [h_1 \quad h_2]\begin{bmatrix} x_1(t) \\ x_2(t) \end{bmatrix}.$$

2.10 Determine the controllability and observability of the dynamic system model given below:

$$\frac{d}{dt}\begin{bmatrix} x_1(t) \\ x_2(t) \end{bmatrix} = \begin{bmatrix} 1 & 0 \\ 1 & 0 \end{bmatrix}\begin{bmatrix} x_1(t) \\ x_2(t) \end{bmatrix} + \begin{bmatrix} 1 & 0 \\ 0 & -1 \end{bmatrix}\begin{bmatrix} u_1 \\ u_2 \end{bmatrix},$$

$$z(t) = [0 \quad 1]\begin{bmatrix} x_1(t) \\ x_2(t) \end{bmatrix}.$$

2.11 Derive the state transition matrix of the time-varying system

$$\dot{x}(t) = \begin{bmatrix} t & 0 \\ 0 & t \end{bmatrix} x(t).$$

2.12 Find the state transition matrix for

$$F = \begin{bmatrix} 0 & 1 \\ 1 & 0 \end{bmatrix}.$$

2.13 For the system of three first-order differential equations

$$\dot{x}_1 = x_2, \qquad \dot{x}_2 = x_3, \qquad \dot{x}_3 = 0$$

(a) What is the companion matrix F?
(b) What is the fundamental solution matrix $\Phi(t)$ such that $(d/dt)\Phi(t) = F\Phi(t)$ and $\Phi(0) = I$?

2.14 Show that the matrix exponential of an antisymmetric matrix is an orthogonal matrix.

2.15 Derive the formula of Equation 2.32 for the observability matrix of a linear dynamic system model in continuous time. (*Hint*: Use the approach of Example 1.2 for estimating the initial state of a system and Equation 2.19 for the state of a system as a linear function of its initial state and its inputs.)

2.16 Derive the formula of Equation 2.33 for the observability matrix of a dynamic system in discrete time. (*Hint*: Use the method of least squares of Example 1.1 for estimating the initial state of a system, and compare the resulting Gramian matrix to the observability matrix of Equation 2.33.)

3

Random Processes and Stochastic Systems

A completely satisfactory definition of random sequence is yet to be discovered.
G. James and R. C. James, Mathematics Dictionary,
D. Van Nostrand Co., Princeton, New Jersey, 1959

3.1 CHAPTER FOCUS

The previous chapter presents methods for representing a class of dynamic systems with relatively small numbers of components, such as a harmonic resonator with one mass and spring. The results are models for *deterministic mechanics*, in which the state of every component of the system is represented and propagated explicitly.

Another approach has been developed for extremely large dynamic systems, such as the ensemble of gas molecules in a reaction chamber. The state-space approach for such large systems would be impractical. Consequently, this other approach focuses on the ensemble *statistical* properties of the system and treats the underlying dynamics as a *random process*. The results are models for *statistical mechanics*, in which only the ensemble statistical properties of the system are represented and propagated explicitly.

In this chapter, some of the basic notions and mathematical models of statistical and deterministic mechanics are combined into a *stochastic system model*, which represents the *state of knowledge* about a dynamic system. These models represent *what we know* about a dynamic system, including a quantitative model for our *uncertainty* about what we know.

In the next chapter, methods will be derived for modifying the state of knowledge, based on observations related to the state of the dynamic system.

3.1.1 Discovery and Modeling of Random Processes

Brownian Motion and Stochastic Differential Equations. The British botanist Robert Brown (1773–1858) reported in 1827 a phenomenon he had observed while studying pollen grains of the herb *Clarkia pulchella* suspended in water and similar observations by earlier investigators. The particles appeared to move about erratically, as though propelled by some unknown force. This phenomenon came to be called *Brownian movement* or *Brownian motion*. It has been studied extensively—both empirically and theoretically—by many eminent scientists (including Albert Einstein [157]) for the past century. Empirical studies demonstrated that no biological forces were involved and eventually established that individual collisions with molecules of the surrounding fluid were causing the motion observed. The empirical results quantified how some statistical properties of the random motion were influenced by such physical properties as the size and mass of the particles and the temperature and viscosity of the surrounding fluid. Mathematical models with these statistical properties were derived in terms of what has come to be called *stochastic differential equations*. P. Langevin (1872–1946) modeled the velocity v of a particle in terms of a differential equation of the form

$$\frac{dv}{dt} = -\beta v + a(t), \tag{3.1}$$

where β is a damping coefficient (due to the viscosity of the suspending medium) and $a(t)$ is called a "random force." This is now called the *Langevin equation*.

Idealized Stochastic Processes. The random forcing function $a(t)$ of the Langevin equation has been idealized in two ways from the physically motivated example of Brownian motion: (1) the velocity changes imparted to the particle have been assumed to be statistically independent from one collision to another and (2) the effective time between collisions has been allowed to shrink to zero, with the magnitude of the imparted velocity change shrinking accordingly. This model transcends the ordinary (Riemann) calculus, because a "white-noise" process is not integrable in the ordinary calculus. A special calculus was developed by Kiyosi Itô (called the *Itô calculus* or the *stochastic calculus*) to handle such functions.

White-Noise Processes and Wiener Processes. A more precise mathematical characterization of white noise was provided by Norbert Weiner, using his generalized harmonic analysis, with a result that is difficult to square with intuition. It has a power spectral density that is uniform over an infinite bandwidth, implying that the noise power is proportional to bandwidth and that the total power is infinite. (If "white light" had this property, would we be able to see?) Wiener preferred to focus on the mathematical properties of $v(t)$, which is now called a *Wiener process*. Its mathematical properties are more benign than those of white-noise processes.

3.1.2 Main Points to Be Covered

The theory of random processes and stochastic systems represents the evolution over time of the uncertainty of our knowledge about physical systems. This representation includes the effects of any measurements (or *observations*) that we make of the physical process and the effects of uncertainties about the measurement processes and dynamic processes involved. The uncertainties in the measurement and dynamic processes are modeled by random processes and stochastic systems.

Properties of uncertain dynamic systems are characterized by statistical parameters such as *means*, *correlations*, and *covariances*. By using only these numerical parameters, one can obtain a finite representation of the problem, which is important for implementing the solution on digital computers. This representation depends upon such statistical properties as orthogonality, stationarity, ergodicity, and Markovianness of the random processes involved and the Gaussianity of probability distributions. Gaussian, Markov, and uncorrelated (white-noise) processes will be used extensively in the following chapters. The autocorrelation functions and power spectral densities (PSDs) of such processes are also used. These are important in the development of frequency-domain and time-domain models. The time-domain models may be either continuous or discrete.

Shaping filters (continuous and discrete) are developed for random-constant, random-walk, and ramp, sinusoidally correlated and exponentially correlated processes. We derive the linear covariance equations for continuous and discrete systems to be used in Chapter 4. The *orthogonality principle* is developed and explained with scalar examples. This principle will be used in Chapter 4 to derive the Kalman filter equations.

3.1.3 Topics Not Covered

It is assumed that the reader is already familiar with the mathematical foundations of probability theory, as covered by Papoulis [39] or Billingsley [53], for example. The treatment of these concepts in this chapter is heuristic and very brief. The reader is referred to textbooks of this type for more detailed background material.

The Itô calculus for the integration of otherwise nonintegrable functions (white noise, in particular) is not defined, although it is used. The interested reader is referred to books on the mathematics of stochastic differential equations (e.g., those by Arnold [51], Baras and Mirelli [52], Itô and McKean [64], Sobczyk [77], or Stratonovich [78]).

3.2 PROBABILITY AND RANDOM VARIABLES

The relationships between unknown physical processes, probability spaces, and random variables are illustrated in Figure 3.1. The behavior of the physical processes is investigated by what is called a *statistical experiment*, which helps to define a model for the physical process as a probability space. Strictly speaking, this is not a

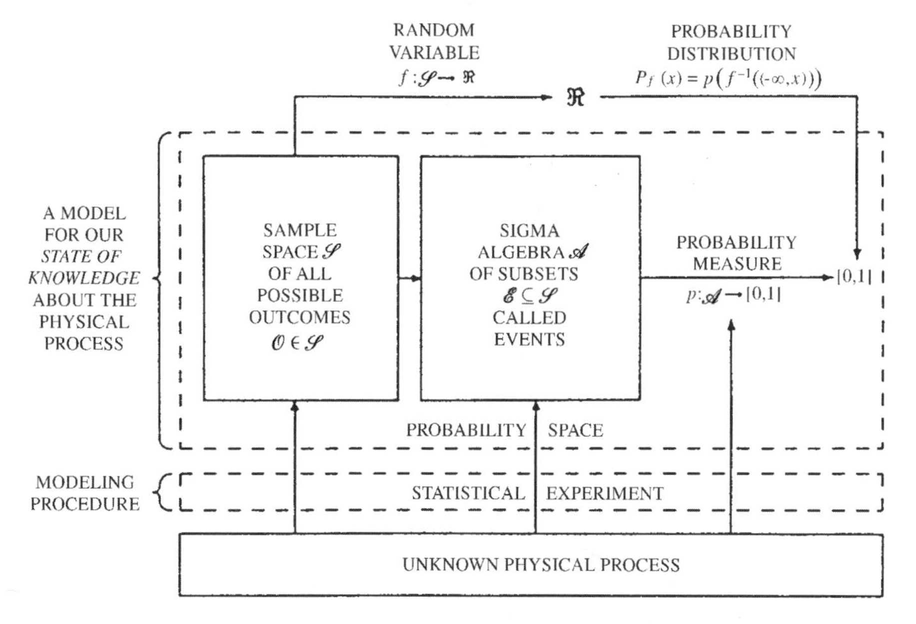

Fig. 3.1 *Conceptual model for a random variable.*

model for the physical process itself, but a model of our own understanding of the physical process. It defines what might be called our "state of knowledge" about the physical process, which is essentially a model for our *uncertainty* about the physical process.

A *random variable* represents a *numerical attribute* of the state of the physical process. In the following subsections, these concepts are illustrated by using the numerical score from tossing dice as an example of a random variable.

3.2.1 An Example of a Random Variable

EXAMPLE 3.1: Score from Tossing a Die A *die* (plural of *dice*) is a cube with its six faces marked by patterns of one to six dots. It is thrown onto a flat surface such that it tumbles about and comes to rest with one of these faces on top. This can be considered an unknown process in the sense that which face will wind up on top is not reliably predictable before the toss. The tossing of a die in this manner is an example of a statistical experiment for defining a statistical model for the process. Each toss of the die can result in but one *outcome*, corresponding to which one of the six faces of the die is on top when it comes to rest. Let us label these outcomes $\mathscr{O}_a$, $\mathscr{O}_b$, $\mathscr{O}_c$, $\mathscr{O}_d$, $\mathscr{O}_e$, $\mathscr{O}_f$. The set of all possible outcomes of a statistical experiment is called a *sample space*. The sample space for the statistical experiment with one die is the set $\mathscr{S} = \{\mathscr{O}_a, \mathscr{O}_b, \mathscr{O}_c, \mathscr{O}_d, \mathscr{O}_e, \mathscr{O}_f\}$.

A random variable assigns real numbers to outcomes. There is an integral number of dots on each face of the die. This defines a "dot function" $d : \mathscr{S} \rightarrow \Re$ on the sample space $\mathscr{S}$, where $d(\mathscr{O})$ is the number of dots showing for the outcome $\mathscr{O}$ of the statistical experiment. Assign the values

$$\begin{array}{lll} d(\mathscr{O}_a) = 1, & d(\mathscr{O}_c) = 3, & d(\mathscr{O}_e) = 5, \\ d(\mathscr{O}_b) = 2, & d(\mathscr{O}_d) = 4, & d(\mathscr{O}_f) = 6. \end{array}$$

This *function* is an example of a *random variable*. The useful statistical properties of this random variable will depend upon the probability space defined by statistical experiments with the die.

Events and sigma algebras. The statistical properties of the random variable d depend on the probabilities of sets of outcomes (called *events*) forming what is called a *sigma algebra*[1] of subsets of the sample space $\mathscr{S}$. Any collection of events that includes the sample space itself, the *empty set* (the set with no elements), and the *set unions* and *set complements* of all its members is called a *sigma algebra* over the sample space. The set of *all subsets* of $\mathscr{S}$ is a sigma algebra with $2^6 = 64$ events.

The probability space for a fair die. A die is considered "fair" if, in a large number of tosses, all outcomes tend to occur with equal frequency. The *relative frequency* of any outcome is defined as the ratio of the number of occurrences of that outcome to the number of occurrences of all outcomes. Relative frequencies of outcomes of a statistical experiment are called *probabilities*. Note that, by this definition, the sum of the probabilities of all outcomes will always be equal to 1. This defines a probability $p(\mathscr{E})$ for every event $\mathscr{E}$ (a set of outcomes) equal to

$$p(\mathscr{E}) = \frac{\#(\mathscr{E})}{\#(\mathscr{S})},$$

where $\#(\mathscr{E})$ is the *cardinality* of $\mathscr{E}$, equal to the *number* of outcomes $\mathscr{O} \in \mathscr{E}$. Note that this assigns probability zero to the empty set and probability one to the sample space.

The *probability distribution* of the random variable d is a nondecreasing function $P_d(x)$ defined for every real number x as the probability of the event for which the score is less than x. It has the formal definition

$$\begin{aligned} P_d(x) &\stackrel{\text{def}}{=} p(d^{-1}((-\infty, x))), \\ d^{-1}((-\infty, x)) &\stackrel{\text{def}}{=} \{\mathscr{O} | d(\mathscr{O}) \leq x\}. \end{aligned}$$

[1]Such a collection of subsets $\mathscr{E}_i$ of a set $\mathscr{S}$ is called an *algebra* because it is a Boolean algebra with respect to the operations of set union ($\mathscr{E}_1 \cup \mathscr{E}_2$), set intersection ($\mathscr{E}_1 \cap \mathscr{E}_2$), and set complement ($\mathscr{S} \backslash \mathscr{E}$)—corresponding to the logical operations *or*, *and*, and *not*, respectively. The "sigma" refers to the summation symbol Σ, which is used for defining the additive properties of the associated probability measure. However, the lowercase symbol σ is used for abbreviating "sigma algebra" to "σ-algebra."

For every real value of x, the set $\{\mathcal{O}|d(\mathcal{O}) < x\}$ is an *event*. For example,

$$\begin{aligned}
P_d(1) &= p(d^{-1}((-\infty, 1))) \\
&= p(\{\mathcal{O}|d(\mathcal{O}) < 1\}) \\
&= p(\{\,\}) \quad \text{(the empty set)} \\
&= 0, \\
P_d(1.0\cdots 01) &= p(d^{-1}((-\infty, 1.0\cdots 01))) \\
&= p(\{\mathcal{O}|d(\mathcal{O}) < 1.0\cdots 01\}) \\
&= p(\{\mathcal{O}_a\}) = \tfrac{1}{6}, \\
&\vdots \\
P_d(6.0\cdots 01) &= p(\mathcal{S}) = 1,
\end{aligned}$$

as plotted in Figure 3.2. Note that P_d *is not a continuous function* in this particular example.

3.2.2 Probability Distributions and Densities

Random variables f are required to have the property that, for every real a and b such that $-\infty \leq a \leq b \leq +\infty$, the outcomes $\mathcal{O}$ such that $a < f(\mathcal{O}) < b$ are an event $\mathcal{E} \in \mathcal{A}$. This property is needed for defining the *probability distribution function* P_f of f as

$$P_f(x) \stackrel{\text{def}}{=} p(f^{-1}((-\infty, x))), \tag{3.2}$$

$$f^{-1}((-\infty, x)) \stackrel{\text{def}}{=} \{\mathcal{O} \in \mathcal{S} \,|\, f(\mathcal{O}) \leq x\}. \tag{3.3}$$

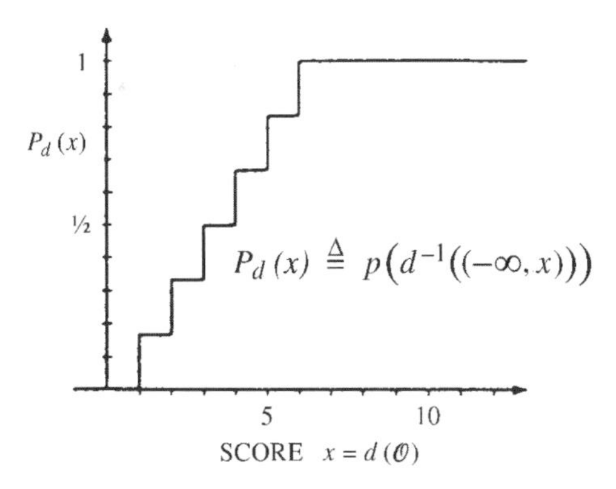

Fig. 3.2 *Probability distribution of scores from a fair die.*

The probability distribution function may not be a differentiable function. However, if it *is* differentiable, then its derivative

$$p_f(x) = \frac{d}{dx} P_f(x) \tag{3.4}$$

is called the *probability density function* of the random variable, f, and the differential

$$p_f(x)\,dx = dP_f(x) \tag{3.5}$$

is the *probability measure* of f defined on a sigma algebra containing the open intervals (called the *Borel*[2] *algebra* over $\Re$).

A *vector-valued random variable* is a vector with random variables as its components. An analogous derivation applies to vector-valued random variables, for which the analogous probability measures are defined on the Borel algebras over $\Re^n$.

3.2.3 Gaussian Probability Densities

The probability distribution of the average score from tossing n dice (i.e., the total number of dots divided by the number of dice) tends toward a particular type of distribution as $n \to \infty$, called a *Gaussian distribution*.[3] It is the limit of many such distributions, and it is common to many models for random phenomena. It is commonly used in stochastic system models for the distributions of random variables.

Univariate Gaussian Probability Distributions. The notation $\mathcal{N}(\bar{x}, \sigma^2)$ is used to denote a probability distribution with density function

$$p(x) = \frac{1}{\sqrt{2\pi}\sigma} \exp\left[-\frac{1}{2} \frac{(x - \bar{x})^2}{\sigma^2} \right], \tag{3.6}$$

where

$$\bar{x} = E\langle x \rangle \tag{3.7}$$

is the *mean* of the distribution (a term that will be defined later on, in Section 3.4.2) and σ^2 is its *variance* (also defined in Section 3.4.2). The "$\mathcal{N}$" stands for "normal,"

[2]Named for the French mathematician Félix Borel (1871–1956).

[3]It is called the *Laplace distribution* in France. It has had many discoverers besides Gauss and Laplace, including the American mathematician Robert Adrian (1775–1843). The physicist Gabriel Lippman (1845–1921) is credited with the observation that "mathematicians think it [the normal distribution] is a law of nature and physicists are convinced that it is a mathematical theorem."

another name for the Gaussian distribution. Because so many other things are called *normal* in mathematics, it is less confusing if we call it *Gaussian*.

Gaussian Expectation Operators and Generating Functions. Because the Gaussian probability density function depends only on the difference $x - \bar{x}$, the expectation operator

$$\underset{x}{E}\langle f(x)\rangle = \int_{-\infty}^{+\infty} f(x)p(x)\,dx \tag{3.8}$$

$$= \frac{1}{\sqrt{2\pi}\sigma}\int_{-\infty}^{+\infty} f(x)e^{-(x-\bar{x})^2/2\sigma^2}\,dx \tag{3.9}$$

$$= \frac{1}{\sqrt{2\pi}\sigma}\int_{-\infty}^{+\infty} f(x+\bar{x})e^{-x^2/2\sigma^2}\,dx \tag{3.10}$$

has the form of a convolution integral. This has important implications for problems in which it must be implemented numerically, because the convolution can be implemented more efficiently as a fast Fourier transform of f, followed by a pointwise product of its transform with the Fourier transform of p, followed by an inverse fast Fourier transform of the result. One does not need to take the numerical Fourier transform of p, because its Fourier transform can be expressed analytically in closed form. Recall that the Fourier transform of p is called its *generating function*. Gaussian generating functions are also (possibly scaled) Gaussian density functions:

$$p(\omega) = \frac{1}{\sqrt{2\pi}}\int_{-\infty}^{\infty} p(x)e^{i\omega x}\,dx \tag{3.11}$$

$$= \frac{1}{\sqrt{2\pi}}\int_{-\infty}^{\infty} \frac{e^{-x^2/2\sigma^2}}{\sqrt{2\pi}\sigma}\,e^{i\omega x}\,dx \tag{3.12}$$

$$= \frac{\sigma}{\sqrt{2\pi}}e^{(-1/2)\omega^2\sigma^2}, \tag{3.13}$$

a Gaussian density function with variance σ^{-2}. Here we have used a probability-preserving form of the Fourier transform, defined with the factor of $1/\sqrt{2\pi}$ in front of the integral. If other forms of the Fourier transform are used, the result is not a probability distribution but a scaled probability distribution.

3.2.3.1 Vector-Valued (Multivariate) Gaussian Distributions. The formula for the n-dimensional Gaussian distribution $\mathcal{N}(\bar{x}, P)$, where the mean $\bar{x}$ is an n-vector and the covariance P is an $n \times n$ symmetric positive-definite matrix, is

$$p(x) = \frac{1}{\sqrt{(2\pi)^n \det P}}e^{(-1/2)(x-\bar{x})^{\mathrm{T}}P^{-1}(x-\bar{x})}. \tag{3.14}$$

The multivariate Gaussian generating function has the form

$$p(\omega) = \frac{1}{\sqrt{(2\pi)^n \det P^{-1}}} e^{(-1/2)\omega^T P \omega}, \tag{3.15}$$

where ω is an n-vector. This is also a multivariate Gaussian probability distribution $\mathcal{N}(0, P^{-1})$ if the scaled form of the Fourier transform shown in Equation 3.11 is used.

3.2.4 Joint Probabilities and Conditional Probabilities

The *joint probability* of two events $\mathscr{E}_a$ and $\mathscr{E}_b$ is the probability of their *set intersection* $p(\mathscr{E}_a \cap \mathscr{E}_b)$, which is the probability that *both* events occur. The joint probability of *independent events* is the *product* of their probabilities.

The *conditional probability* of event $\mathscr{E}$, given that event $\mathscr{E}_c$ has occurred, is defined as the probability of $\mathscr{E}$ in the "conditioned" probability space with sample space $\mathscr{E}_c$. This is a probability space defined on the sigma algebra

$$\mathscr{A}|\mathscr{E}_c = \{\mathscr{E} \cap \mathscr{E}_c | \mathscr{E} \in \mathscr{A}\} \tag{3.16}$$

of the set intersections of all events $\mathscr{E} \in \mathscr{A}$ (the original sigma algebra) with the conditioning event $\mathscr{E}_c$. The probability measure on the "conditioned" sigma algebra $\mathscr{A}|\mathscr{E}_c$ is defined in terms of the joint probabilities in the original probability space by the rule

$$p(\mathscr{E}|\mathscr{E}_c) = \frac{p(\mathscr{E} \cap \mathscr{E}_c)}{p(\mathscr{E}_c)}, \tag{3.17}$$

where $p(\mathscr{E} \cap \mathscr{E}_c)$ is the joint probability of $\mathscr{E}$ and $\mathscr{E}_c$. Equation 3.17 is called *Bayes' rule*[4].

EXAMPLE 3.2: Experiment with Two Dice Consider a toss with two dice in which one die has come to rest before the other and just enough of its face is visible to show that it contains either four or five dots. The question is: What is the probability distribution of the score, given that information?

The probability space for two dice. This example illustrates just how rapidly the sizes of probability spaces grow with the "problem size" (in this case, the number of dice). For a single die, the sample space has 6 outcomes and the sigma algebra has 64 events. For two dice, the sample space has 36 possible outcomes (6 independent outcomes for each of two dice) and $2^{36} = 68,719,476,736$ possible events. If each

[4]Discovered by the English clergyman and mathematician Thomas Bayes (1702–1761). Conditioning on impossible events is not defined. Note that the conditional probability is based on the assumption that $\mathscr{E}_c$ *has occurred*. This would seem to imply that $\mathscr{E}_c$ is an event with *nonzero probability*, which one might expect from practical applications of Bayes' rule.

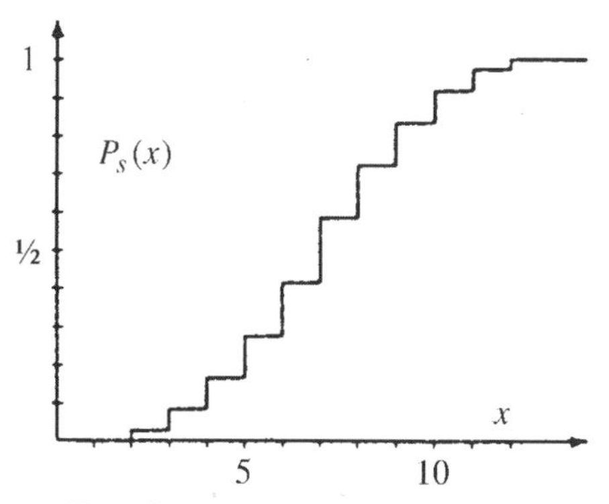

(a) Two dice without conditioning

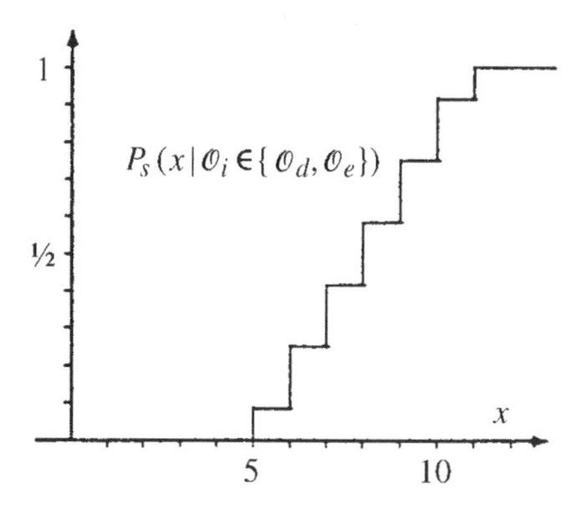

(b) Two dice with conditioning

Fig. 3.3 Probability distributions of dice scores.

die is fair and their outcomes are independent, then all outcomes with two dice have probability $(\frac{1}{6}) \times (\frac{1}{6}) = \frac{1}{36}$ and the probability of any event is the number of outcomes in the event divided by 36 (the number of outcomes in the sample space). Using the same notation as the previous (one-die) example, let the outcome from tossing a pair of dice be represented by an ordered pair (in parentheses) of the outcomes of the first and second die, respectively. Then the score $s((\mathcal{O}_i, \mathcal{O}_j)) = d(\mathcal{O}_i) + d(\mathcal{O}_j)$, where $\mathcal{O}_i$ represents the outcome of the first die and $\mathcal{O}_j$ represents the outcome of the second die. The corresponding probability distribution function of the score x for two dice is shown in Figure 3.3a.

The event corresponding to the condition that the first die have either four or five dots showing contains all outcomes in which $\mathcal{O}_i = \mathcal{O}_d$ or $\mathcal{O}_e$, which is the set

$$\begin{aligned}\mathcal{E}_c = \{&(\mathcal{O}_d, \mathcal{O}_a), (\mathcal{O}_d, \mathcal{O}_b), (\mathcal{O}_d, \mathcal{O}_c), (\mathcal{O}_d, \mathcal{O}_d), (\mathcal{O}_d, \mathcal{O}_e), (\mathcal{O}_d, \mathcal{O}_f)\\ &(\mathcal{O}_e, \mathcal{O}_a), (\mathcal{O}_e, \mathcal{O}_b), (\mathcal{O}_e, \mathcal{O}_c), (\mathcal{O}_e, \mathcal{O}_d), (\mathcal{O}_e, \mathcal{O}_e), (\mathcal{O}_e, \mathcal{O}_f)\},\end{aligned}$$

of 12 outcomes. It has probability $p(\mathcal{E}_c) = \frac{12}{36} = \frac{1}{3}$.

By applying Bayes' rule, the conditional probabilities of all events corresponding to unique scores can be calculated as shown in Figure 3.4. The corresponding probability distribution function for two dice with this conditioning is shown in Figure 3.3*b*.

3.3 STATISTICAL PROPERTIES OF RANDOM VARIABLES

3.3.1 Expected Values of Random Variables

Expected values. The symbol E is used as an operator on random variables. It is called the *expectancy*, *expected value*, or *average* operator, and the expression $\underset{x}{E}\langle f(x)\rangle$ is used to denote the expected value of the function f applied to the ensemble of possible values of the random variable x. The symbol under the E indicates the random variable (RV) over which the expected value is to be evaluated. When the RV in question is obvious from context, the symbol underneath the E will be eliminated. If the argument of the expectancy operator is also obvious from context, the angular brackets can also be disposed with, using Ex instead of E$\langle x\rangle$, for example.

Moments. The nth moment of a scalar RV x with probability density $p(x)$ is defined by the formula

$$\eta_n(x) \stackrel{\text{def}}{=} \underset{x}{E}\langle x^n\rangle \stackrel{\text{def}}{=} \int_{\infty}^{\infty} x^n p(x)\, dx. \tag{3.18}$$

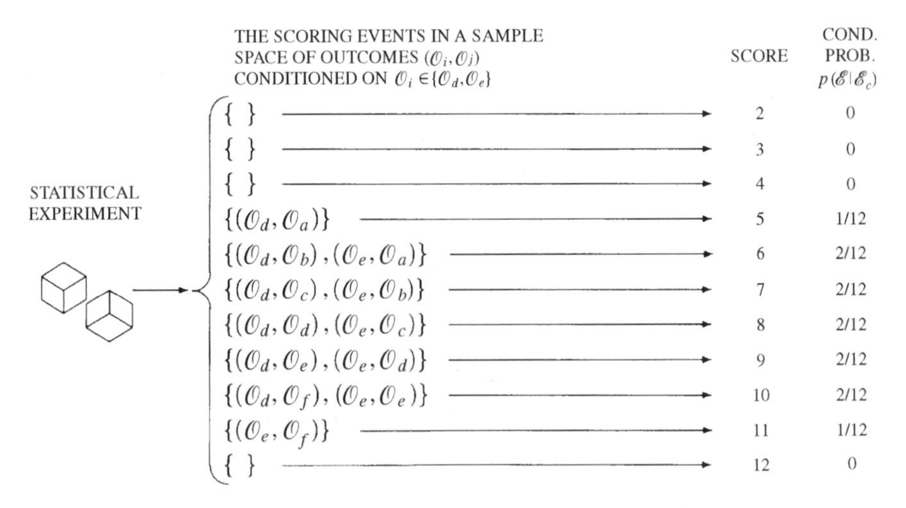

Fig. 3.4 *Conditional scoring probabilities for two dice.*

The nth *central moment* of x is defined as

$$\mu_n(x) \stackrel{\text{def}}{=} E\langle x - Ex\rangle^n \tag{3.19}$$

$$= \int_{-\infty}^{\infty} (x - Ex)^n p(x)\, dx. \tag{3.20}$$

The first moment of x is called its mean[5]:

$$\eta_1 = Ex = \int_{-\infty}^{\infty} xp(x)\, dx. \tag{3.21}$$

In general, a function of several arguments such as $f(x,y,z)$ has first moment

$$Ef(x, y, z) = \int\int_{-\infty}^{\infty}\int f(x, y, z)p(x, y, z)\, dx\, dy\, dz. \tag{3.22}$$

Array Dimensions of Moments. The first moment will be a scalar or a vector, depending on whether the function $f(x, y, z)$ is scalar or vector valued. Higher order moments have tensorlike properties, which we can characterize in terms of the number of subscripts used in defining them as data structures. Vectors are singly subscripted data structures. The higher order moments of vector-valued variates are successively higher order data structures. That is, the second moments of vector-valued RVs are matrices (doubly subscripted data structures), and the third-order moments will be triply subscripted data structures.

These definitions of a moment apply to discrete-valued random variables if we simply substitute summations in place of integrations in the definitions.

3.3.2 Functions of Random Variables

A function of RV x is the operation of assigning to each value of x another value, for example y, according to rule or function. This is represented by

$$y = f(x), \tag{3.23}$$

where x and y are usually called input and output, respectively. The statistical properties of y in terms of x are, for example,

$$\begin{aligned} Ey &= \int_{-\infty}^{\infty} f(x)p(x)\, dx, \\ &\vdots \\ Ey^n &= \int_{-\infty}^{\infty} [f(x)]^n p(x)\, dx \end{aligned} \tag{3.24}$$

when y is scalar. For vector-valued functions y, similar expressions can be shown.

[5]We here restrict the order of the moment to the positive integers. The zeroth-order moment would otherwise always evaluate to 1.

The probability density of y can be obtained from the density of x. If equation 3.23 can be solved for x, yielding the unique solution

$$x = g(y). \tag{3.25}$$

Then we have

$$p_y(y) = \frac{p_x(g(y))}{\left.\dfrac{\partial f(x)}{\partial x}\right|_{x=g(y)}} \tag{3.26}$$

where $p_y(y)$ and $p_x(x)$ are the density functions of y and x, respectively. A function of two RVs, x, y is the process of assigning to each pair of x, y another value, for example, z, according to the same rule,

$$z = f(y, x), \tag{3.27}$$

and similarly functions of n RVs. When x and y in Equation 3.23 are n-dimensional vectors and if a unique solution for x in terms of y exists,

$$x = g(y), \tag{3.28}$$

Equation 3.26 becomes

$$p_y(y) = \frac{p_x[g(y)]}{|J|_{x=g(y)}}, \tag{3.29}$$

where the Jacobian $|J|$ is defined as the determinant of the array of partial derivatives $\partial f_i / \partial x_j$:

$$|J| = \det \begin{bmatrix} \dfrac{\partial f_1}{\partial x_1} & \dfrac{\partial f_1}{\partial x_2} & \cdots & \dfrac{\partial f_1}{\partial x_n} \\ \dfrac{\partial f_2}{\partial x_1} & \dfrac{\partial f_2}{\partial x_2} & \cdots & \dfrac{\partial f_2}{\partial x_n} \\ \vdots & \vdots & \ddots & \vdots \\ \dfrac{\partial f_n}{\partial x_1} & \dfrac{\partial f_n}{\partial x_2} & \cdots & \dfrac{\partial f_n}{\partial x_n} \end{bmatrix}. \tag{3.30}$$

3.4 STATISTICAL PROPERTIES OF RANDOM PROCESSES

3.4.1 Random Processes (RPs)

A RV was defined as a function $x(s)$ defined for each outcome of an experiment identified by s. Now if we assign to each outcome s a time function $x(t, s)$, we obtain a family of functions called random processes or stochastic processes. A random process is called discrete if its argument is a discrete variable set as

$$x(k, s), \quad k = 1, 2 \ldots. \tag{3.31}$$

It is clear that the value of a random process $x(t)$ at any particular time $t = t_0$, namely $x(t_0, s)$, is a random variable [or a random vector if $x(t_0, s)$ is vector valued].

3.4.2 Mean, Correlation, and Covariance

Let $x(t)$ be an n-vector random process. Its mean

$$Ex(t) = \int_{-\infty}^{\infty} x(t)p[x(t)]\,dx(t), \tag{3.32}$$

which can be expressed elementwise as

$$Ex_i(t) = \int_{-\infty}^{\infty} x_i(t)p[x_i(t)]\,dx(t), \quad i = 1 \ldots n.$$

For a random sequence, the integral is replaced by a sum.

The *correlation* of the vector-valued process $x(t)$ is defined by

$$E\langle x(t_1)x^{\mathrm{T}}(t_2)\rangle = \begin{bmatrix} E\langle x(t_1)x_1(t_2)\rangle & \cdots & E\langle x_1(t_1)x_n(t_2)\rangle \\ \vdots & \ddots & \vdots \\ E\langle x_n(t_1)x_1(t_2)\rangle & \cdots & E\langle x_n(t_1)x_n(t_2)\rangle \end{bmatrix}, \tag{3.33}$$

where

$$Ex_i(t_1)x_j(t_2) = \int_{-\infty}^{\infty}\int x_i(t_1)x_j(t_2)p[x_i(t_1), x_j(t_2)]\,dx_i(t_1)\,dx_j(t_2). \tag{3.34}$$

The *covariance* of $x(t)$ is defined by

$$\begin{aligned} &E\langle [x(t_1) - Ex(t_1)][x(t_2) - Ex(t_2)]^{\mathrm{T}}\rangle \\ &\quad = E\langle x(t_1)x^{\mathrm{T}}(t_2)\rangle - E\langle x(t_1)\rangle E\langle x^{\mathrm{T}}(t_2)\rangle. \end{aligned} \tag{3.35}$$

When the process $x(t)$ has zero mean (i.e., $Ex(t) = 0$ for all t), its correlation and covariance are equal.

The correlation matrix of two RPs $x(t)$, an n-vector, and $y(t)$, an m-vector, is given by an $n \times m$ matrix

$$Ex(t_1)y^{\mathrm{T}}(t_2), \tag{3.36}$$

where

$$Ex_i(t_1)y_j(t_2) = \int_{-\infty}^{\infty}\int x_i(t_1)y_j(t_2)p[x_i(t_1), y_j(t_2)]\, dx_i(t_1)\, dy_j(t_2) \tag{3.37}$$

Similarly, the cross-covariance $n \times m$ matrix is

$$E\langle[x(t_1) - Ex(t_1)][y(t_2) - Ey(t_2)]^{\mathrm{T}}\rangle. \tag{3.38}$$

3.4.3 Orthogonal Processes and White Noise

Two RPs $x(t)$ and $y(t)$ are called *uncorrelated* if their cross-covariance matrix is identically zero for all t_1 and t_2:

$$E\langle[x(t_1) - E\langle x(t_1)\rangle][y(t_2) - E\langle y(t_2)\rangle]^{\mathrm{T}}\rangle = 0. \tag{3.39}$$

The processes $x(t)$ and $y(t)$ are called *orthogonal* if their correlation matrix is identically zero:

$$E\langle x(t_1)y^{\mathrm{T}}(t_2)\rangle = 0. \tag{3.40}$$

The random process $x(t)$ is called uncorrelated if

$$E\langle[x(t_1) - E\langle x(t_1)\rangle][x(t_2) - E\langle x(t_2)\rangle]^{\mathrm{T}}\rangle = Q(t_1, t_2)\delta(t_1 - t_2) \tag{3.41}$$

where $\delta(t)$ is the Dirac delta "function"[6] (actually, a generalized function), defined by

$$\int_a^b \delta(t)\, dt = \begin{cases} 1 & \text{if } a \le 0 \le b, \\ 0 & \text{otherwise.} \end{cases} \tag{3.42}$$

Similarly, a random sequence x_k is called uncorrelated if

$$E\langle[x_k - E\langle x_k\rangle][x_j - E\langle x_j\rangle]^{\mathrm{T}}\rangle = Q(k, j)\,\Delta(k - j), \tag{3.43}$$

where $\Delta(\cdot)$ is the Kronecker delta function[7], defined by

$$\Delta(k) = \begin{cases} 1 & \text{if } k = 0 \\ 0 & \text{otherwise.} \end{cases} \tag{3.44}$$

A white-noise process or sequence is an example of an uncorrelated process or sequence.

[6]Named for the English physicist Paul Adrien Maurice Dirac (1902–1984).
[7]Named for the German mathematician Leopold Kronecker (1823–1891).

A process $x(t)$ is considered independent if for any choice of distinct times $t_1, t_2, \ldots t_n$, the random variables $x(t_1), x(t_2), \ldots, x(t_n)$ are independent. That is,

$$p_{x(t_1)}, \ldots, p_{x(t_n)}(s_1, \ldots, s_n) = \prod_{i=1}^{n} p_{x(t_i)}(s_i). \tag{3.45}$$

Independence (all of the moments) implies no correlation (which restricts attention to the second moments), but the opposite implication is not true, except in such special cases as Gaussian processes (see Section 3.2.3). Note that *whiteness* means *uncorrelated* in time rather than *independent* in time (i.e., including all moments), although this distinction disappears for the important case of white Gaussian processes (see Chapter 4).

3.4.4 Strict-Sense and Wide-Sense Stationarity

The random process $x(t)$ (or random sequence x_k) is called *strict-sense stationary* if all its statistics (meaning $p[x(t_1),\ x(t_2), \ldots]$) are invariant with respect to shifts of the time origin:

$$\begin{aligned} p(x_1, x_2, \ldots, x_n, t_1, \ldots, t_n) \\ = p(x_1, x_2, \ldots, x_n,\ t_1 + \varepsilon, t_2 + \varepsilon, \ldots, t_n + \varepsilon) \end{aligned} \tag{3.46}$$

The random process $x(t)$ (or x_k) is called *wide-sense stationary* (WSS) (or "weak-sense" stationary) if

$$E\langle x(t)\rangle = c \quad \text{(a constant)} \tag{3.47}$$

and

$$E\langle x(t_1)x^{\mathrm{T}}(t_2)\rangle = Q(t_2 - t_1) = Q(\tau), \tag{3.48}$$

where Q is a matrix with each element depending only on the difference $t_2 - t_1 = \tau$. Therefore, when $x(t)$ is stationary in the weak sense, it implies that its first- and second-order statistics are independent of time origin, while strict stationarity by definition implies that statistics of all orders are independent of the time origin.

3.4.5 Ergodic Random Processes

A process is considered ergodic[8] if all of its statistical parameters, mean, variance, and so on, can be determined from arbitrarily chosen member functions. A sampled function $x(t)$ is ergodic if its time-averaged statistics equal the ensemble averages.

[8]The term *ergodic* came originally from the development of statistical mechanics for thermodynamic systems. It is taken from the Greek words for *energy* and *path*. The term was applied by the American physicist Josiah Willard Gibbs (1839–1903) to the time history (or path) of the state of a thermodynamic system of constant energy. Gibbs had assumed that a thermodynamic system would eventually take on all possible states consistent with its energy. It was shown to be impossible from function-theoretic considerations in the nineteenth century. The so-called ergodic hypothesis of James Clerk Maxwell (1831–1879) is that the temporal means of a stochastic system are equivalent to the ensemble means. The concept was given firmer mathematical foundations by George David Birkhoff and John von Neumann around 1930 and by Norbert Wiener in the 1940s.

3.4.6 Markov Processes and Sequences

An RP $x(t)$ is called a *Markov process*[9] if its future state distribution, conditioned on knowledge of its present state, is not improved by knowledge of previous states:

$$p\{x(t_i)|x(\tau);\ \tau < t_{i-1}\} = p\{x(t_i)|x(t_{i-1})\}, \tag{3.49}$$

where the times $t_1 < t_2 < t_3 < \cdots < t_i$.

Similarly, a random sequence (RS) x_k is called a *Markov sequence* if

$$p(x_i|x_k;\ k \le i-1) = p\{x_i|x_{i-1}\}. \tag{3.50}$$

The solution to a general first-order differential or difference equation with an independent process (uncorrelated normal RP) as a forcing function is a Markov process. That is, if $x(t)$ and x_k are n-vectors satisfying

$$\dot{x}(t) = F(t)x(t) + G(t)w(t) \tag{3.51}$$

or

$$x_k = \Phi_{k-1}x_{k-1} + G_{k-1}w_{k-1}, \tag{3.52}$$

where $w(t)$ and w_{k-1} are r-dimensional independent random processes and sequences, the solutions $x(t)$ and x_k are then vector Markov processes and sequences, respectively.

3.4.7 Gaussian Processes

An n-dimensional RP $x(t)$ is called Gaussian (or normal) if its probability density function is Gaussian, as given by the formulas of Section 3.2.3, with covariance matrix

$$P = E\langle |x(t) - E\langle x(t)\rangle|[x(t) - E\langle x(t)\rangle]^{\mathrm{T}}\rangle \tag{3.53}$$

for the random variable x.

Gaussian random processes have some useful properties:

1. A Gaussian RP $x(t)$ is WSS—and stationary in the strict sense.
2. Orthogonal Gaussian RPs are independent.
3. Any linear function of jointly Gaussian RP results in another Gaussian RP.
4. All statistics of a Gaussian RP are completely determined by its first- and second-order statistics.

[9]Defined by Andrei Andreevich Markov (1856–1922).

3.4.8 Simulating Multivariate Gaussian Processes

Cholesky decomposition methods are discussed in Chapter 6 and Appendix B. We show here how these methods can be used to generate uncorrelated pseudorandom vector sequences with zero mean (or any specified mean) and a specified covariance P.

There are many programs that will generate pseudorandom sequences of uncorrelated Gaussian scalars $\{s_i | i = 1, 2, 3, \ldots\}$ with zero mean and unit variance:

$$E\langle s_i \rangle \in \mathcal{N}(0, 1) \quad \text{for all } i, \tag{3.54}$$

$$E\langle s_i s_j \rangle = \begin{cases} 0 & \text{if} \quad i \neq j, \\ 1 & \text{if} \quad i = j \end{cases} \tag{3.55}$$

These can be used to generate sequences of Gaussian n-vectors x_k with mean zero and covariance I_m:

$$u_k = [s_{nk+1} \quad s_{nk+2} \quad s_{nk+3} \quad \cdots \quad s_{n(k+1)}]^{\mathrm{T}}, \tag{3.56}$$

$$E\langle u_k \rangle = 0, \tag{3.57}$$

$$E\langle u_k u_k^{\mathrm{T}} \rangle = I_n. \tag{3.58}$$

These vectors, in turn, can be used to generate a sequence of n-vectors w_k with zero mean and covariance P. For that purpose, let

$$CC^{\mathrm{T}} = P \tag{3.59}$$

be a Cholesky decomposition of P, and let the sequence of n-vectors w_k be generated according to the rule

$$w_k = Cu_k. \tag{3.60}$$

Then the sequence of vectors $\{w_0, w_1, w_2, \ldots\}$ will have mean

$$E\langle w_k \rangle = CE\langle u_k \rangle \tag{3.61}$$

$$= 0 \tag{3.62}$$

(an n-vector of zeros) and covariance

$$E\langle w_k w_k^{\mathrm{T}} \rangle = E\langle Cu_k (Cu_k)^{\mathrm{T}} \rangle \tag{3.63}$$

$$= CI_n C^{\mathrm{T}} \tag{3.64}$$

$$= P. \tag{3.65}$$

The same technique can be used to obtain pseudorandom Gaussian vectors with a given mean v by adding v to each w_k. These techniques are used in simulation and Monte Carlo analysis of stochastic systems.

3.4.9 Power Spectral Density

Let $x(t)$ be a zero-mean scalar stationary RP with autocorrelation $\psi_x(\tau)$,

$$E\langle x(t)x(t+\tau)\rangle = \psi_x(\tau) \tag{3.66}$$

The power spectral density (PSD) is defined as

$$\Psi_x(\omega) = \int_{-\infty}^{\infty} \psi_x(\tau)e^{-j\omega\tau}\,d\tau \tag{3.67}$$

and the inverse transform as

$$\psi_x(\tau) = \frac{1}{2\pi}\int_{-\infty}^{\infty} \Psi_x(\omega)e^{j\omega\tau}\,d\omega. \tag{3.68}$$

The following are properties of autocorrelation functions:

1. Autocorrelation functions are symmetrical ("even" functions).
2. An autocorrelation function attains its maximum value at the origin.
3. Its Fourier transform is nonnegative (greater than or equal to zero).

These properties are satisfied by valid autocorrelation functions.

Setting $\tau = 0$ in Equation 3.68 gives

$$\mathop{E}_{t}\langle x^2(t)\rangle = \psi_x(0) = \frac{1}{2\pi}\int_{-\infty}^{\infty} \Psi_x(\omega)\,d\omega. \tag{3.69}$$

Because of property 1 of the autocorrelation function,

$$\Psi_x(\omega) = \Psi_x(-\omega); \tag{3.70}$$

that is, the PSD is a symmetric function of frequency.

EXAMPLE 3.3 If $\psi_x(\tau) = \sigma^2 e^{-\alpha|\tau|}$, find the associated PSD:

$$\begin{aligned}\Psi_x(\omega) &= \int_{-\infty}^{0} \sigma^2 e^{\alpha\tau} e^{-j\omega\tau}\, d\tau + \int_{0}^{\infty} \sigma^2 e^{-\alpha\tau} e^{-j\omega\tau}\, d\tau \\ &= \sigma^2\left(\frac{1}{\alpha - j\omega} + \frac{1}{\alpha + j\omega}\right) \\ &= \frac{2\sigma^2\alpha}{\omega^2 + \alpha^2}.\end{aligned}$$

EXAMPLE 3.4 This is an example of a second-order Markov process generated by passing WSS white noise with zero mean and unit variance through a second-order "shaping filter" with the dynamic model of a harmonic resonator. (This is the same example introduced in Chapter 2 and will be used again in Chapters 4 and 5.)

The transfer function of the dynamic system is

$$H(s) = \frac{as + b}{s^2 + 2\zeta w_n s + w_n^2}.$$

Definitions of ζ, w_n, and s are the same as in Example 2.7. The state-space model of $H(s)$ is given as

$$\begin{bmatrix}\dot{x}_1(t) \\ \dot{x}_2(t)\end{bmatrix} = \begin{bmatrix} 0 & 1 \\ -w_n^2 & -2\zeta w_n\end{bmatrix}\begin{bmatrix} x_{1(t)} \\ x_2(t)\end{bmatrix} + \begin{bmatrix} a \\ b - 2a\zeta w_n\end{bmatrix} w(t),$$
$$z(t) = x_1(t) = x(t).$$

The general form of the autocorrelation is

$$\psi_x(\tau) = \frac{\sigma^2}{\cos\theta} e^{-\zeta w_n|\tau|} \cos\left(\sqrt{1-\zeta^2}\, w_n|\tau| - \theta\right).$$

In practice, σ^2, θ, ζ, and w_n are chosen to fit empirical data (see Problem 3.13). The PSD corresponding to the $\psi_x(\tau)$ will have the form

$$\Psi_x(w) = \frac{a^2 w^2 + b^2}{w^4 + 2w_n^2(2\zeta^2 - 1)w^2 + w_n^4}.$$

(The peak of this PSD will not be at the "natural" (undamped) frequency ω_n, but at the "resonant" frequency defined in Example 2.6.)

The block diagram corresponding to the state-space model is shown in Figure 3.5.

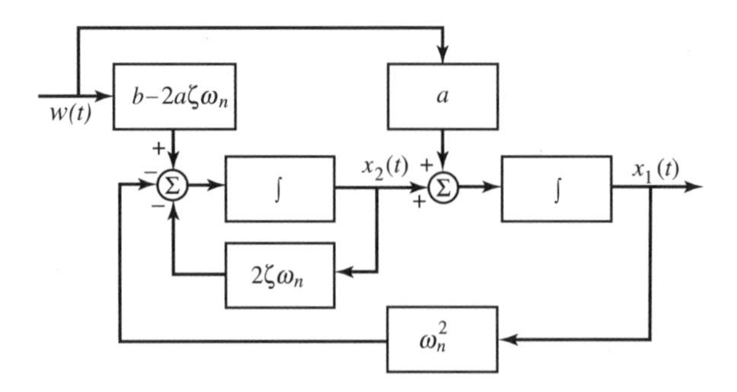

Fig. 3.5 Diagram of a second-order Markov process.

The *mean power* of a scalar random process is given by the equations

$$\underset{t}{E}\langle x^2(t)\rangle = \lim_{T\to\infty}\int_{-T}^{\mathrm{T}} x^2(t)\,dt \tag{3.71}$$

$$= \frac{1}{2\pi}\int_{-\infty}^{\infty} \Psi_x(\omega)\,d\omega \tag{3.72}$$

$$= \sigma^2. \tag{3.73}$$

The *cross power spectral density* between an RP x and an RP y is given by the formula

$$\Psi_{xy}(\omega) = \int_{-\infty}^{\infty} \psi_{xy}(\tau)e^{-j\omega\tau}\,d\tau \tag{3.74}$$

3.5 LINEAR SYSTEM MODELS OF RANDOM PROCESSES AND SEQUENCES

Assume that a linear system is given by

$$y(t) = \int_{-\infty}^{\infty} x(\tau)h(t,\tau)\,d\tau, \tag{3.75}$$

where $x(t)$ is input and $h(t,\tau)$ is the system weighting function (see Figure 3.6). If the system is time invariant, then Equation 3.75 becomes

$$y(t) = \int_{0}^{\infty} h(\tau)x(t-\tau)d\tau. \tag{3.76}$$

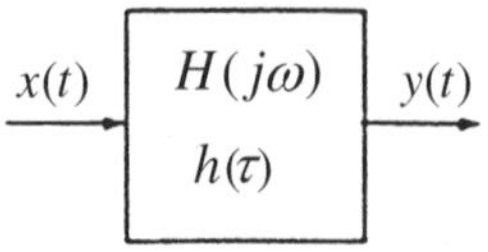

$x(t)$ → [$H(j\omega)$ / $h(\tau)$] → $y(t)$

Fig. 3.6 Block diagram representation of a linear system.

This type of integral is called a *convolution integral*. Manipulation of Equation 3.76 leads to relationships between autocorrelation functions of $x(t)$ and $y(t)$,

$$\psi_y(\tau) = \int_0^\infty d\tau_1\, h(\tau_1) \int_0^\infty d\tau_2\, h(\tau_2)\psi_x(\tau + \tau_1 - \tau_2), \tag{3.77}$$

$$\psi_{xy}(\tau) = \int_0^\infty h(\tau_1)\psi_x(\tau - \tau_1)d\tau_1 \tag{3.78}$$

and PSD relationships

$$\Psi_{xy}(\omega) = H(j\omega)\Psi_x(\omega), \tag{3.79}$$

$$\Psi_y(\omega) = |H(j\omega)|^2\Psi_x(\omega), \tag{3.80}$$

where H is the system transfer function shown in Figure 3.6, defined in Laplace transform notation as

$$H(s) = \int_0^\infty h(\tau)e^{s\tau}\, d\tau, \tag{3.81}$$

where $s = j\omega$.

3.5.1 Stochastic Differential Equations for Random Processes

A Note on the Calculus of Stochastic Differential Equations. Differential equations involving random processes are called stochastic differential equations. Introducing random processes as inhomogeneous terms in ordinary differential equations has ramifications beyond the level of rigor that will be followed here, but the reader should be aware of them. The problem is that random processes are not integrable functions in the conventional (Riemann) calculus. The resolution of this problem requires foundational modifications of the calculus to obtain many of the results presented. The Riemann integral of the "ordinary" calculus must be modified to what is called the *Itô calculus*. The interested reader will find these issues treated more rigorously in the books by Bucy and Joseph [15] and Itô [113].

A linear stochastic differential equation as a model of an RP with initial conditions has the form

$$\begin{aligned} \dot{x}(t) &= F(t)x(t) + G(t)w(t) + C(t)u(t), \\ z(t) &= H(t)x(t) + v(t) + D(t)u(t), \end{aligned} \tag{3.82}$$

where the variables are defined as

$$
\begin{aligned}
x(t) &= n \times 1 \text{ state vector,}\\
z(t) &= \ell \times 1 \text{ measurement vector,}\\
u(t) &= r \times 1 \text{ deterministic input vector,}\\
F(t) &= n \times n \text{ time-varying dynamic coefficient matrix,}\\
C(t) &= n \times r \text{ time-varying input coupling matrix,}\\
H(t) &= \ell \times n \text{ time-varying measurement sensitivity matrix,}\\
D(t) &= \ell \times r \text{ time-varying output coupling matrix,}\\
G(t) &= n \times r \text{ time-varying process noise coupling matrix,}\\
w(t) &= r \times 1 \text{ zero-mean uncorrelated "plant noise" process,}\\
v(t) &= \ell \times 1 \text{ zero-mean uncorrelated "measurement noise" process}
\end{aligned}
$$

and the expected values as

$$
\begin{aligned}
E\langle w(t)\rangle &= 0,\\
E\langle v(t)\rangle &= 0,\\
E\langle w(t_1)w^{\mathrm{T}}(t_2)\rangle &= Q(t_1)\delta(t_2 - t_1),\\
E\langle v(t_1)v^{\mathrm{T}}(t_2)\rangle &= R(t_1)\delta(t_2 - t_1).\\
E\langle w(t_1)v^{\mathrm{T}}(t_2)\rangle &= M(t_1)\delta(t_2 - t_1).
\end{aligned}
$$

The symbols Q, R, and M represent $r \times r$, $\ell \times \ell$, and $r \times \ell$ matrices, respectively, and δ represents the Dirac delta "function" (a measure). The values over time of the variable $x(t)$ in the differential equation model define vector-valued Markov processes. This model is a fairly accurate and useful representation for many real-world processes, including stationary Gaussian and nonstationary Gaussian processes, depending on the statistical properties of the random variables and the temporal properties of the deterministic variables. [The function $u(t)$ usually represents a known control input. For the rest of the discussion in this chapter, we will assume that $u(t) = 0$.]

EXAMPLE 3.5 Continuing with Example 3.3, let the RP $x(t)$ be a zero-mean stationary normal RP having autocorrelation

$$\psi_x(\tau) = \sigma^2 e^{-\alpha|\tau|}. \tag{3.83}$$

The corresponding power spectral density is

$$\Psi_x(\omega) = \frac{2\sigma^2\alpha}{\omega^2 + \alpha^2}. \tag{3.84}$$

This type of RP can be modeled as the output of a linear system with input $w(t)$, a zero-mean white Gaussian noise with PSD equal to unity. Using Equation 3.80, one can derive the transfer function $H(j\omega)$ for the following model:

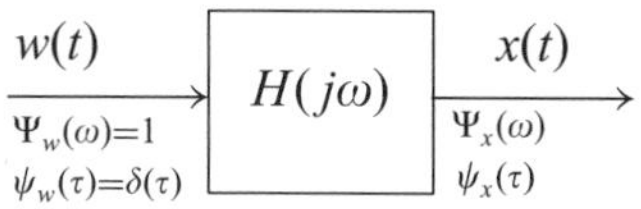

$w(t)$ → [$H(j\omega)$] → $x(t)$

$\Psi_w(\omega)=1$, $\psi_w(\tau)=\delta(\tau)$; $\Psi_x(\omega)$, $\psi_x(\tau)$

$$H(j\omega)H(-j\omega) = \frac{\sqrt{2\alpha\sigma}}{\alpha + j\omega} \cdot \frac{\sqrt{2\alpha\sigma}}{\alpha - j\omega}.$$

Take the stable portion of this system transfer function as

$$H(s) = \frac{\sqrt{2\alpha\sigma}}{s + \alpha}, \tag{3.85}$$

which can be represented as

$$\frac{x(s)}{w(s)} = \frac{\sqrt{2\alpha\sigma}}{s + \alpha}, \tag{3.86}$$

By taking the inverse Laplace transform of both sides of this last equation, one can obtain the following sequence of equations:

$$\begin{aligned} \dot{x}(t) + \alpha x(t) &= \sqrt{2\alpha\sigma} w(t), \\ \dot{x}(t) &= -\alpha x(t) + \sqrt{2\alpha\sigma} w(t), \\ z(t) &= x(t), \end{aligned}$$

with $\sigma_x^2(0) = \sigma^2$. The parameter $1/\alpha$ is called the *correlation time* of the process.

The block diagram representation of the process in Example 3.5 is shown in Table 3.1. This is called a *shaping filter*. Some other examples of differential equation models are also given in Table 3.1.

3.5.2 Discrete Model of a Random Sequence

A vector discrete-time recursive equation for modeling a random sequence (RS) with initial conditions can be given in the form

$$\begin{aligned} x_k &= \Phi_{k-1} x_{k-1} + G_{k-1} w_{k-1} + \Gamma_{k-1} u_{k-1}, \\ z_k &= H_k x_k + v_k + D_k u_k. \end{aligned} \tag{3.87}$$

TABLE 3.1 System Models of Random Processes

Random Process	Autocorrelation Function and Power Spectral Density	Shaping Filter Diagram	State-Space Formulation
White noise	$\psi_x(\tau) = \sigma^2\delta^2(\tau)$ $\psi_x(\omega) = \sigma^2$	None	Always treated as measurement noise
Random walk	$\psi_x(\tau) =$ (undefined) $\psi_x(\omega) \propto \sigma^2/\omega^2$	 $w(t)$, s^{-1}, $x(t)$	$\dot{x} = w(t)$ $\sigma_x^2(0) = 0$
Random constant	$\psi_x(\tau) = \sigma^2$ $\psi_x(\omega) = 2\pi\sigma^2\delta(\omega)$	None	$\dot{x} = 0$ $\sigma_x^2(0) = \sigma^2$
Sinusoid	$\psi_x(\tau) = \sigma^2\cos(\omega_0\tau)$ $\Psi_x(\omega) = \pi\sigma^2\delta(\omega - \omega_0) + \pi\sigma^2\delta(\omega + \omega_0)$	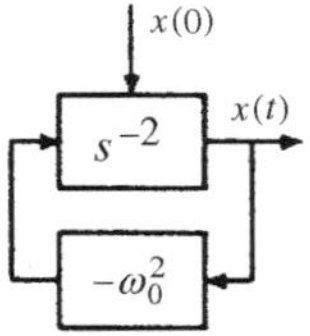 $x(0)$, s^{-2}, $x(t)$, $-\omega_0^2$	$\dot{x} = \begin{bmatrix} 0 & 1 \\ -\omega_0^2 & 0 \end{bmatrix} x$ $P(0) = \begin{bmatrix} \sigma^2 & 0 \\ 0 & 0 \end{bmatrix}$
Exponentially correlated	$\psi_x(\tau) = \sigma^2 e^{-\alpha\|\tau\|}$ $\Psi_x(\omega) = \dfrac{2\sigma^\alpha\alpha}{\omega^2 + \alpha^2}$ $\dfrac{1}{\alpha} =$ correlation time	$w(t)$, $\sigma\sqrt{2\alpha}$, $x(0)$, s^{-1}, $x(t)$, $-\alpha$	$\dot{x} = -\alpha x + \sigma\sqrt{2\alpha}w(t)$ $\sigma_x^2(0) = \sigma^2$

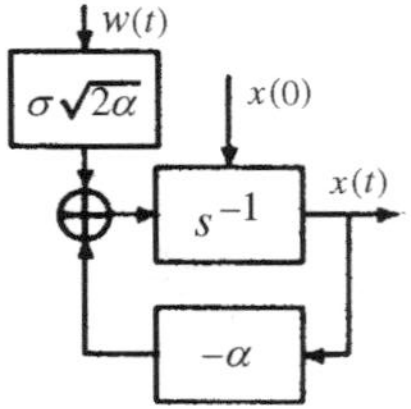

This is the complete model with deterministic inputs u_k as discussed in Chapter 2, Equations 2.28 and 2.29, and random sequence noise w_k and v_k as described in Chapter 4 equations:

$$\begin{aligned}
x_k &= n \times 1 \text{ state vector} \\
z_k &= \ell \times 1 \text{ measurement vector} \\
u_k &= r \times 1 \text{ deterministic input vector} \\
\Phi_{k-1} &= n \times n \text{ time varying matrix} \\
G_{k-1} &= n \times r \text{ time varying matrix} \\
H_k &= \ell \times n \text{ time varying matrix} \\
D_k &= \ell \times r \text{ time varying matrix} \\
\Gamma_{k-1} &= n \times r \text{ time varying matrix} \\
E\langle w_k \rangle &= 0 \\
E\langle v_k \rangle &= 0 \\
E\langle w_{k_1} w_{k_2}^{\mathrm{T}} \rangle &= Q_{k_1}\ \Delta(k_2 - k_1) \\
E\langle v_{k_1} v_{k_2}^{\mathrm{T}} \rangle &= R_{k_1}\ \Delta(k_2 - k_1) \\
E\langle w_{k_1} v_{k_2}^{\mathrm{T}} \rangle &= M_{k_1}\ \Delta(k_2 - k_1)
\end{aligned}$$

EXAMPLE 3.6 Let the $\{x_k\}$ be a zero-mean stationary Gaussian RS with autocorrelation

$$\psi_x(k_2 - k_1) = \sigma^2 e^{-\alpha|k_2 - k_1|}.$$

This type of RS can be modeled as the output of a linear system with input w_k being zero-mean white Gaussian noise with PSD equal to unity.

A difference equation model for this type of process can be defined as

$$x_k = \Phi x_{k-1} + G w_{k-1}, \quad z_k = x_k. \tag{3.88}$$

In order to use this model, we need to solve for the unknown parameters Φ and G as functions of the parameter α. To do so, we first multiply Equation 3.88 by x_{k-1} on both sides and take the expected values to obtain the equations

$$\begin{aligned}
E\langle x_k x_{k-1} \rangle &= \Phi E\langle x_{k-1} x_{k-1} \rangle + G E\langle w_{k-1} x_{k-1} \rangle, \\
\sigma^2 e^{-\alpha} &= \Phi \sigma^2,
\end{aligned}$$

assuming the w_k are uncorrelated and $E\langle w_K \rangle = 0$, so that $E\langle w_{k-1} x_{k-1} \rangle = 0$. One obtains the solution

$$\Phi = e^{-\alpha}. \tag{3.89}$$

Next, square the state variable defined by Equation 3.88 and take its expected value:

$$E\langle x_k^2\rangle = \Phi^2 E\langle x_{k-1}x_{k-1}\rangle + G^2 E\langle w_{k-1}w_{k-1}\rangle, \tag{3.90}$$

$$\sigma^2 = \sigma^2\Phi^2 + G^2, \tag{3.91}$$

because the variance $E\langle w_{k-1}^2\rangle = 1$ and the parameter $G = \sigma\sqrt{1 - e^{-2\alpha}}$.

The complete model is then

$$x_k = e^{-\alpha}x_{k-1} + \sigma\sqrt{1 - e^{-2\alpha}}w_{k-1}$$

with $E\langle w_k\rangle = 0$ and $E\langle w_{k_1}w_{k_2}\rangle = \Delta(k_2 - k_1)$.

The dynamic process model derived in Example 3.6 is called a shaping filter. Block diagrams of this and other shaping filters are given in Table 3.2, along with their difference equation models.

3.5.3 Autoregressive Processes and Linear Predictive Models

A *linear predictive model* for a signal is a representation in the form

$$\acute{x}_{k+1} = \sum_{i=1}^{n} a_i \acute{x}_{k-i+1} + \acute{u}_k, \tag{3.92}$$

where $\acute{u}_k$ is the *prediction error*. Successive samples of the signal are predicted as linear combinations of the n previous values.

An *autoregressive process* has the same formulation, except that $\acute{u}_k$ is a white Gaussian noise process. Note that this formula for an autoregressive process can be rewritten in state transition matrix (STM) form as

$$\begin{bmatrix} \acute{x}_{k+1} \\ \acute{x}_k \\ \acute{x}_{k_1} \\ \vdots \\ \acute{x}_{k-n+2} \end{bmatrix} = \begin{bmatrix} a_1 & a_2 & \cdots & a_{n-1} & a_n \\ 1 & 0 & \cdots & 0 & 0 \\ 0 & 1 & \cdots & 0 & 0 \\ \vdots & \vdots & \ddots & \vdots & \vdots \\ 0 & 0 & \cdots & 1 & 0 \end{bmatrix} \begin{bmatrix} \acute{x}_k \\ \acute{x}_{k-1} \\ \acute{x}_{k-2} \\ \vdots \\ \acute{x}_{k-n+1} \end{bmatrix} + \begin{bmatrix} \acute{u} \\ 0 \\ 0 \\ \vdots \\ 0 \end{bmatrix}, \tag{3.93}$$

$$x_{k+1} = \Phi x_k + u_k, \tag{3.94}$$

where the "state" is the n-vector of the last n samples of the signal and the covariance matrix Q_k of the associated process noise u_k will be filled with zeros, except for the term $Q_{11} = E\langle \acute{u}_k^2\rangle$.

TABLE 3.2 Stochastic System Models of Discrete Random Sequences

Process Type	Autocorrelation	Block Diagram	State-Space Model
Random constant	$\psi_x(k_1 - k_2) = \sigma^2$	x_k Delay x_{k-1}	$x_k = x_{k-1}$ $\sigma_x^2(0) = \sigma^2$
Random walk	$\to +\infty$	w_{k-1} x_k Delay x_{k-1}	$x_k = x_{k-1} + w_{k-1}$ $\sigma_x^2(0) = 0$
Exponentially correlated noise	$\psi_x(k_2 - k_1) = \sigma^2 e^{-\alpha \lvert k_2 - k_1 \rvert}$	w_{k-1} $\sigma\sqrt{1-e^{-2\alpha}}$ x_k Delay x_{k-1} $e^{-\alpha}$	$x_k = e^{-\alpha} x_{k-1} + \sigma\sqrt{1 - e^{-2\alpha}}\, w_{k-1}$ $\sigma_x^2(0) = \sigma^2$

3.6 SHAPING FILTERS AND STATE AUGMENTATION

Shaping Filters. The focus of this section is nonwhite models for stationary processes. For many physical systems encountered in practice, it may not be justified to assume that all noises are white Gaussian noise processes. It can be useful to generate an autocorrelation function or PSD from real data and then develop an appropriate noise model using differential or difference equations. These models are called shaping filters. They are driven by noise with a flat spectrum (white-noise processes), which they shape to represent the spectrum of the actual system. It was shown in the previous section that a linear time-invariant system (shaping filter) driven by WSS white Gaussian noise provides such a model. The state vector can be "augmented" by appending to it the state vector components of the shaping filter, with the resulting model having the form of a linear dynamic system driven by white noise.

3.6.1 Correlated Process Noise Models

Shaping Filters for Process Noise. Let a system model be given by

$$\dot{x}(t) = F(t)x(t) + G(t)w_1(t), \qquad z(t) = H(t)x(t) + v(t) \tag{3.95}$$

where $w_1(t)$ is nonwhite, for example, correlated Gaussian noise. As given in the previous section, $v(t)$ is a zero-mean white Gaussian noise. Suppose that $w_1(t)$ can be modeled by a linear shaping filter[10]:

$$\begin{aligned} \dot{x}_{\mathrm{SF}}(t) &= F_{\mathrm{SF}}(t)x_{\mathrm{SF}}(t) + G_{\mathrm{SF}}(t)w_2(t) \\ w_1(t) &= H_{\mathrm{SF}}(t)x_{\mathrm{SF}}(t) \end{aligned} \tag{3.96}$$

where SF denotes the shaping filter and $w_2(t)$ is zero mean white Gaussian noise. Now define a new augmented state vector

$$X(t) = [x(t)\ x_{\mathrm{SF}}(t)]^{\mathrm{T}}. \tag{3.97}$$

Equations 3.95 and 3.96 can be combined into the matrix form

$$\begin{bmatrix} \dot{x}(t) \\ \dot{x}_{\mathrm{SF}}(t) \end{bmatrix} = \begin{bmatrix} F(t) & G(t)H_{\mathrm{SF}}(t) \\ 0 & F_{\mathrm{SF}}(t) \end{bmatrix} \begin{bmatrix} x(t) \\ x_{\mathrm{SF}}(t) \end{bmatrix} + \begin{bmatrix} 0 \\ G_{\mathrm{SF}}(t) \end{bmatrix} w_2(t), \tag{3.98}$$

$$\dot{X}(t) = F_T(t)X(t) + G_T(t)w_2(t), \tag{3.99}$$

[10] See Example in Section 3.7 for WSS processes.

and the output equation can be expressed in compatible format as

$$z(t) = [H(t) \quad 0] \begin{bmatrix} x(t) \\ x_{\text{SF}}(t) \end{bmatrix} + v(t) \tag{3.100}$$

$$= H_T(t)X(t) + v(t). \tag{3.101}$$

This total system given by Equations 3.99 and 3.101 is a linear differential equation model driven by white Gaussian noise. (See Figure 3.7 for a nonwhite-noise model.)

3.6.2 Correlated Measurement Noise Models

Shaping Filters for Measurement Noise. A similar development is feasible for the case of time-correlated measurement noise $v_1(t)$:

$$\begin{aligned} \dot{x}(t) &= F(t)x(t) + G(t)w(t), \\ z(t) &= H(t)x(t) + v_1(t). \end{aligned} \tag{3.102}$$

In this case, let $v_2(t)$ be zero-mean white Gaussian noise and let the measurement noise $v_1(t)$ be modeled by

$$\begin{aligned} \dot{x}_{\text{SF}}(t) &= F_{\text{SF}}(t)x_{\text{SF}}(t) + G_{\text{SF}}(t)v_2(t), \\ v_1(t) &= H_{\text{SF}}(t)x_{\text{SF}}(t). \end{aligned} \tag{3.103}$$

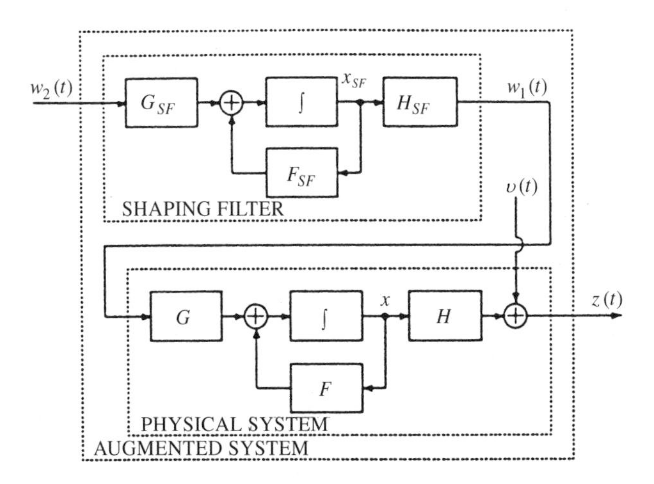

Fig. 3.7 Shaping filter model for nonwhite noise.

The total augmented system is given by

$$\begin{bmatrix} \dot{x}(t) \\ \dot{x}_{SF}(t) \end{bmatrix} = \begin{bmatrix} F(t) & 0 \\ 0 & F_{SF}(t) \end{bmatrix} \begin{bmatrix} x(t) \\ x_{SF}(t) \end{bmatrix} + \begin{bmatrix} G(t) & 0 \\ 0 & G_{SF}(t) \end{bmatrix} \begin{bmatrix} w(t) \\ v_2(t) \end{bmatrix}, \tag{3.104}$$

$$z(t) = [H(t)\ H_{SF}(t)] \begin{bmatrix} x(t) \\ x_{SF}(t) \end{bmatrix}.$$

This is in the form of a linear system model driven by white Gaussian noise and output equation with no input noise.

These systems can be specialized to the WSS process for continuous and discrete cases as by shaping filters shown in Tables 3.1 and 3.2.

EXAMPLE 3.7 The "midpoint acceleration" error for an acceleration sensor (accelerometer) is defined as the effective acceleration error at the midpoint of the sampling period. The associated error model for an accelerometer in terms of unknown parameters of the sensor is as follows:

$$\Delta_{\beta m} = \beta_m \otimes \zeta + b_A + h_A \beta_m + \beta_m^2 (FI1 - FX1) + \delta\beta,$$

where

$\Delta_{\beta m}$ = the midpoint acceleration error

$\otimes$ = the cross product (for 3 − vectors)

ζ = a 3×1 vector representing attitude alignment errors between "platform" axes and computational axes

b_A = a 3×1 vector of unknown accelerometer biases, normalized to the magnitude of gravity

$$h_A = \begin{bmatrix} S_1 & \delta_{12} & \delta_{13} \\ 0 & S_2 & \delta_{23} \\ 0 & 0 & S_3 \end{bmatrix}$$

and

S_i = unknown accelerometer scale factor errors ($i = 1, 2, 3$)
δ_{ij} = unknown accelerometer axes nonorthogonalities
$\delta\beta$ = other error terms, some of which are observable; for reason of practicality in our example they are not estimated, only compensated with factory-calibrated values
$FI1$ = a 3×1 unknown acceleration-squared nonlinearity for acceleration along the accelerometer input axis
$FX1$ = a 3×1 unknown acceleration-squared nonlinearity for acceleration normal to the accelerometer input axis
β_m = a 3×1vector $(\beta_1, \beta_2, \beta_3)^{\mathrm{T}}$ of midpoint components of acceleration in platform coordinates

$$\beta_m^2 = \begin{bmatrix} \beta_m^2 & 0 & 0 \\ 0 & \beta_2^2 & 0 \\ 0 & 0 & \beta_3^2 \end{bmatrix}$$

The 12×1 accelerometer state vector x^A is composed of the components

$$(x^A) = \Bigg[\underbrace{b_A}_{3\times 1} \quad S_1 \quad \delta_{12} \quad S_2 \quad \delta_{13} \quad \delta_{23} \quad S_3 \quad \underbrace{(FX1 - FI1)}_{3\times 1}\Bigg]^{\mathrm{T}}.$$

The 12 unknown parameters will be modeled as random walks (see Table 3.1) for the parameter identification problem to be discussed in Chapter 5 (Example 5.4).

EXAMPLE 3.8 A gyroscope drift error model is given as follows:

$$\varepsilon = b_g + h_g\omega + U_g\beta + K_g\beta^1 + [|\omega|]T_g + b_{gt}t + U_{gt}t\beta,$$

where

b_g = a 3×1 vector of unknown gyroscope fixed drift parameters
h_g = a 3×3 matrix containing unknown scale factor (S_{gi}) and linear axes alignment errors (Δ_{ij}) as components ($i, j = 1, 2, 3$)

$$\begin{bmatrix} S_{g1} & \Delta_{12} & \Delta_{13} \\ \Delta_{21} & S_{g2} & \Delta_{23} \\ \Delta_{31} & \Delta_{32} & S_{g3} \end{bmatrix}$$

T_g = a 3×1 vector of unknown nonlinear gyroscope torquer scale factor errors, with elements δS_{gi}

$[|\omega|]$ = a 3×3 diagonal matrix composed of absolute values of the components of ω (platform inertial angular rate) on the corresponding diagonal element

U_g = a 3×3 matrix of unknown gyroscope mass unbalance parameters $(d_{k,j})$

$$\begin{bmatrix} d_{I1} & d_{01} & d_{S1} \\ d_{S2} & d_{I2} & d_{02} \\ d_{03} & d_{S3} & d_{I3} \end{bmatrix}$$

indices $I, 0$, and S denoting input, output, and spin axes, respectively, for each gyroscope 1, 2 and 3.

K_g = a 3×6 matrix of unknown gyroscope compliance (g-squared) errors k_{kji}

$$\begin{bmatrix} k_{II1} & k_{001} & k_{SS1} & I_{I01} & k_{IS1} & k_{S01} \\ k_{SS2} & k_{II2} & k_{002} & k_{IS2} & k_{S02} & k_{I02} \\ k_{003} & k_{SS3} & k_{II3} & k_{S03} & k_{I03} & k_{IS3} \end{bmatrix}$$

b_{gt} = 3×1 vector of unknown gyroscope fixed-drift trend parameters

U_{gt} = 3×6 matrix of unknown gyroscope mass unbalance trend parameters

β = 3×1 vector of vertical direction cosines (normalized gravity) $(\beta_1, \beta_2, \beta_3)^{\mathrm{T}}$

β^1 = 6×1 vector with components $(\beta_1^2, \beta_2^2, \beta_3^2, \beta_1\beta_2, \beta_1\beta_3, \beta_2\beta_3)^{\mathrm{T}}$

$$x^g(t) = \begin{bmatrix} 3\times 1 & 9\times 1 & 9\times 1 & 15\times 1 & 3\times 1 & 3\times 1 & 6\times 1 \\ b_g & h_g^1 & U_g^1 & K_g^1 & T_g & b_{gt} & U_{gt}^1 \end{bmatrix}^{\mathrm{T}}$$

The 48 unknown parameters will be modeled as random walks and random ramps (see Table 3.1) for the parameter identification problem to be discussed in Chapter 5 (Example 5.4).

3.7 COVARIANCE PROPAGATION EQUATIONS

The second moments of the state $x(t)$ (a random process) and forcing function $w(t)$ (another random process) can be described in terms of covariance matrices. Let us define the $n \times n$ covariance matrix

$$P(t) = E\langle[x(t) - \mathrm{E}\langle x(t)\rangle][x(t) - \mathrm{E}\langle x(t)\rangle]^{\mathrm{T}}\rangle. \tag{3.105}$$

If we replace $E\langle x(t)\rangle$ with the estimate of $x(t)$ defined by $\hat{x}(t)$ in Chapter 4, then $P(t)$ will be called the error covariance matrix.

3.7.1 Propagation in Continuous Time

Let

$$\begin{aligned} \dot{x} &= F(t)x + G(t)w(t), \quad E\langle w(t)\rangle = 0, \\ E\langle w(t_1)w^{\mathrm{T}}(t_2)\rangle &= Q(t_2, t_1)\delta(t_2 - t_1). \end{aligned} \tag{3.106}$$

The solution of this equation with $x(t_0)$ as initial condition and $\Phi(t, t_0)$ as state transition matrix has been discussed in Chapter 2,

$$x(t) = \Phi(t, t_0)x(t_0) + \int_{t_0}^{t} \Phi(t, \tau)G(\tau)w(\tau)\, d\tau. \tag{3.107}$$

Take the expected value

$$E\langle x(t)\rangle = \Phi(t, t_0)E\langle x(t_0)\rangle + \int_{t_0}^{t} \Phi(t, \tau)G(\tau)E\langle w(\tau)\rangle\, d\tau. \tag{3.108}$$

Then

$$[x(t) - E\langle x(t)\rangle] = \Phi(t, t_0)[x(t_0) - E\langle x(t_0)\rangle] + \int_{t_0}^{t} \Phi(t, \tau)G(\tau)w(\tau)\, d\tau \tag{3.109}$$

The covariance matrix $P(t)$ is given by

$$\begin{aligned} P(t) &= E\langle [x(t) - \mathrm{E}\langle x(t)\rangle][x(t) - E\langle x(t)\rangle]^{\mathrm{T}}\rangle \\ &= E\left\langle \left[\Phi(t, t_0)[x(t_0) - E\langle x(t_0)\rangle] + \int_{t_0}^{t} \Phi(t, \tau)G(\tau)w(\tau)\, d\tau \right] \right. \\ &\quad \left. \times \left[\Phi(t, t_0)[x(t_0) - E\langle x(t_0)\rangle] + \int_{t_0}^{t} \Phi(t, \tau)G(\tau)w(\tau)\, d\tau \right]^{\mathrm{T}} \right\rangle \\ &= \Phi(t, t_0)E\langle [x(t_0) - E\langle x(t_0)\rangle][x(t_0) - E\langle x(t_0)\rangle]^{\mathrm{T}}\rangle \Phi^{\mathrm{T}}(t, t_0) \\ &\quad + \Phi(t, t_0)\int_{t_0}^{t} \mathrm{E}\langle \underline{x(t_0) - E\langle x(t_0)\rangle [w(\tau)]^{\mathrm{T}}}\rangle G^{\mathrm{T}}(\tau)\Phi^{\mathrm{T}}(t, \tau)\, d\tau \\ &\quad + \int_{t_0}^{t} \Phi(t, \tau)G(\tau)E\langle \underline{[w(\tau)][x(t_0) - E\langle x(t_0)\rangle]^{\mathrm{T}}}\rangle \Phi^{\mathrm{T}}(t, t_0)\, d\tau \\ &\quad + \int_{t_0}^{t}\int_{t_0}^{t} \Phi(t, \tau_1)G(\tau_1)E\langle w(\tau_1)w^{\mathrm{T}}(\tau_2)\rangle G^{\mathrm{T}}(\tau_2)\Phi^{\mathrm{T}}(t, \tau_2)\, d\tau_1 d\tau_2. \end{aligned}$$

The underlined quantities are zero since $w(t)$ and $x(t_0)$ are uncorrelated for $t > t_0$,

$$\begin{aligned}
P(t) &= \Phi(t, t_0)P(t_0)\Phi^{\mathrm{T}}(t, t_0) \\
&\quad + \int_{t_0}^{t}\int_{t_0}^{t} \Phi(t, \tau_1)G(\tau_1)Q\delta(\tau_2 - \tau_1)G^{\mathrm{T}}(\tau_2)\Phi^{\mathrm{T}}(t, \tau_2)\, d\tau_1\, d\tau_2 \\
&= \Phi(t, t_0)P(t_0)\Phi^{\mathrm{T}}(t, t_0) \\
&\quad + \int_{t_0}^{t} \Phi(t, \tau)G(\tau)QG^{\mathrm{T}}(\tau)\Phi^{\mathrm{T}}(t, \tau)\, d\tau.
\end{aligned}$$

The double integral reduces to a single integral, due to the δ-function (see Equation 3.42). We can use the fact that $\dot{\Phi}(t, t_0) = F(t)\Phi(t, t_0)$ and apply Leibniz's rule to differentiate the integral:

$$\begin{aligned}
\dot{P}(t) &= \dot{\Phi}(t, t_0)P(t_0)\Phi^{\mathrm{T}}(t, t_0) + \Phi(t, t_0)P(t_0)\dot{\Phi}^{\mathrm{T}}(t, t_0) \\
&\quad + \frac{d}{dt}\int_{t_0}^{t} \Phi(t, \tau)G(\tau)QG^{\mathrm{T}}(\tau)\Phi^{\mathrm{T}}(t, \tau)\, d\tau \\
&= F(t)\Phi(t, t_0)P(t_0)\Phi^{\mathrm{T}}(t, t_0) + \Phi(t, t_0)P(t_0)\Phi^{\mathrm{T}}(t, t_0)F^{\mathrm{T}}(t) \\
&\quad + \int_{t_0}^{t} [\dot{\Phi}(t, \tau)G(\tau)QG^{\mathrm{T}}(\tau)\Phi^{\mathrm{T}}(t, \tau) + \Phi(t, \tau)G(\tau)QG^{\mathrm{T}}(\tau)\dot{\Phi}^{\mathrm{T}}(t, \tau)]\, d\tau \\
&\quad + G(t)QG^{\mathrm{T}}(t).
\end{aligned}$$

When terms containing F or F^{T} as a factor are collected in square brackets, as shown below, they can be recognized as being equal to $P(t)$:

$$\begin{aligned}
\dot{P}(t) &= F(t)\left[\Phi(t, t_0)P(t_0)\Phi^{\mathrm{T}}(t, t_0) + \int_{t_0}^{t} \Phi(t, \tau)G(\tau)QG^{\mathrm{T}}(\tau)\Phi^{\mathrm{T}}(t, \tau)\, d\tau\right] \\
&\quad + \left[\Phi(t, t_0)P(t_0)\Phi^{\mathrm{T}}(t, t_0) + \int_{t_0}^{t} \Phi(t, \tau)G(\tau)QG^{\mathrm{T}}(\tau)\Phi^{\mathrm{T}}(t, \tau)\, d\tau\right]F^{\mathrm{T}}(t) + G(t)QG^{\mathrm{T}}(t) \\
&= F(t)P(t) + P(t)F^{\mathrm{T}}(t) + G(t)QG^{\mathrm{T}}(t).
\end{aligned}$$

Furthermore, if F and G are constant, the so-called steady-state equation $\dot{P} = 0$ is an *algebraic* matrix equation:

$$0 = FP(\infty) + P(\infty)F^{\mathrm{T}} + GQG^{\mathrm{T}}, \tag{3.110}$$

which may fail to have a nonnegative definite solution. (See Problem 3.35.)

EXAMPLE 3.9: Steady-State Solution of the State Covariance Equation for the Harmonic Resonator Model Consider the stochastic system model

$$
\begin{aligned}
\dot{x}(t) &= Fx(t) + w(t), \\
E\langle w(t_1)w^{\mathrm{T}}(t_2)\rangle &= \delta(t_1 - t_2)Q, \\
Q &= \begin{bmatrix} 0 & 0 \\ 0 & q \end{bmatrix}
\end{aligned}
$$

for a harmonic resonator driven by white *acceleration* noise $w(t)$. That is, the additive process noise on the resonator *velocity* is zero.

It is of interest to find the covariance of the process $x(t)$ at *steady state*. (It could be infinite, but in this example it will be finite.)

Recall that the state-space model for the harmonic resonator from Examples 2.2, 2.3, 2.6, and 2.7 has as its dynamic coefficient matrix the 2×2 matrix

$$
\begin{aligned}
f &= \begin{bmatrix} 0 & 1 \\ -\dfrac{k_s}{m} & -\dfrac{k_d}{m} \end{bmatrix} \\
&= \begin{bmatrix} 0 & 1 \\ -\omega^2 - \xi^2 & -2\xi \end{bmatrix},
\end{aligned}
$$

where the alternate model parameters

$$
\begin{aligned}
\omega &= \sqrt{\frac{k_s}{m} - \frac{k_d^2}{4m^2}}, \\
\tau &= \frac{2m}{k_d}, \\
\xi &= \frac{1}{\tau}
\end{aligned}
$$

are the resonant frequency, the damping time constant, and the damping frequency, respectively. The state covariance equation

$$
\frac{d}{dt}P(t) = FP(t) + P(t)F^{\mathrm{T}} + Q
$$

has the steady-state form

$$
\begin{aligned}
0 &= \lim_{t \to \infty}\left(\frac{d}{dt}X(t)\right) \\
&= FP(\infty) + P(\infty)F^{\mathrm{T}} + Q,
\end{aligned}
$$

which is a *linear* equation in the unknown elements $p_{11}, p_{12}, p_{21}, p_{22}$ of $P(\infty)$.

Because this is a *symmetric* 2×2 matrix equation, it is equivalent to three scalar linear equations:

$$0 = \sum_{k=1}^{2} f_{1k} p_{k1} + \sum_{k=1}^{2} p_{1k} f_{k1} + q_{11},$$
$$0 = \sum_{k=1}^{2} f_{1k} p_{k2} + \sum_{k=1}^{2} p_{1k} f_{k2} + q_{12},$$
$$0 = \sum_{k=1}^{2} f_{2k} p_{k2} + \sum_{k=1}^{2} p_{2k} f_{k2} + q_{22},$$

with known parameters

$$\begin{aligned} q_{11} &= 0, & q_{12} &= 0, & q_{22} &= q, \\ f_{11} &= 0, & f_{12} &= 1, & & \\ f_{21} &= -\omega^2 - \xi^2, & f_{22} &= -2\xi. & & \end{aligned}$$

However, because the unknown matrix $P(\infty)$ is a symmetric 2×2 matrix, it has only three independent elements ($p_{12} = p_{21}$). Therefore, the above linear system of equations can be reduced to the nonsingular 3×3 system of equations

$$\begin{bmatrix} 0 \\ 0 \\ q \end{bmatrix} = - \begin{bmatrix} 0 & 2 & 0 \\ -(\omega^2 + \tau^{-2}) & -\dfrac{2}{\tau} & 1 \\ 0 & -2(\omega^2 + \tau^{-2}) & -\dfrac{4}{\tau} \end{bmatrix} \begin{bmatrix} p_{11} \\ p_{12} \\ p_{22} \end{bmatrix}$$

with solution

$$\begin{bmatrix} p_{11} \\ p_{12} \\ p_{22} \end{bmatrix} = \begin{bmatrix} \dfrac{q\tau}{4(\omega^2 + \tau^{-2})} \\ 0 \\ \dfrac{q\tau}{4} \end{bmatrix},$$

$$P(\infty) = \begin{bmatrix} p_{11} & p_{12} \\ p_{12} & p_{22} \end{bmatrix} = \frac{q\tau}{4(\omega^2 + \tau^{-2})} \begin{bmatrix} 1 & 0 \\ 0 & (\omega^2 + \tau^{-2}) \end{bmatrix}.$$

[11]The dimensionless quantity $2\pi\omega\tau$ is called the *quality factor*, *Q-factor*, or simply the Q of a resonator. It equals the number of cycles that the unforced resonator will go through before its amplitude falls to $1/e \approx 37\,\%$ of its initial amplitude.

Note that the steady-state state covariance depends linearly on the process noise covariance q. The steady-state covariance of velocity also depends linearly on the damping time constant τ.[11]

3.7.2 Covariance Propagation in Discrete Time

Recall that the equations for the state and its first moment (expected value) with $G_{k-1} = I$ are

$$\begin{aligned} x_k &= \Phi_{k-1}x_{k-1} + w_{k-1}, \\ E\langle x_k\rangle &= \Phi_{k-1}E\langle x_{k-1}\rangle + E\langle w_{k-1}\rangle, \\ &= \Phi_{k-1}E\langle x_{k-1}\rangle, \end{aligned}$$

respectively (because $E\langle w_{k-1}\rangle = 0$). The corresponding equations for the second moment (covariance) of the state can then be derived as follows:

$$\begin{aligned} P_k &= E\langle [x_k - E\langle x_k\rangle][x_k - E\langle x_k\rangle]^{\mathrm{T}}\rangle \\ &= E\langle [\Phi_{k-1}[x_{k-1} - E\langle x_{k-1}\rangle] + w_{k-1}] \\ &\quad \times [\Phi_{k-1}[x_{k-1} - E\langle x_{k-1}\rangle] + w_{k-1}]^{\mathrm{T}}\rangle \\ &= E\langle \Phi_{k-1}[x_{k-1} - E\langle x_{k-1}\rangle][x_{k-1} - E\langle x_{k-1}\rangle]^{\mathrm{T}}\Phi_{k-1}^{\mathrm{T}} \\ &\quad + \Phi_{k-1}[x_{k-1} - E\langle x_{k-1}\rangle]w_{k-1}^{\mathrm{T}} \\ &\quad + w_{k-1}[x_{k-1} - E\langle x_{k-1}\rangle]^{\mathrm{T}}\Phi_{k-1}^{\mathrm{T}} + w_{k-1}w_{k-1}^{\mathrm{T}}\rangle \\ &= \Phi_{k-1}\underbrace{E\langle [x_{k-1} - E\langle x_{k-1}\rangle][x_{k-1} - E\langle x_{k-1}\rangle]^{\mathrm{T}}\rangle}_{P_{k-1}}\Phi_{k-1}^{\mathrm{T}} \\ &\quad + \Phi_{k-1}E\langle [x_{k-1} - E\langle x_{k-1}\rangle]w_{k-1}^{\mathrm{T}}\rangle \\ &\quad + E\langle w_{k-1}[x_{k-1} - E\langle x_{k-1}\rangle]^{\mathrm{T}}\rangle\Phi_{k-1}^{\mathrm{T}} + \underbrace{E\langle w_{k-1}w_{k-1}^{\mathrm{T}}\rangle}_{Q_{k-1}} \\ &= \Phi_{k-1}P_{k-1}\Phi_{k-1}^{\mathrm{T}} + Q_{k-1}, \end{aligned} \tag{3.111}$$

which is the *evolution equation* for P. When Φ and Q are constant, the corresponding steady-state equation $P_k = P_{k-1}$ is also an *algebraic* matrix equation

$$P_\infty = \Phi P_\infty \Phi^{\mathrm{T}} + Q, \tag{3.112}$$

which can fail to have a finite solution. (See Problem 3.36.) An example with a finite solution is provided in the next subsection (Example 3.11).

3.7.3 Dependence of Q_k on $Q(t)$

The process noise covariance $Q(t)$ in continuous time is related to the equivalent discrete-time process noise covariance Q_k by the matrix differential equation

$$\frac{d}{dt}Q_k(t) = F(t)Q_k(t) + Q_k(t)F^{\mathrm{T}}(t) + G(t)Q(t)G^{\mathrm{T}}(t)$$

with initial condition

$$Q_k(t_{k-1}) = 0$$

and final condition

$$Q_k = Q_k(t_k)$$

defining the equivalent discrete-time process noise covariance. See Section 4.10.

Initial-Condition Solution. If

$$\dot{x} = F(t)x + G(t)w(t), \tag{3.113}$$

then the covariance equation is

$$\dot{P} = F(t)P + PF^{\mathrm{T}}(t) + G(t)QG^{\mathrm{T}}(t), \tag{3.114}$$

$$P(t_{k-1}) = P_{k-1}, \tag{3.115}$$

and solution of the above differential equation with $t = t_k$ and $t_0 = t_{k-1}$ is

$$x(t_k) = \Phi(t_k, t_{k-1})x(t_{k-1}) + \int_{t_{k-1}}^{t_k} \Phi(t_k, \tau)G(\tau)w(\tau)\, d\tau. \tag{3.116}$$

Then the difference equation is

$$x_k = \Phi_{k-1}x_{k-1} + w_{k-1}. \tag{3.117}$$

Observe that x_k is equal to w_{k-1} if $x_{k-1} = 0$. Hence, the covariance of w_{k-1} is equal to the covariance of x_k when the covariance of x_{k-1} is zero. This covariance is determined by the stated procedure.

EXAMPLE 3.10: Discrete-Time Process Noise Covariance for the Harmonic Resonator Model Consider the problem of determining the covariance matrix Q_{k-1} for the equivalent *discrete-time model*

$$x_k = \Phi_{k-1}x_{k-1} + w_{k-1},$$

$$E\langle w_{k-1}w_{k-1}^{\mathrm{T}}\rangle = Q_{k-1}$$

for a harmonic resonator driven by white acceleration noise, given the variance q of the process noise in its *continuous-time model*

$$\frac{d}{dt}x(t) = Fx(t) + w(t),$$

$$E\langle w(t_1)w^{\mathrm{T}}(t_2)\rangle = \delta(t_1 - t_2)Q,$$

$$Q = \begin{bmatrix} 0 & 0 \\ 0 & q \end{bmatrix},$$

where ω is the resonant frequency, τ is the damping time constant, ξ is the corresponding damping "frequency" (i.e., has frequency units), and q is the process noise covariance in continuous time. [This stochastic system model for a harmonic resonator driven by white acceleration noise $w(t)$ is derived in Examples 3.4 and 3.9.]

Following the derivation of Example 2.6, the fundamental solution matrix for the *unforced* dynamic system model can be expressed in the form

$$\begin{aligned}\Phi(t) &= e^{Ft} \\ &= e^{-t/\tau}\begin{bmatrix} \dfrac{S(t) + C(t)\omega\tau}{\omega\tau} & \dfrac{S(t)}{\omega} \\ -\dfrac{S(t)(1+\omega^2\tau^2)}{\omega\tau^2} & \dfrac{-S(t) + C(t)\omega\tau}{\omega\tau} \end{bmatrix}, \\ S(t) &= \sin(\omega t), \\ C(t) &= \cos(\omega t),\end{aligned}$$

in terms of its resonant frequency ω and damping time-constant τ. Its matrix inverse

$$\Phi^{-1}(s) = \frac{e^{s/\tau}}{\omega\tau^2}\begin{bmatrix} \tau[\omega\tau C(s) - S(s)] & -\tau^2 S(s) \\ (1+\omega^2\tau^2)S(s) & \tau[\omega\tau C(s) + S(s)] \end{bmatrix}$$

at time $t = s$. Consequently, the indefinite integral matrix

$$\begin{aligned}\Psi(t) &= \int_0^t \Phi^{-1}(s)\begin{bmatrix} 0 & 0 \\ 0 & \mathrm{q} \end{bmatrix}\Phi^{\mathrm{T}-1}(s)\, ds \\ &= \frac{q}{\omega^2\tau^2}\int_0^t \begin{bmatrix} \tau^2 S(s)^2 & -tS(s)[\omega\tau C(s) + S(s)] \\ -\tau S(s)[\omega\tau C(s) + S(s)] & [\omega\tau C(s) + S(s)]^2 \end{bmatrix} e^{2s/\tau}\, ds\end{aligned}$$

$$
= \begin{bmatrix} \dfrac{q\tau\{-\omega^2\tau^2 + [2S(t)^2 - 2C(t)\omega S(t)\tau + \omega^2\tau^2]\zeta^2\}}{4\omega^2(1+\omega^2\tau^2)} & \dfrac{-qS(t)^2\zeta^2}{2\omega^2} \\ \dfrac{-qS(t)^2\zeta^2}{2\omega^2} & \dfrac{q\{-\omega^2\tau^2 + [2S(t)^2 + 2C(t)\omega S(t)\tau + \omega^2\tau^2]\zeta^2\}}{4\omega^2\tau} \end{bmatrix},
$$

$\zeta = e^{t/\tau}$.

The discrete-time covariance matrix Q_{k-1} can then be evaluated as (see Section 4.10.)

$$
\begin{aligned}
Q_{k-1} &= \Phi(\Delta t)\Psi(\Delta t)\Phi^{\mathrm{T}}(\Delta t) \\
&= \begin{bmatrix} q_{11} & q_{12} \\ q_{21} & q_{22} \end{bmatrix}, \\
q_{11} &= \frac{q\tau\{\omega^2\tau^2(1 - e^{-2\Delta t/\tau}) - 2S(\Delta t)e^{-2\Delta t/\tau}[S(\Delta t) + \omega\tau C(\Delta t)]\}}{4\omega^2(1+\omega^2\tau^2)}, \\
q_{12} &= \frac{qe^{-2\Delta t/\tau}S(\Delta t)^2}{2\omega^2}, \\
q_{21} &= q_{12}, \\
q_{22} &= \frac{q\{\omega^2\tau^2(1 - e^{-2\Delta t/\tau}) - 2S(\Delta t)e^{-2\Delta t/\tau}[S(\Delta t) - \omega\tau C(\Delta t)]\}}{4\omega^2\tau}
\end{aligned}
$$

Note that the *structure* of the discrete-time process noise covariance Q_{k-1} for this example is quite different from the continuous-time process noise Q. In particular, Q_{k-1} is a full matrix, although Q is a sparse matrix.

First-order approximation of Q_k for constant F and G. The justification of a truncated power series expansion for Q_k when F and G are constant is as follows:

$$
Q_k = \sum_{i=1}^{\infty} \frac{Q^i \, \Delta t^i}{i!} \tag{3.118}
$$

Consider the Taylor series expansion of Q_k about t_{k-1}, where

$$
\begin{aligned}
Q^i &= \left.\frac{d^iQ}{dt^i}\right|_{t=t_{k-1}}, \\
\dot{Q} &= FQ_k + Q_kF^{\mathrm{T}} + GQ(t)G^{\mathrm{T}}, \\
Q^{(1)} &= \dot{Q}(t_{k-1}) = GQ(t)G^{\mathrm{T}} \text{ since } Q(t_{k-1}) = 0, \\
Q^{(2)} &= \ddot{Q}(t_{k-1}) = F\dot{Q}(t_{k-1}) + \dot{Q}(t_{k-1})F^{\mathrm{T}}, \\
&= FQ^{(1)} + Q^{(1)}F^{\mathrm{T}}, \\
&\vdots \\
Q^{(i)} &= FQ^{(i-1)} + Q^{(i-1)}F^{\mathrm{T}}, \qquad i = 1, 2, 3, \ldots.
\end{aligned}
$$

Taking only first-order terms in the above series,

$$Q_k \approx GQ(t)G^{\mathrm{T}}\Delta t \tag{3.119}$$

This is not always a good approximation, as is shown in the following example.

EXAMPLE 3.11: First-Order Approximation of Q_k for the Harmonic Resonator Let us see what happens if this first-order approximation

$$Q_k \approx Q\,\Delta t$$

is applied to the previous example of the harmonic resonator with acceleration noise.

The solution to the steady-state "state covariance" equation (i.e., the equation of covariance of the state vector itself, not the estimation error)

$$P_\infty = \Phi P_\infty \Phi^{\mathrm{T}} + Q\,\Delta t$$

has the solution (for $\theta = 2\pi f_{\text{resonance}}/f_{\text{sampling}}$)

$$\begin{aligned}\{P_\infty\}_{11} &= q\,\Delta t\, e^{-2\Delta t/\tau}\sin(\theta)^2(e^{-2\Delta t/\tau}+1)/D,\\ D &= \omega^2(e^{-2\Delta t/\tau}-1)\\ &\quad\times(e^{-2\Delta t/\tau}-2e^{-2\Delta t/\tau}\cos(\theta)+1)(e^{-2\Delta t/\tau}+2e^{-2\Delta t/\tau}\cos(\theta)+1)\end{aligned}$$

for its upper-left element, which is the *steady-state mean-squared resonator displacement*. Note, however, that

$$\{P_\infty\}_{11} = 0 \quad \text{if} \quad \sin(\theta) = 0,$$

which would imply that there is *no* displacement if the sampling frequency is twice the resonant frequency. This is absurd, of course. This proves by contradiction that

$$Q_k \neq Q\,\Delta t$$

in general—even though it may be a reasonable approximation in some instances.

3.8 ORTHOGONALITY PRINCIPLE

3.8.1 Estimators Minimizing Expected Quadratic Loss Functions

A block diagram representation of an estimator of the state of a system represented by Equation 3.82 is shown in Figure 3.8. The estimate $\hat{x}(t)$ of $x(t)$ will be the output of a Kalman filter.

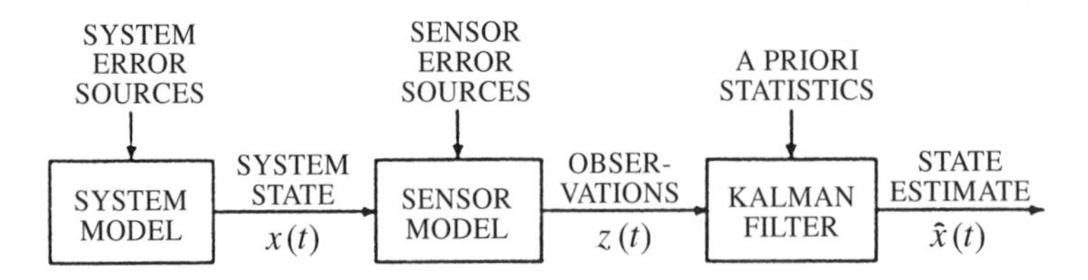

Fig. 3.8 Block diagram of estimator model.

The *estimation error* is defined as the difference between the "true" value of a random variable $x(t)$ and an "estimate" $\hat{x}(t)$.

A quadratic "loss" function of the estimation error has the form

$$[x(t) - \hat{x}(t)]^{\mathrm{T}} M [x(t) - \hat{x}(t)], \tag{3.120}$$

where M is an $n \times n$ symmetric, positive-definite "weighting matrix."

An "optimal" estimator for a particular quadratic loss function is defined as that estimate $\hat{x}(t)$ minimizing the *expected value* of the loss, with the probabilities conditioned on the observation $z(t)$. It will be shown that the optimal estimate of $x(t)$ (minimizing the average of a quadratic cost function) is the conditional expectation of $x(t)$ given the observation $z(t)$:

$$\begin{aligned} \hat{x} = E\langle x(t) | z(t)\rangle \text{ minimizes} \\ E\langle [x(t) - \hat{x}(t)]^{\mathrm{T}} M [x(t) - \hat{x}(t)] | z(t)\rangle. \end{aligned} \tag{3.121}$$

Let $z(t), 0 \le t \le t_1$, be the observed quantity and it is desired to estimate $x(t)$ at $t = t_2$. Then Equation 3.121 assumes the form

$$\hat{x}(t_2) = E\langle x(t_2) | z(t), 0 \le t \le t_1\rangle \tag{3.122}$$

and the corresponding equation for a similar discrete model is

$$\hat{x}_{k_2} = E\langle \hat{x}_{k_2} | z_1, z_2, \ldots, z_{k_1}\rangle, \quad 1 \le k_2 \le k_1. \tag{3.123}$$

Let

$$J = E\langle [x(t) - \hat{x}(t)]^{\mathrm{T}} M [x(t) - \hat{x}(t)] | z(t)\rangle. \tag{3.124}$$

Recall that $\hat{x}(t)$ is a nonrandom function of the observations

$$0 = \frac{dJ}{d\hat{x}} \tag{3.125}$$

$$= -2ME\langle [x(t) - \hat{x}(t)] | z(t)\rangle, \tag{3.126}$$

$$E\langle \hat{x}(t) | z(t)\rangle = \hat{x}(t) = E\langle x(t) | z(t)\rangle. \tag{3.127}$$

This proves the result of Equation 3.121. If $x(t)$ and $z(t)$ are jointly normal (Gaussian), the nonlinear minimum variance and linear minimum variance estimators coincide:

$$E\langle x_{k_2}|z_1, z_2, \ldots, z_{k_1}\rangle = \sum_{i=1}^{k_1} \alpha_i z_i \tag{3.128}$$

and

$$E\langle x(t_2)|z(t), 0 \leq t \leq t_1\rangle = \int_0^{t_1} \alpha(t, \tau)z(\tau)\, d\tau. \tag{3.129}$$

Proof for the Discrete Case: Recall the properties of jointly Gaussian processes from Section 3.2.3. Let the probability density

$$p[x_{k_2}|z_{k_1}] \tag{3.130}$$

be Gaussian and let $\alpha_1, \alpha_2, \ldots, \alpha_{k_1}$ satisfy

$$E\left\langle \left[x_{k_2} - \sum_{i=1}^{k_1} \alpha_i z_i\right] z_j^{\mathrm{T}}\right\rangle = 0, \quad j = 1, \ldots, k_1, \tag{3.131}$$

and

$$k_1 < k_2, k_1 = k_2, \quad \text{or} \quad k_1 > k_2. \tag{3.132}$$

The existence of vectors α_i satisfying this equation is guaranteed because the covariance $[z_i, z_j]$ is nonsingular.

The vectors

$$\left[x_{k_2} - \sum \alpha_i z_i\right] \tag{3.133}$$

and z_i are independent. Then it follows from the zero-mean property of the sequence x_k that

$$\begin{aligned} E\left\langle \left[x_{k_2} - \sum_{i=1}^{k_1} \alpha_i z_i\right] \middle| z_1, \ldots, z_{k_1}\right\rangle &= E\left\langle x_{k_2} - \sum_{i+1}^{k_1} \alpha_i z_i\right\rangle \\ &= 0, \\ E\langle x_{k_2}|z_1, z_2, \ldots, z_{k_1}\rangle &= \sum_{i=1}^{k_1} \alpha_i z_i. \end{aligned}$$

The proof of the continuous case is similar.

The linear minimum variance estimator is unbiased, that is,

$$E\langle x(t) - \hat{x}(t)\rangle = 0, \tag{3.134}$$

where

$$\hat{x}(t) = E\langle x(t)|z(t)\rangle. \tag{3.135}$$

In other words, an unbiased estimate is one whose expected value is the same as that of the quantity being estimated.

3.8.2 Orthogonality Principle

The nonlinear solution $E\langle x|z\rangle$ of the estimation problem is not simple to evaluate. If x and z are jointly normal, then $E\langle x|z\rangle = \alpha_1 z + \alpha_0$.

Let x and z be scalars and M be a 1×1 weighting matrix. The constants α_0 and α_1 that minimize the *mean-squared* (MS) *error*

$$e = E\langle[x - (\alpha_0 + \alpha_1 z)]^2\rangle = \int_{\infty}^{\infty}\int_{\infty}^{\infty} [x - (\alpha_0 + \alpha_1 z)]^2 p(x, z)\, dx\, dz \tag{3.136}$$

are given by

$$\alpha_1 = \frac{r\sigma_x}{\sigma_z},$$
$$\alpha_0 = E\langle x\rangle - \alpha_1 E\langle z\rangle,$$

and the resulting minimum mean-squared error $e_{\min}$ is

$$e_{\min} = \sigma_x^2(1 - r^2), \tag{3.137}$$

where the ratio

$$r = \frac{E\langle xz\rangle}{\sigma_x \sigma_z} \tag{3.138}$$

is called the *correlation coefficient* of x and z, and σ_x, σ_z are standard deviations of x and z, respectively.

Suppose α_1 is specified. Then

$$\frac{d}{d\alpha_0} E\langle[x - \alpha_0 - \alpha_1 z]^2\rangle = 0 \tag{3.139}$$

and

$$\alpha_0 = E\langle x\rangle - \alpha_1 E\langle z\rangle. \tag{3.140}$$

Substituting the value of α_0 in $E\langle[x - \alpha_0 - \alpha_1 z]^2\rangle$ yields

$$\begin{aligned}E\langle[x - \alpha_0 - \alpha_1 z]^2\rangle &= E\langle[x - E\langle x\rangle - \alpha_1(z - E\langle z\rangle)]^2\rangle \\ &= E\langle[(x - E\langle x\rangle) - \alpha_1(z - E\langle z\rangle)]^2\rangle \\ &= E\langle[x - E\langle x\rangle]^2\rangle + \alpha_1^2 E\langle[z - E\langle z\rangle]^2\rangle \\ &\quad - 2\alpha_1 E\langle(x - E\langle x\rangle)(z - E\langle z\rangle)\rangle,\end{aligned}$$

and differentiating with respect to α_1 as

$$\begin{aligned}0 &= \frac{d}{d\alpha_1} E\langle[x - \alpha_0 - \alpha_1 z]^2\rangle \\ &= 2\alpha_1 E\langle(z - E\langle z\rangle)^2\rangle - 2E\langle(x - E\langle x\rangle)(z - E\langle z\rangle)\rangle, \qquad (3.141) \\ \alpha_1 &= \frac{E\langle(x - E\langle x\rangle)(z - E\langle z\rangle)\rangle}{E\langle(z - E\langle z\rangle)^2\rangle} \\ &= \frac{r\sigma_x\sigma_z}{\sigma_z^2} \\ &= \frac{r\sigma_x}{\sigma_z}, \qquad (3.142) \\ e_{\min} &= \sigma_x^2 - 2r^2\sigma_x^2 + r^2\sigma_x^2 \\ &= \sigma_x^2(1 - r^2).\end{aligned}$$

Note that, if one assumes that x and z have zero means,

$$E\langle x\rangle = E\langle z\rangle = 0; \qquad (3.143)$$

then we have the solution

$$\alpha_0 = 0. \qquad (3.144)$$

Orthogonality Principle. The constant α_1 that minimizes the mean-squared error

$$e = E\langle[x - \alpha_1 z]^2\rangle \qquad (3.145)$$

is such that $x - \alpha_1 z$ is *orthogonal* to z. That is,

$$E\langle[x - \alpha_1 z]z\rangle = 0, \qquad (3.146)$$

and the value of the minimum mean-squared error is given by the formula

$$e_m = E\langle(x - \alpha_1 z)x\rangle. \qquad (3.147)$$

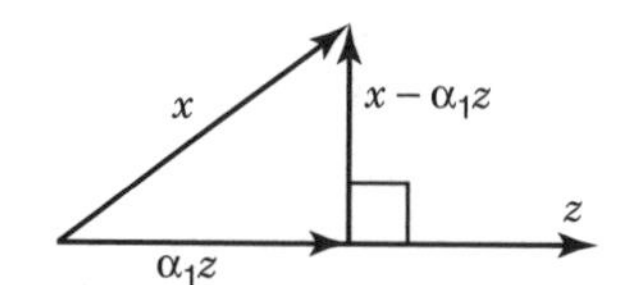

Fig. 3.9 Orthogonality diagram.

3.8.3 A Geometric Interpretation of Orthogonality

Consider all random variables as vectors in abstract vector spaces. The inner product of x and z is taken as the second moment $E\langle xz\rangle$. Thus

$$E\langle x^2\rangle = E\langle x^{\mathrm{T}}x\rangle \tag{3.148}$$

is the square of the length of x. The vectors x, z, $\alpha_1 z$, and $x - \alpha_1 z$ are as shown in Figure 3.9.

The mean-squared error $E\langle(x - \alpha_1 z)^2\rangle$ is the square of the length of $x - \alpha_1 z$. This length is minimum if $x - \alpha_1 z$ is orthogonal (perpendicular) to z,

$$E\langle(x - \alpha_1 z)z\rangle = 0. \tag{3.149}$$

We will apply the orthogonality principle to derive Kalman estimators in Chapter 4.

3.9 SUMMARY

3.9.1 Important Points to Remember

Probabilities are measures. That is, they are functions whose arguments are *sets* of points, not individual points. The *domain* of a probability measure P is a *sigma algebra* of subsets of a given set S, called the *measurable sets* of S. The sigma algebra of measurable sets has an algebraic structure under the operations of *set union* and *set complement*. The measurable sets always include the empty set $\{\,\}$ and the set S, and the probability $P(S) = 1$, $P(\{\,\}) = 0$, $P(A \cup B) + P(A \cap B) = P(A) + P(B)$ for all measurable sets A and B. A *probability space* is characterized by a set S, a sigma algebra of its measurable subsets, and a probability measure defined on the measurable subsets.

Events Form a Sigma Algebra of Outcomes of an Experiment. A *statistical experiment* is an undertaking with an uncertain outcome. The set of *all* possible outcomes of an experiment is called a *sample space*. An event is said to *occur* if the outcome of an experiment is an element of the event.

Independent Events. A collection of events is called *mutually independent* if the occurrence or nonoccurrence of any finite number of them has no influence on the possibilities for occurrence or nonoccurrence of the others.

Random Variables Are Functions. A *scalar random variable* is a real-valued function defined on the sample space of a probability space such that, for every open interval (a, b), $-\infty < a \leq b < +\infty$, the set

$$f^{-1}((a, b)) = \{s \in S | a < f(s) < b\}$$

is an event (i.e., is in the sigma algebra of events). A vector-valued random variable has scalar random variables as its components. A random variable is also called a *variate*.

Random processes (RPs) are functions of time with random variables as their values. A *process* is the evolution over time of a system. If the future state of the system can be predicted from its initial state and its inputs, then the process is considered *deterministic*. Otherwise, it is called *nondeterministic*. If the possible states of a nondeterministic system at any time can be represented by a random variable, then the evolution of the state of the system is a random process, or a *stochastic process*. Formally, a random or stochastic process is a function f defined on a time interval with *random variables* as its values $f(t)$.

A random process is called:

A *Bernoulli process*, or *independent, identically distributed (i.i.d.)* process if the probability distribution of its values at any time is independent of its values at any other time.

A *Markov process* if, for any time t, the probability distribution of its state at any time $\tau > t$, given its state at time t, is the same as its probability distribution given its state at all times $s \leq t$.

A *Gaussian process* if the probability distribution of its possible values at any time is a Gaussian distribution.

Stationary if certain statistics of its probability distributions are invariant under shifts of the time origin. If only its first and second moments are invariant, it is called *wide-sense stationary* or *weak-sense stationary*. If all its statistics are invariant, it is called *strict sense stationary*.

Ergodic if the probability distribution of its values at any one time, over the ensemble of sample functions, equals the probability distribution over all time of the values of randomly chosen member functions.

Orthogonal to another random process if the expected value of their pointwise product is zero.

3.9.2 Important Equations to Remember

The density function of an n-vector-valued (or *multivariate*) *Gaussian probability distribution* $\mathcal{N}(\bar{x}, P)$ has the functional form

$$p(x) = \frac{1}{\sqrt{(2\pi)^n \det P}} e^{-(1/2)(x-\bar{x})^{\mathrm{T}} P^{-1}(x-\bar{x})},$$

where $\bar{x}$ is the *mean* of the distribution and P is the covariance matrix of deviations from the mean.

A *linear stochastic process in continuous time* with state x and state covariance P has the model equations

$$\begin{aligned}
\dot{x}(t) &= F(t)x(t) + G(t)w(t), \\
z(t) &= H(t)x(t) + v(t), \\
\dot{P}(t) &= F(t)P(t) + P(t)F^{\mathrm{T}}(t) + G(t)Q(t)G^{\mathrm{T}}(t),
\end{aligned}$$

where $Q(t)$ is the covariance of zero-mean *plant noise* $w(t)$. A *discrete-time linear stochastic process* has the model equations

$$\begin{aligned}
x_k &= \Phi_{k-1}x_{k-1} + G_{k-1}w_{k-1}, \\
z_k &= H_k x_k + v_k, \\
P_k &= \Phi_{k-1}P_{k-1}\Phi_{k-1}^{\mathrm{T}} + G_{k-1}Q_{k-1}G_{k-1}^{\mathrm{T}},
\end{aligned}$$

where x is the system state, z is the system output, w is the zero-mean uncorrelated *plant noise*, Q_{k-1} is its covariance of w_{k-1}, and v is the zero-mean uncorrelated *measurement noise*. Plant noise is also called *process noise*. These models may also have known inputs. *Shaping filters* are models of these types that are used to represent random processes with certain types of spectral properties or temporal correlations.

PROBLEMS

3.1 Let a deck of 52 cards be divided into four piles (labeled North, South, East, West). Find the probability that each pile contains exactly one ace. (There are four aces in all.)

3.2 Show that

$$\binom{n+1}{k+1} = \binom{n}{k+1} + \binom{n}{k}.$$

3.3 How many ways are there to divide a deck of 52 cards into four piles of 13 each?

3.4 If a hand of 13 cards are drawn from a deck of 52, what is the probability that exactly 3 cards are spades? (There are 13 spades in all.)

3.5 If the 52 cards are divided into four piles of 13 each, and if we are told that North has exactly three spades, find the probability that South has exactly three spades.

3.6 A hand of 13 cards is dealt from a well-randomized bridge deck. (The deck contains 13 spades, 13 hearts, 13 diamonds, and 13 clubs.)

(a) What is the probability that the hand contains exactly 7 hearts?

(b) During the deal, the face of one of the cards is inadvertently exposed and it is seen to be a heart. What is now the probability that the hand contains exactly 7 hearts?

You may leave the above answers in terms of factorials.

3.7 The random variables $X_1, X_2, \ldots, X_n$ are independent with mean zero and the same variance σ_X^2. We define the new random variables $Y_1, Y_2, \ldots, Y_n$ by

$$Y_n = \sum_{j=1}^{n} X_j.$$

Find the correlation coefficient r between Y_{n-1} and Y_n.

3.8 The random variables X and Y are independent and uniformly distributed between 0 and 1 (rectangular distribution). Find the probability density function of $Z = |X - Y|$.

3.9 Two random variables X and Y have the density function

$$p_{XY}(x, y) = \begin{cases} C(y - x + 1), & 0 \le y \le x \le 1, \\ 0, & \text{elsewhere}, \end{cases}$$

where the constant $C < 0$ is chosen to normalize the distribution.

(a) Sketch the density function in the x, y *plane.*

(b) Determine the value of C for normalization.

(c) Obtain two marginal density functions.

(d) Obtain $E\langle Y|x\rangle$.

(e) Discuss the nature and use of the relation $y = E\langle Y|x\rangle$.

3.10 The random variable X has the probability density function

$$f_X(x) = \begin{cases} 2x, & 0 \le x \le 1, \\ 0, & \text{elsewhere}. \end{cases}$$

Find the following:

(a) The cumulative function $F_X(x)$.

(b) The median.

(c) The mode.

(d) The mean, $E\langle X \rangle$.

(e) The mean-square value $E\langle X^2 \rangle$.

(f) The variance $\sigma^2[X]$.

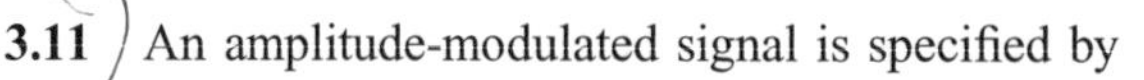

3.11 An amplitude-modulated signal is specified by

$$y(t) = [1 + mx(t)]\cos(\Omega t + \lambda).$$

Here $x(t)$ is a wide sense stationary random process independent of λ, which is a random variable uniformly distributed over $[0, 2\pi]$. We are given that

$$\psi_x(\tau) = \frac{1}{\tau^2 + 1}.$$

(a) Verify that $\psi_x(\tau)$ is an autocorrelation.

(b) Let $x(t)$ have the autocorrelation given above. Using the direct method for computing the spectral density, calculate Ψ_y.

3.12 Let $R(T)$ be an arbitrary autocorrelation function for a mean-square continuous stochastic process $x(t)$ and let $\Psi(\omega)$ be the power spectral density for the process $x(t)$. Is it true that

$$\lim_{|\omega| \to \infty} \Psi(\omega) = 0?$$

Justify your answer.

3.13 Find the state-space models for longitudinal, vertical, and lateral turbulence for the following PSD of the "Dryden" turbulence model:

$$\Psi(\omega) = \sigma^2 \left(\frac{2L}{\pi V} \right) \left(\frac{1}{1 + (L\omega/V)^2} \right)$$

where

$\omega =$ frequency in radians per second
$\sigma =$ root-mean-square (RMS) turbulence intensity
$L =$ scale length in feet
$V =$ airplane velocity in feet per second (290 ft/sec)

(a) For longitudinal turbulence:

$$L = 600\ \text{ft}$$
$$\sigma_u = 0.15 \text{ mean head wind or tail wind (knots)}$$

(b) For vertical turbulence:

$$L = 300\ \text{ft}$$
$$\sigma_w = 1.5\ \text{knots}$$

(c) For lateral turbulence:

$$L = 600\ \text{ft}$$
$$\sigma_v = 0.15 \text{ mean cross-wind (knots)}$$

3.14 Consider the random process

$$x(t) = \cos(\omega_0 t + \theta_1)\cos(\omega_0 t + \theta_2),$$

where θ_1 and θ_2 are independent random variables uniformly distributed between 0 and 2π.

(a) Show that $x(t)$ is wide-sense stationary.

(b) Calculate $\psi_x(\tau)$ and $\Psi_x(\omega)$.

(c) Discuss the ergodicity of $x(t)$.

3.15 Let $\psi_x(\tau)$ be the autocorrelation of a wide-sense stationary random process. Is the real part of $\psi_x(\tau)$ necessarily also an autocorrelation? If your answer is affirmative, prove it; if negative, give a counterexample.

3.16 Assume $x(t)$ is wide-sense stationary:

$$y(t) = x(t)\cos(\omega t + \theta),$$

where ω is a constant and θ is a uniformly distributed $[0, 2\pi]$ random phase. Find $\psi_{xy}(\tau)$.

3.17 The random process $x(t)$ has mean zero and autocorrelation function

$$\psi_x(\tau) = e^{-|\tau|}.$$

Find the autocorrelation function for

$$y(t) = \int_0^t x(u)\, du, \quad t > 0.$$

3.18 Assume $x(t)$ is wide-sense stationary with power spectral density

$$\Psi_x(\omega) = \begin{cases} 1, & -a \leq \omega \leq a, \\ 0, & \text{otherwise}. \end{cases}$$

Sketch the spectral density of the process

$$y(t) = x(t)\cos(\Omega t + \theta),$$

where θ is a uniformly distributed random phase and $\Omega > a$.

3.19 **(a)** Define a wide-sense stationary random process.

(b) Define a strict-sense stationary random process.

(c) Define a realizable linear system.

(d) Is the following an autocorrelation function?

$$\psi(\tau) = \begin{cases} 1 - |\tau|, & |\tau| < 1, \\ 0 & \text{otherwise}, \end{cases}$$

Explain.

3.20 Assume $x(t)$ is a stationary random process with autocorrelation function

$$\psi_x(\tau) = \begin{cases} 1 - |\tau|, & -1 \leq \tau \leq 1, \\ 0 & \text{otherwise}. \end{cases}$$

Find the spectral density $\Psi_y(\omega)$ for

$$y(t) = x(t)\cos(\omega_0 t + \lambda)$$

when ω_0 is a constant and λ is a random variable uniformly distributed on the interval $[0, 2\pi]$.

3.21 A random process $x(t)$ is defined by

$$x(t) = \cos(t + \theta),$$

where θ is a random variable uniformly distributed on the interval $[0, 2\pi]$. Calculate the autocorrelation function $\psi_y(t, s)$ for

$$y(t) = \int_0^t x(u)\, du.$$

3.22 Let ψ_1 and ψ_2 be two arbitrary continuous, absolutely integrable autocorrelation functions. Are the following necessarily autocorrelation functions?

Briefly explain your answer.

(a) $\psi_1 \cdot \psi_2$

(b) $\psi_1 + \psi_2$

(c) $\psi_1 - \psi_2$

(d) $\psi_1 * \psi_2$ (the convolution of ψ_1 with ψ_2)

3.23 Give a short reason for each answer:

(a) If $F(T)$ and $G(T)$ are autocorrelation functions, $f^2(t) + g(t)$ is (necessarily, perhaps, never) an autocorrelation function.

(b) As in (a), $f^2(t) - g(t)$ is (necessarily, perhaps, never) an autocorrelation function.

(c) If $x(t)$ is a strictly stationary process, $x^2(t) + 2x(t-1)$ is (necessarily, perhaps, never) strictly stationary.

(d) The function

$$\omega(\tau) = \begin{bmatrix} \cos\tau, & -\frac{9}{2}\pi \le \tau \le \frac{9}{2}\pi, \\ 0 & \text{otherwise}, \end{bmatrix}$$

(is, is not) an autocorrelation function.

(e) Let $x(t)$ be strictly stationary and ergodic and α be a Gaussian random variable with mean zero and variance one and α is independent of $x(t)$. Then $y(t) = \alpha x(t)$ is (necessarily, perhaps, never) ergodic.

3.24 Which of the following functions is an autocorrelation function of a wide-sense stationary process? Give a brief reason for each answer.

(a) $e^{-|\tau|}$

(b) $e^{-|\tau|}\cos\tau$

(c) $\Gamma(t) = \begin{cases} 1, & |t| < a, \\ 0 & |t| \ge a \end{cases}$

(d) $e^{-|\tau|}\sin\tau$

(e) $\frac{3}{2}e^{-|\tau|} - e^{-2|\tau|}$

(f) $2e^{-2|\tau|} - e^{-|\tau|}$

3.25 Discuss each of the following:

(a) The distinction between stationarity and wide-sense stationarity.

(b) The periodic character of the cross-correlation function of two processes that are themselves periodic with periods mT and nT, respectively.

3.26 A system transfer function can sometimes be experimentally determined by injecting white noise $n(t)$ and measuring the cross correlation between the system output and the white noise. Here we consider the following system:

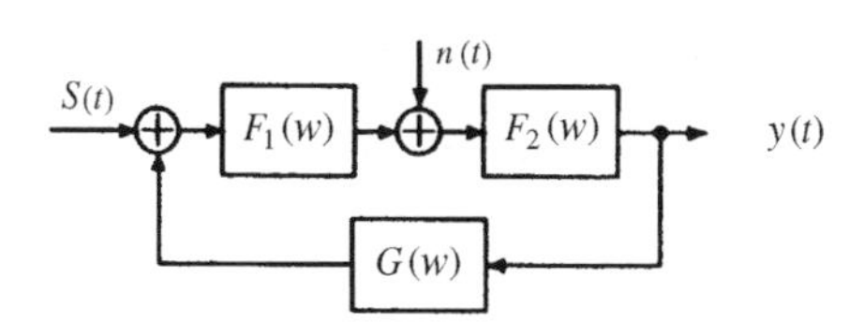

We assume $\Psi_S(\omega)$ known, $S(t)$ and $n(t)$ independent, and $\Psi_n(\omega) = 1$. Find $\Psi_{yn}(\omega)$. *Hint:* Write $y(t) = y_S(t) + y_n(t)$, where y_S and y_n are the parts of the output due to S and n, respectively.

3.27 Let $S(t)$ and $n(t)$ be real stationary uncorrelated random processes, each with mean zero.

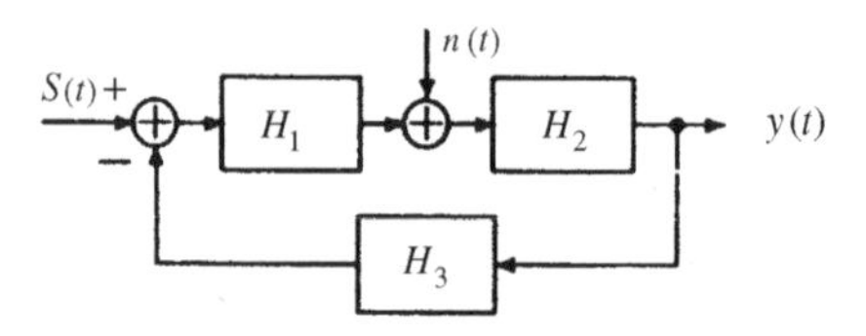

Here, $H_1(j2\pi\omega)$, $H_2(j2\pi\omega)$, and $H_3(j2\pi\omega)$ are transfer functions of time-invariant linear systems and $S_0(t)$ is the output when $n(t)$ is zero and $n_0(t)$ is the output when $S(t)$ is zero. Find the output signal-to-noise ratio, defined as $E\langle S_0^2(t)\rangle / E\langle n_0^2(t)\rangle$.

3.28 A single random data source is measured by two different transducers, and their outputs are suitably combined into a final measurement $y(t)$. The system is as pictured below:

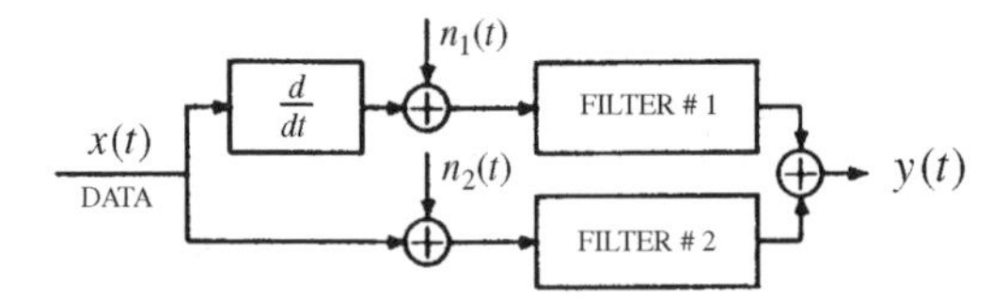

Assume that $n_1(t)$ and $n_2(t)$ are uncorrelated random processes, data and noises are uncorrelated, filter 1 has transfer function $Y(s)/s$, and filter 2 has transfer function $1 - Y(s)$. Suppose that it is desired to determine the mean square error of measurement, where the error is defined by $e(t) = x(t) - y(t)$. Calculate the mean-square value of the error in terms of $Y(s)$ and the spectral densities Ψ_x, Ψ_{n_1}, and Ψ_{n_2}.

3.29 Let $x(t)$ be the solution of

$$\dot{x} + x = n(t),$$

where $n(t)$ is white noise with spectral density 2π.

(a) Assuming that the above system has been operating since $t = -\infty$, find $\psi_x(t_1 t_2)$. Investigate whether $x(t)$ is wide-sense stationary, and if so, express ψ_x accordingly.

(b) Instead of the system in (a), consider

$$\dot{x} + x = \begin{cases} n(t), & t \geq 0, \\ 0, & t < 0, \end{cases}$$

where $x(0) = 0$. Again, compute $\psi_x(t_1, t_2)$.

(c) Let $y(t) = \int_0^t x(\tau)\, d\tau$. Find $\psi_{xy}(t_1, t_2)$ for both of the systems described in (a) and (b).

(d) It is desired to predict $x(t + \alpha)$ from $x(t)$, that is, a future value of the process from its present value. A possible predictor $\hat{x}(t + \alpha)$ is of the form

$$\hat{x}(t + \alpha) = ax(t).$$

Find that a that will give the smallest mean-square prediction error, that is, that minimizes

$$E\langle|\hat{x}(t + \alpha) - x(t + \alpha)|^2\rangle,$$

where $x(t)$ is as in part (a).

3.30 Let $x(t)$ be the solution of

$$\dot{x} + x = n(t)$$

with initial condition $x(0) = x_0$. It is assumed that $n(t)$ is white noise with spectral density 2π and is turned on at $t = 0$. The initial condition x_0 is a random variable independent of $n(t)$ and with zero mean.

(a) If x_0 has variance σ^2, what is $\psi_x(t_1, t_2)$? Derive the result.

(b) Find that value of σ (call it σ_0) for which $\psi_x(t_1, t_2)$ is the same for all $t \geq 0$. Determine whether, with $\sigma = \sigma_0$, $\psi_x(t_1, t_2)$ is a function only of $t_1 - t_2$.

(c) If the white noise had been turned on at $t = -\infty$ and the initial condition has zero mean and variance σ_0^2 as above, is $x(t)$ wide-sense stationary? Justify your answer by appropriate reasoning and/or computation.

3.31 Let

$$\dot{x}(t) = F(t)x(t) + w(t),$$
$$x(a) = x_a, \qquad t \geq a,$$

where x_a is a zero-mean random variable with covariance matrix P_a and

$$E\langle w(t)\rangle = 0 \qquad \forall t,$$
$$E\langle w(t)w^{\mathrm{T}}(s)\rangle = Q(t)\delta(t - s) \qquad \forall t, s$$
$$E\langle x(a)w^{\mathrm{T}}(t)\rangle = 0 \qquad \forall t.$$

(a) Determine the mean $m(t)$ and covariance $P(t, t)$ for the process $x(t)$.

(b) Derive a differential equation for $P(t, t)$.

3.32 Find the covariance matrix $P(t)$ and its steady-state value $P(\infty)$ for the following continuous systems:

(a) $\dot{x} = \begin{bmatrix} -1 & 0 \\ -1 & 0 \end{bmatrix} x + \begin{bmatrix} 1 \\ 1 \end{bmatrix} w(t), \qquad P(0) = \begin{bmatrix} 1 & 0 \\ 0 & 1 \end{bmatrix}$

(b) $\dot{x} = \begin{bmatrix} -1 & 0 \\ 0 & -1 \end{bmatrix} x + \begin{bmatrix} 5 \\ 1 \end{bmatrix} w(t), \qquad P(0) = \begin{bmatrix} 1 & 0 \\ 0 & 1 \end{bmatrix}$

where $w \in \mathcal{N}(0, 1)$ and white.

3.33 Find the covariance matrix P_k and its steady-state value P_∞ for the following discrete system:

$$x_{k+1} = \begin{bmatrix} 0 & \frac{1}{2} \\ -\frac{1}{2} & 2 \end{bmatrix} x_k + \begin{bmatrix} 1 \\ 1 \end{bmatrix} w_k, \quad P_0 = \begin{bmatrix} 1 & 0 \\ 0 & 1 \end{bmatrix},$$

where $w_k \in \mathcal{N}(0, 1)$ and white.

3.34 Find the steady-state covariance for the state-space model given in Example 3.4.

3.35 Show that the continuous-time steady-state algebraic equation

$$0 = FP(\infty) + P(\infty)F^{\mathrm{T}} + GQG^{\mathrm{T}}$$

has no nonnegative solution for the scalar case with $F = Q = G = 1$. (See Equation 3.110.)

3.36 Show that the discrete-time steady-state algebraic equation

$$P_\infty = \Phi P_\infty \Phi^{\mathrm{T}} + Q$$

has no solution for the scalar case with $\Phi = Q = 1$. (See Equation 3.112.)

3.37 Find the covariance of x_k as a function of k and its steady-state value for the system

$$x_k = -2x_{k-1} + w_{k-1}$$

where $Ew_{k-1} = 0$ and $E(w_k w_j) = e^{-|k-j|}$. Assume the initial value of the covariance (P_0) is 1.

3.38 Find the covariance of $x(t)$ as a function of t and its steady-state value for the system

$$\dot{x}(t) = -2x(t) + w(t),$$

where $Ew(t) = 0$ and $E\big(w(t_1)w(t_2)\big) = e^{-|t_1 - t_2|}$. Assume the initial value of the covariance (P_0) is 1.

3.39 Suppose that $x(t)$ has autocorrelation function $\psi_x(\tau) = e^{-c|\tau|}$. It is desired to predict $x(t+\alpha)$ on the basis of the past and present of $x(t)$, that is, the predictor may use $x(s)$ for all $s \leq t$.

(a) Show that the minimum mean-square error linear prediction is

$$\hat{x}(t+\alpha) = e^{-c\alpha} x(t).$$

(b) Find the mean-square error corresponding to the above. *Hint:* Use the orthogonality principle.

4

Linear Optimal Filters and Predictors

Prediction is difficult—especially of the future.
Attributed to Niels Henrik David Bohr (1885–1962)

4.1 CHAPTER FOCUS

4.1.1 Estimation Problem

This is the problem of estimating the state of a linear stochastic system by using measurements that are linear functions of the state.

We suppose that stochastic systems can be represented by the types of plant and measurement models (for continuous and discrete time) shown as Equations 4.1–4.5 in Table 4.1, with dimensions of the vector and matrix quantities as shown in Table 4.2. The symbols $\Delta(k-\ell)$ and $\delta(t-s)$ stand for the *Kronecker delta function* and the *Dirac delta function* (actually, a *generalized* function), respectively.

TABLE 4.1 Linear Plant and Measurement Models

Model	Continuous Time	Discrete Time	Equation Number
Plant	$\dot{x}(t) = F(t)x(t) + w(t)$	$x_k = \Phi_{k-1}x_{k-1} + w_{k-1}$	(4.1)
Measurement	$z(t) = H(t)x(t) + v(t)$	$z_k = H_k x_k + v_k$	(4.2)
Plant noise	$E\langle w(t)\rangle = 0$	$E\langle w_k\rangle = 0$	(4.3)
	$E\langle w(t)w^{\mathrm{T}}(s)\rangle = \delta(t-s)Q(t)$	$E\langle w_k w_i^{\mathrm{T}}\rangle = \Delta(k-i)Q_k$	(4.4)
Observation noise	$E\langle v(t)\rangle = 0$	$E\langle v_k\rangle = 0$	
	$E\langle v(t)v^{\mathrm{T}}(s)\rangle = \delta(t-s)R(t)$	$E\langle v_k v_i^{\mathrm{T}}\rangle = \Delta(k-i)R_k$	(4.5)

TABLE 4.2 Dimensions of Vectors and Matrices in Linear Model

Symbol	Dimensions	Symbol	Dimensions
x, w	$n \times 1$	Φ, Q	$n \times n$
z, v	$\ell \times 1$	H	$\ell \times n$
R	$\ell \times \ell$	Δ, δ	scalar

The measurement and plant noise v_k and w_k are assumed to be zero-mean Gaussian processes, and the initial value x_0 is a Gaussian variate with known mean x_0 and known covariance matrix P_0. Although the noise sequences w_k and v_k are assumed to be uncorrelated, the derivation in Section 4.5 will remove this restriction and modify the estimator equations accordingly.

The objective will be to find an estimate of the n state vector x_k represented by $\hat{x}_k$, a linear function of the measurements $z_i, \ldots, z_k$, that minimizes the weighted mean-squared error

$$E[x_k - \hat{x}_k]^{\mathrm{T}} M [x_k - \hat{x}_k], \tag{4.6}$$

where M is any *symmetric nonnegative-definite weighting matrix.*

4.1.2 Main Points to Be Covered

Linear Quadratic Gaussian Estimation Problem. We are now prepared to derive the mathematical forms of optimal linear estimators for the states of linear stochastic systems defined in the previous chapters. This is called the *linear quadratic Gaussian* (LQG) estimation problem. The dynamic systems are linear, the performance cost functions are quadratic, and the random processes are Gaussian.

Filtering, Prediction, and Smoothing. There are three general types of estimators for the LQG problem:

- *Predictors* use observations *strictly prior* to the time that the state of the dynamic system is to be estimated:

$$t_{\text{obs}} < t_{\text{est}}.$$

- *Filters* use observations *up to and including* the time that the state of the dynamic system is to be estimated:

$$t_{\text{obs}} \leq t_{\text{est}}.$$

- *Smoothers* use observations *beyond* the time that the state of the dynamic system is to be estimated:

$$t_{\mathrm{obs}} > t_{\mathrm{est}}.$$

Orthogonality Principle. A straightforward and simple approach using the orthogonality principle is used in the derivation[1] of *estimators*. These estimators will have *minimum variance* and be *unbiased* and *consistent*.

Unbiased Estimators. The Kalman filter can be characterized as an algorithm for computing the conditional mean and covariance of the probability distribution of the state of a linear stochastic system with uncorrelated Gaussian process and measurement noise. The conditional mean is the unique *unbiased* estimate. It is propagated in feedback form by a system of linear differential equations or by the corresponding discrete-time equations. The conditional covariance is propagated by a nonlinear differential equation or its discrete-time equivalent. This implementation automatically minimizes the expected risk associated with any quadratic loss function of the estimation error.

Performance Properties of Optimal Estimators. The statistical performance of the estimator can be predicted a priori (that is, before it is actually used) by solving the nonlinear differential (or difference) equations used in computing the optimal feedback gains of the estimator. These are called *Riccati equations*,[2] and the behavior of their solutions can be shown analytically in the most trivial cases. These equations also provide a means for verifying the proper performance of the actual estimator when it is running.

4.2 KALMAN FILTER

Observational Update Problem for System State Estimator. Suppose that a measurement has been made at time t_k and that the information it provides is to be

[1] For more mathematically oriented derivations, consult any of the references such as Anderson and Moore [1], Bozic [9], Brammer and Siffling [10], Brown [11], Bryson and Ho [14], Bucy and Joseph [15], Catlin [16], Chui and Chen [18], Gelb et al. [21], Jazwinski [23], Kailath [24], Maybeck [30, 31], Mendel [34, 35], Nahi [36], Ruymgaart and Soong [42], and Sorenson [47].

[2] Named in 1763 by Jean le Rond D'Alembert (1717–1783) for Count Jacopo Francesco Riccati (1676–1754), who had studied a second-order scalar differential equation [213], although not the form that we have here [54, 210]. Kalman gives credit to Richard S. Bucy for showing him that the Riccati differential equation is analogous to spectral factorization for defining optimal gains. The Riccati equation also arises naturally in the problem of separation of variables in ordinary differential equations and in the transformation of two-point boundary-value problems to initial-value problems [155].

applied in updating the estimate of the state x of a stochastic system at time t_k. It is assumed that the measurement is linearly related to the state by an equation of the form $z_k = Hx_k + v_k$, where H is the *measurement sensitivity matrix* and v_k is the *measurement noise*.

Estimator in Linear Form. The optimal linear estimate is equivalent to the general (nonlinear) optimal estimator if the variates x and z are jointly Gaussian (see Section 3.8.1). Therefore, it suffices to seek an updated estimate $\hat{x}_k(+)$—based on the observation z_k—that is a *linear* function of the a priori estimate and the measurement z:

$$\hat{x}_k(+) = K_k^1 \hat{x}_k(-) + \overline{K}_k z_k, \tag{4.7}$$

where $\hat{x}_k(-)$ is the a priori estimate of x_k and $\hat{x}_k(+)$ is the a posteriori value of the estimate.

Optimization Problem. The matrices K_k^1 and $\overline{K}_k$ are as yet unknown. We seek those values of K_k^1 and $\overline{K}_k$ such that the new estimate $\hat{x}_k(+)$ will satisfy the orthogonality principle of Section 3.8.2. This orthogonality condition can be written in the form

$$E\langle [x_k - \hat{x}_k(+)] z_i^{\mathrm{T}} \rangle = 0, \quad i = 1, 2, \ldots, k-1, \tag{4.8}$$

$$E\langle [x_k - \hat{x}_k(+)] z_k^{\mathrm{T}} \rangle = 0. \tag{4.9}$$

If one substitutes the formula for x_k from Equation 4.1 (in Table 4.1) and for $\hat{x}_k(+)$ from Equation 4.7 into Equation 4.8, then one will observe from Equations 4.1 and 4.2 that the data $z_1, \ldots, z_k$ do not involve the noise term w_k. Therefore, because the random sequences w_k and v_k are uncorrelated, it follows that $E w_k z_i^{\mathrm{T}} = 0$ for $1 \le i \le k$. (See Problem 4.5.)

Using this result, one can obtain the following relation:

$$E[\langle \Phi_{k-1} x_{k-1} + w_{k-1} - K_k^1 \hat{x}_k(-) - \overline{K}_k z_k \rangle z_i^{\mathrm{T}}] = 0, \quad i = 1, \ldots, k-1. \tag{4.10}$$

But because $z_k = H_k x_k + v_k$, Equation 4.10 can be rewritten as

$$E[\Phi_{k-1} x_{k-1} - K_k^1 \hat{x}_k(-) - \overline{K}_k H_k x_k - \overline{K}_k v_k] z_i^{\mathrm{T}} = 0, \quad i = 1, \ldots, k-1. \tag{4.11}$$

We also know that Equations 4.8 and 4.9 hold at the previous step, that is,

$$E\langle [x_{k-1} - \hat{x}_{k-1}(+)] z_i^{\mathrm{T}} \rangle = 0, \quad i = 1, \ldots, k-1,$$

and

$$E\langle v_k z_i^{\mathrm{T}} \rangle = 0, \quad i = 1, \ldots, k-1.$$

Then Equation 4.11 can be reduced to the form

$$\begin{aligned}
\Phi_{k-1}Ex_{k-1}z_i^{\mathrm{T}} - K_k^1 E\hat{x}_k(-)z_i^{\mathrm{T}} - \overline{K}_k H_k \Phi_{k-1} E x_{k-1} z_i^{\mathrm{T}} - \overline{K}_k E v_k z_i^{\mathrm{T}} &= 0,\\
\Phi_{k-1}Ex_{k-1}z_i^{\mathrm{T}} - K_k^1 E\hat{x}_k(-)z_i^{\mathrm{T}} - \overline{K}_k H_k \Phi_{k-1} E x_{k-1} z_i^{\mathrm{T}} &= 0,\\
E\langle [x_k - \overline{K}_k H_k x_k - K_k^1 x_k] - K_k^1(\hat{x}_k(-) - x_k)\rangle z_i^{\mathrm{T}} &= 0,\\
[I - K_k^1 - \overline{K}_k H_k] E x_k z_i^{\mathrm{T}} &= 0.
\end{aligned} \tag{4.12}$$

Equation 4.12 can be satisfied for any given x_k if

$$K_k^1 = I - \overline{K}_k H_k. \tag{4.13}$$

Clearly, this choice of K_k^1 causes Equation 4.7 to satisfy a portion of the condition given by Equation 4.8, which was derived in Section 3.8. The choice of $\overline{K}_k$ is such that Equation 4.9 is satisfied.

Let the errors

$$\tilde{x}_k(+) \stackrel{\Delta}{=} \hat{x}_k(+) - x_k, \tag{4.14}$$

$$\tilde{x}_k(-) \stackrel{\Delta}{=} \hat{x}_k(-) - x_k, \tag{4.15}$$

$$\begin{aligned}
\tilde{z}_k &\stackrel{\Delta}{=} \hat{z}_k(-) - z_k\\
&= H_k \hat{x}_k(-) - z_k.
\end{aligned} \tag{4.16}$$

Vectors $\tilde{x}_k(+)$ and $\tilde{x}_k(-)$ are the estimation errors after and before updates, respectively.[3]

The parameter $\hat{x}_k$ depends linearly on x_k, which depends linearly on z_k. Therefore, from Equation 4.9

$$E[x_k - \hat{x}_k(+)]z_k^{\mathrm{T}}(-) = 0 \tag{4.17}$$

and also (by subtracting Equation 4.9 from Equation 4.17)

$$E[x_k - \hat{x}_k(+)]\tilde{z}_k^{\mathrm{T}} = 0. \tag{4.18}$$

Substitute for x_k, $\hat{x}_k(+)$ and $\tilde{z}_k$ from Equations 4.1, 4.7, and 4.16, respectively. Then

$$E[\Phi_{k-1}x_{k-1} + w_{k-1} - K_k^1(-) - \overline{K}_k z_k][H_k \hat{x}_k(-) - z_k]^{\mathrm{T}} = 0.$$

However, by the system structure

$$\begin{aligned}
E w_k z_k^{\mathrm{T}} = E w_k \hat{x}_k^{\mathrm{T}}(+) &= 0,\\
E[\Phi_{k-1}x_{k-1} - K_k^1 \hat{x}_k(-) - \overline{K}_k z_k][H_k \hat{x}_k(-) - z_k]^{\mathrm{T}} &= 0.
\end{aligned}$$

[3]The symbol ~ is officially called a *tilde* but often called a "squiggle."

Substituting for K_k^1, z_k, and $\tilde{x}_k(-)$ and using the fact that $E\tilde{x}_k(-)v_k^{\mathrm{T}} = 0$, this last result can be modified as follows:

$$\begin{aligned}
0 &= E\langle[\Phi_{k-1}x_{k-1} - \hat{x}_k(-) + \overline{K}_k H_k \hat{x}_k(-) - \overline{K}_k H_k x_k - \overline{K}_k v_k] \\
&\quad [H_k \hat{x}_k(-) - H_k x_k - v_k]^{\mathrm{T}}\rangle \\
&= E\langle[(x_k - \hat{x}_k(-)) - \overline{K}_k H_k(x_k - \hat{x}_k(-)) - \overline{K}_k v_k][H_k \tilde{x}_k(-) - v_k]^{\mathrm{T}}\rangle \\
&= E\langle[(-\tilde{x}_k(-) + \overline{K}_k H_k \tilde{x}_k(-) - \overline{K}_k v_k][H_k \tilde{x}_k(-) - v_k]^{\mathrm{T}}\rangle.
\end{aligned}$$

By definition, the a priori covariance (the error covariance matrix before the update) is

$$P_k(-) = E\langle\tilde{x}_k(-)\tilde{x}_k^{\mathrm{T}}(-)\rangle.$$

It satisfies the equation

$$[I - \overline{K}_k H_k]P_k(-)H_k^{\mathrm{T}} - \overline{K}_k R_k = 0,$$

and therefore the gain can be expressed as

$$\overline{K}_k = P_k(-)H_k^{\mathrm{T}}[H_k P_k(-)H_k^{\mathrm{T}} + R_k]^{-1}, \tag{4.19}$$

which is the solution we seek for the gain as a function of the a priori covariance.

One can derive a similar formula for the a posteriori covariance (the error covariance matrix after update), which is defined as

$$P_k(+) = E\langle[\tilde{x}_k(+)\tilde{x}_k^{\mathrm{T}}(+)]\rangle. \tag{4.20}$$

By substituting Equation 4.13 into Equation 4.7, one obtains the equations

$$\begin{aligned}
\hat{x}_k(+) &= (I - \overline{K}_k H_k)\hat{x}_k(-) + \overline{K}_k z_k, \\
\hat{x}_k(+) &= \hat{x}_k(-) + \overline{K}_k[z_k - H_k \hat{x}_k(-)].
\end{aligned} \tag{4.21}$$

Subtract x_k from both sides of the latter equation to obtain the equations

$$\begin{aligned}
\hat{x}_k(+) - x_k &= \hat{x}_k(-) + \overline{K}_k H_k x_k + \overline{K}_k v_k - \overline{K}_k H_k \hat{x}_k(-) - x_k, \\
\tilde{x}_k(+) &= \tilde{x}_k(-) - \overline{K}_k H_k \tilde{x}_k(-) + \overline{K}_k v_k, \\
\tilde{x}_k(+) &= (I - \overline{K}_k H_k)\tilde{x}_k(-) + \overline{K}_k v_k.
\end{aligned} \tag{4.22}$$

By substituting Equation 4.22 into Equation 4.20 and noting that $E\tilde{x}_k(-)v_k^{\mathrm{T}} = 0$, one obtains

$$\begin{aligned}
P_k(+) &= E\{[I - \overline{K}_k H_k]\tilde{x}_k(-)\tilde{x}_k^{\mathrm{T}}(-)[I - \overline{K}_k H_k]^{\mathrm{T}} + \overline{K}_k v_k v_k^{\mathrm{T}} \overline{K}_k^{\mathrm{T}}\} \\
&= (I - \bar{K}_k H_k)P_k(-)(I - \overline{K}_k H_k)^{\mathrm{T}} + \overline{K}_k R_k \overline{K}_k^{\mathrm{T}}.
\end{aligned} \tag{4.23}$$

This last equation is the so-called "Joseph form" of the covariance update equation derived by P. D. Joseph [15]. By substituting for $\overline{K}_k$ from Equation 4.19, it can be put into the following forms:

$$\begin{aligned}
P_k(+) &= P_k(-) - \overline{K}_k H_k P_k(-) \\
&\quad - P_k(-) H_k^{\mathrm{T}} \overline{K}_k^{\mathrm{T}} + \overline{K}_k H_k P_k(-) H_k^{\mathrm{T}} \overline{K}_k^{\mathrm{T}} + \overline{K}_k R_k \overline{K}_k^{\mathrm{T}} \\
&= (I - \overline{K}_k H_k) P_k(-) - P_k(-) H_k^{\mathrm{T}} \overline{K}_k^{\mathrm{T}} \\
&\quad + \underbrace{\overline{K}_k (H_k P_k(-) H_k^{\mathrm{T}} + R_k)}_{P_k(-) H_k^{\mathrm{T}}} \overline{K}_k^{\mathrm{T}} \\
&= (I - \overline{K}_k H_k) P_k(-),
\end{aligned} \tag{4.24}$$

the last of which is the one most often used in computation. This implements the effect that *conditioning on the measurement* has on the covariance matrix of estimation uncertainty.

Error covariance extrapolation models the effects of time on the covariance matrix of estimation uncertainty, which is reflected in the a priori values of the covariance and state estimates,

$$\begin{aligned}
P_k(-) &= E[\tilde{x}_k(-) \tilde{x}_k^{\mathrm{T}}(-)], \\
\hat{x}_k(-) &= \Phi_{k-1} \hat{x}_{k-1}(+),
\end{aligned} \tag{4.25}$$

respectively. Subtract x_k from both sides of the last equation to obtain the equations

$$\begin{aligned}
\hat{x}_k(-) - x_k &= \Phi_{k-1} \hat{x}_{k-1}(+) - x_k, \\
\tilde{x}_k(-) &= \Phi_{k-1} [\hat{x}_{k-1}(+) - x_{k-1}] - w_{k-1} \\
&= \Phi_{k-1} \tilde{x}_{k-1}(+) - w_{k-1}
\end{aligned}$$

for the propagation of the estimation error, $\tilde{x}$. Postmultiply it by $\tilde{x}_k^{\mathrm{T}}(-)$ (on both sides of the equation) and take the expected values. Use the fact that $E\tilde{x}_{k-1} w_{k-1}^{\mathrm{T}} = 0$ to obtain the results

$$\begin{aligned}
P_k(-) &\overset{\text{def}}{=} E[\tilde{x}_k(-) \tilde{x}_k^{\mathrm{T}}(-)] \\
&= \Phi_{k-1} E[\tilde{x}_{k-1}(+) \tilde{x}_{k-1}^{\mathrm{T}}(+)] \Phi_{k-1}^{\mathrm{T}} + E[w_{k-1} w_{k-1}^{\mathrm{T}}] \\
&= \Phi_{k-1} P_{k-1}^{(+)} \Phi_{k-1}^{\mathrm{T}} + Q_{k-1},
\end{aligned} \tag{4.26}$$

which gives the a priori value of the covariance matrix of estimation uncertainty as a function of the previous a posteriori value.

4.2.1 Summary of Equations for the Discrete-Time Kalman Estimator

The equations derived in the previous section are summarized in Table 4.3. In this formulation of the filter equations, G has been combined with the plant covariance by multiplying G_{k-1} and G_{k-1}^{T}, for example,

$$
\begin{aligned}
Q_{k-1} &= G_{k-1}E(w_{k-1}w_{K-1}^{\mathrm{T}})G_{k-1}^{\mathrm{T}} \\
&= G_{k-1}\bar{Q}_{k-1}G_{k-1}^{\mathrm{T}}.
\end{aligned}
$$

The relation of the filter to the system is illustrated in the block diagram of Figure 4.1. The basic steps of the computational procedure for the discrete-time Kalman estimator are as follows:

1. Compute $P_k(-)$ using $P_{k-1}(+)$, Φ_{k-1}, and Q_{k-1}.
2. Compute $\overline{K}_k$ using $P_k(-)$ (computed in step 1), H_k, and R_k.
3. Compute $P_k(+)$ using $\overline{K}_k$ (computed in step 2) and $P_k(-)$ (from step 1).
4. Compute successive values of $\hat{x}_k(+)$ recursively using the computed values of $\overline{K}_k$ (from step 3), the given initial estimate $\hat{x}_0$, and the input data z_k.

TABLE 4.3 Discrete-Time Kalman Filter Equations

System dynamic model:

$$x_k = \Phi_{k-1}x_{k-1} + w_{k-1}$$
$$w_k \sim \mathcal{N}(0, Q_k)$$

Measurement model:

$$z_k = H_k x_k + v_k$$
$$v_k \sim \mathcal{N}(0, R_k)$$

Initial conditions:

$$E\langle x_0\rangle = \hat{x}_0$$
$$E\langle \tilde{x}_0\tilde{x}_0^{\mathrm{T}}\rangle = P_0$$

Independence assumption:

$$E\langle w_k v_j^{\mathrm{T}}\rangle = 0 \quad \text{for all } k \text{ and } j$$

State estimate extrapolation (Equation 4.25):

$$\hat{x}_k(-) = \Phi_{k-1}\hat{x}_{k-1}(+)$$

Error covariance extrapolation (Equation 4.26):

$$P_k(-) = \Phi_{k-1}P_{k-1}(+)\Phi_{k-1}^{\mathrm{T}} + Q_{k-1}$$

State estimate observational update (Equation 4.21):

$$\hat{x}_k(+) = \hat{x}_k(-) + \overline{K}_k[z_k - H_k\hat{x}_k(-)]$$

Error covariance update (Equation 4.24):

$$P_k(+) = [I - \overline{K}_k H_k]P_k(-)$$

Kalman gain matrix (Equation 4.19):

$$\overline{K}_k = P_k(-)H_k^{\mathrm{T}}[H_k P_k(-)H_k^{\mathrm{T}} + R_k]^{-1}$$

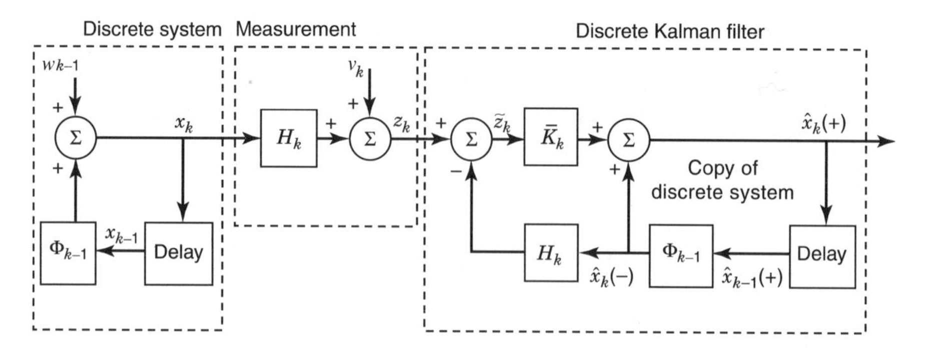

Fig. 4.1 *Block diagram of system, measurement model, and discrete-time Kalman filter.*

Step 4 of the Kalman filter implementation [computation of $\hat{x}_k(+)$] can be implemented only for state vector propagation where simulator or real data sets are available. An example of this is given in Section 4.12.

In the design trade-offs, the covariance matrix update (steps 1 and 3) should be checked for symmetry and positive definiteness. Failure to attain either condition is a sign that something is wrong—either a program "bug" or an ill-conditioned problem. In order to overcome ill-conditioning, another equivalent expression for $P_k(+)$ is called the "Joseph form,"[4] as shown in Equation 4.23:

$$P_k(+) = [I - \overline{K}_k H_k] P_k(-) [I - \overline{K}_k H_k]^{\mathrm{T}} + \overline{K}_k R_k \overline{K}_k^{\mathrm{T}}.$$

Note that the right-hand side of this equation is the summation of two symmetric matrices. The first of these is positive definite and the second is nonnegative definite, thereby making $P_k(+)$ a positive definite matrix.

There are many other forms[5] for $\overline{K}_k$ and $P_k(+)$ that might not be as useful for robust computation. It can be shown that state vector update, Kalman gain, and error covariance equations represent an asymptotically stable system, and therefore, the estimate of state $\hat{x}_k$ becomes independent of the initial estimate $\hat{x}_0$, P_0 as k is increased.

Figure 4.2 shows a typical time sequence of values assumed by the ith component of the estimated state vector (plotted with solid circles) and its corresponding variance of estimation uncertainty (plotted with open circles). The arrows show the successive values assumed by the variables, with the annotation (in parentheses) on the arrows indicating which input variables define the indicated transitions. Note that each variable assumes two distinct values at each discrete time: its a priori value

[4]after Bucy and Joseph [15].

[5]Some of the alternative forms for computing $\overline{K}_k$ and $P_k(+)$ can be found in Jazwinski [23], Kailath [24], and Sorenson [46].

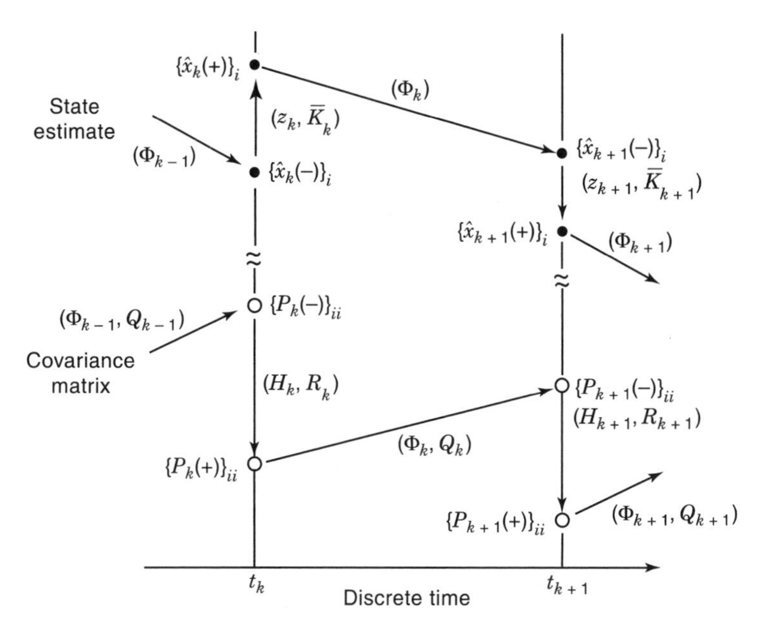

Fig. 4.2 *Representative sequence of values of filter variables in discrete time.*

corresponding to the value *before* the information in the measurement is used, and the a posteriori value corresponding to the value *after* the information is used.

EXAMPLE 4.1 Let the system dynamics and observations be given by the following equations:

$$
\begin{aligned}
x_k &= x_{k-1} + w_{k-1}, \qquad z_k = x_k + v_k, \\
E\langle v_k \rangle &= Ew_k = 0, \\
E\langle v_{k_1} v_{k_2} \rangle &= 2\Delta(k_2 - k_1), \quad E\langle w_{k_1} w_{k_2} \rangle = \Delta(k_2 - k_1), \\
z_1 &= 2, \qquad z_2 = 3, \\
E\langle x(0) \rangle &= \hat{x}_0 = 1, \\
E\langle [x(0) - \hat{x}_0][x(0) - \hat{x}_o]^{\mathrm{T}} \rangle &= P_0 = 10.
\end{aligned}
$$

The objective is to find $\hat{x}_3$ and the steady-state covariance matrix P_∞. One can use the equations in Table 4.3 with

$$\Phi = 1 = H, \qquad Q = 1, \qquad R = 2,$$

for which

$$\boxed{P_k(-) = P_{k-1}(+) + 1},$$

$$\overline{K}_k = \frac{P_k(-)}{P_k(-)+2} = \frac{P_{k-1}(+)+1}{P_{k-1}(+)+3},$$

$$P_k(+) = \left[1 - \frac{P_{k-1}(+)+1}{P_{k-1}(+)+3}\right](P_{k-1}(+)+1),$$

$$\boxed{P_k(+) = \frac{2(P_{k-1}(+)+1)}{P_{k-1}(+)+3}},$$

$$\boxed{\hat{x}_k(+) = \hat{x}_{k-1}(+) + \overline{K}_k(z_k - \hat{x}_{k-1}(+))}.$$

Let

$$P_k(+) = P_{k-1}(+) = P \quad \text{(steady-state covariance)},$$

$$P = \frac{2(P+1)}{P+3},$$

$$P^2 + P - 2 = 0,$$

$$P = 1, \quad \text{positive-definite solution.}$$

For $k = 1$

$$\hat{x}_1(+) = \hat{x}_0 + \frac{P_0+1}{P_0+3}(2 - \hat{x}_0) = 1 + \frac{11}{13}(2-1) = \frac{24}{13}$$

Following is a table for the various values of the Kalman filter:

k	$P_k(-)$	$P_k(+)$	$\overline{K}_k$	$\hat{x}_k(+)$
1	11	$\frac{22}{13}$	$\frac{11}{13}$	$\frac{24}{13}$
2	$\frac{35}{13}$	$\frac{70}{61}$	$\frac{35}{61}$	$\frac{153}{61}$

4.2.2 Treating Vector Measurements with Uncorrelated Errors as Scalars

In many (if not most) applications with vector-valued measurement z, the corresponding matrix R of measurement noise covariance is a diagonal matrix, meaning that the individual components of v_k are uncorrelated. For those applications, it is advantageous to consider the components of z as independent scalar measurements, rather than as a vector measurement. The principal advantages are as follows:

1. **Reduced Computation Time**. The number of arithmetic computations required for processing an ℓ-vector z as ℓ successive scalar measurements is significantly less than the corresponding number of operations for vector measurement processing. (It is shown in Chapter 6 that the number of computations for the vector implementation grows as ℓ^3, whereas that of the scalar implementation grows only as ℓ.)
2. **Improved Numerical Accuracy**. Avoiding matrix inversion in the implementation of the covariance equations (by making the expression $HPH^{\mathrm{T}} + R$ a scalar) improves the robustness of the covariance computations against roundoff errors.

The filter implementation in these cases requires ℓ iterations of the observational update equations using the rows of H as measurement "matrices" (with row dimension equal to 1) and the diagonal elements of R as the corresponding (scalar) measurement noise covariance. The updating can be implemented iteratively as the following equations:

$$\overline{K}_k^{[i]} = \frac{1}{H_k^{[i]} P_k^{[i-1]} H_k^{[i]\mathrm{T}} + R_k^{[i]}} P_k^{[i-1]} H_k^{[i]\mathrm{T}},$$

$$P_k^{[i]} = P_k^{[i-1]} - \overline{K}_k^{[i]} H_k^{[i]} P_k^{[i-1]},$$

$$\hat{x}_k^{[i]} = \hat{x}_k^{[i-1]} + \overline{K}_k^{[i]}[\{z_k\}_i - H_k^{[i]} \hat{x}_k^{[i-1]}],$$

for $i = 1, 2, 3, \ldots, \ell$, using the initial values

$$P_k^{[0]} = P_k(-), \qquad \hat{x}_k^{[0]} = \hat{x}_k(-);$$

intermediate variables

$$R_k^{[i]} = i\text{th diagonal element of the } \ell \times \ell \text{ diagonal matrix } R_k,$$

$$H_k^{[i]} = i\text{th row of the } \ell \times n \text{ matrix } H_k;$$

and final values

$$P_k^{[\ell]} = P_k(+), \qquad \hat{x}_k^{[\ell]} = \hat{x}_k(+).$$

4.2.3 Using the Covariance Equations for Design Analysis

It is important to remember that the Kalman gain and error covariance equations are independent of the actual observations. The covariance equations alone are all that is required for characterizing the performance of a proposed sensor system before it is actually built. At the beginning of the design phase of a measurement and estimation system, when neither real nor simulated data are available, just the covariance calculations can be used to obtain preliminary indications of estimator performance. Covariance calculations consist of solving the estimator equations with steps 1–3 of the previous subsection, repeatedly. These covariance calculations will involve the plant noise covariance matrix Q, measurement noise covariance matrix R, state transition matrix Φ, measurement sensitivity matrix H, and initial covariance matrix P_0—all of which must be known for the designs under consideration.

4.3 KALMAN–BUCY FILTER

Analogous to the discrete-time case, the continuous-time random process $x(t)$ and the observation $z(t)$ are given by

$$\dot{x}(t) = F(t)x(t) + G(t)w(t), \tag{4.27}$$
$$z(t) = H(t)x(t) + v(t), \tag{4.28}$$
$$Ew(t) = Ev(t) = 0,$$
$$Ew(t_1)w^{\mathrm{T}}(t_2) = Q(t)\delta(t_2 - t_1), \tag{4.29}$$
$$Ev(t_1)v^{\mathrm{T}}(t_2) = R(t)\delta(t_2 - t_1), \tag{4.30}$$
$$Ew(t)v^{\mathrm{T}}(\eta) = 0, \tag{4.31}$$

where $F(t)$, $G(t)$, $H(t)$, $Q(t)$, and $R(t)$ are $n \times n$, $n \times n$, $l \times n$, $n \times n$, and $l \times l$ matrices, respectively. The term $\delta(t_2 - t_1)$ is the Dirac delta. The covariance matrices Q and R are positive definite.

It is desired to find the estimate of n state vector $x(t)$ represented by $\hat{x}(t)$ which is a linear function of the measurements $z(t)$, $0 \leq t \leq T$, which minimizes the scalar equation

$$E[x(t) - \hat{x}(t)]^{\mathrm{T}} M[x(t) - \hat{x}(t)], \tag{4.32}$$

where M is a symmetric positive-definite matrix.

The initial estimate and covariance matrix are $\hat{x}_0$ and P_0.

This section provides a formal derivation of the continuous-time Kalman estimator. A rigorous derivation can be achieved by using the orthogonality principle as in the discrete-time case. In view of the main objective (to obtain efficient and practical estimators), less emphasis is placed on continuous-time estimators.

Let Δt be the time interval $[t_k - t_{k-1}]$. As shown in Chapters 2 and 3, the following relationships are obtained:

$$\Phi(t_k, t_{k-1}) = \Phi_k = I + F(t_{k-1})\Delta t + 0(\Delta t^2),$$

where $0(\Delta t^2)$ consists of terms with powers of Δt greater than or equal to two. For measurement noise

$$R_k = \frac{R(t_k)}{\Delta t},$$

and for process noise

$$Q_k = G(t_k)Q(t_k)G^{\mathrm{T}}(t_k)\,\Delta t.$$

Equations 4.24 and 4.26 can be combined. By substituting the above relations, one can get the result

$$\begin{aligned} P_k(-) &= [I + F(t)\Delta t][I - \overline{K}_{k-1}H_{k-1}]P_{k-1}(-) \\ &\quad \times [I + F(t)\Delta t]^{\mathrm{T}} + G(t)Q(t)G^{\mathrm{T}}(t)\Delta t, \end{aligned} \tag{4.33}$$

$$\begin{aligned} \frac{P_k(-) - P_{k-1}(-)}{\Delta t} &= F(t)P_{k-1}(-) + P_{k-1}(-)F^{\mathrm{T}}(t) \\ &\quad + G(t)Q(t)G^{\mathrm{T}}(t) - \frac{\overline{K}_{k-1}H_{k-1}P_{k-1}(-)}{\Delta t} \\ &\quad - F(t)\overline{K}_{k-1}H_{k-1}P_{k-1}(-)F^{\mathrm{T}}(t)\,\Delta t \\ &\quad + \text{higher order terms}. \end{aligned} \tag{4.34}$$

The Kalman gain of Equation 4.19 becomes, in the limit,

$$\begin{aligned} \lim_{\Delta t \to 0}\left[\frac{\overline{K}_{k-1}}{\Delta t}\right] &= \lim_{\Delta t \to 0}\left\{P_{k-1}(-)H_{k-1}^{\mathrm{T}}[H_{k-1}P_{k-1}(-)H_{k-1}^{\mathrm{T}}\,\Delta t + R(t)]^{-1}\right\} \\ &= PH^{\mathrm{T}}R^{-1} = \overline{K}(t). \end{aligned} \tag{4.35}$$

Substituting Equation 4.35 in 4.34 and taking the limit as $\Delta t \to 0$, one obtains the desired result

$$\begin{aligned} \dot{P}(t) &= F(t)P(t) + P(t)F^{\mathrm{T}}(t) + G(t)Q(t)G^{\mathrm{T}}(t) \\ &\quad - P(t)H^{\mathrm{T}}(t)R^{-1}(t)H(t)P(t) \end{aligned} \tag{4.36}$$

with $P(t_0)$ as the initial condition. This is called the *matrix Riccati differential equation*. Methods for solving it will be discussed in Section 4.8. The differential equation can be rewritten by using the identity

$$P(t)H^{\mathrm{T}}(t)R^{-1}(t)R(t)R^{-1}(t)H(t)P(t) = \overline{K}(t)R(t)\overline{K}^{\mathrm{T}}(t)$$

to transform Equation 4.36 to the form

$$\dot{P}(t) = F(t)P(t) + P(t)F^{\mathrm{T}}(t) + G(t)Q(t)G^{\mathrm{T}}(t) - \overline{K}(t)R(t)\overline{K}^{\mathrm{T}}(t). \tag{4.37}$$

In similar fashion, the state vector update equation can be derived from Equations 4.21 and 4.25 by taking the limit as $\Delta t \to 0$ to obtain the differential equation for the estimate:

$$\dot{\hat{x}}(t) = F(t)\hat{x}(t) + \overline{K}(t)[z(t) - H(t)\hat{x}(t)] \tag{4.38}$$

with initial condition $\hat{x}(0)$. Equations 4.35, 4.37, and 4.38 define the continuous-time Kalman estimator, which is also called the *Kalman–Bucy filter* [27, 179, 181, 182].

4.4 OPTIMAL LINEAR PREDICTORS

4.4.1 Prediction as Filtering

Prediction is equivalent to filtering when the measurement data are not available or are unreliable. In such cases, the Kalman gain matrix $\overline{K}_k$ is forced to be zero. Hence, Equations 4.21, 4.25, and 4.38 become

$$\hat{x}_k(+) = \Phi_{k-1}\hat{x}_{k-1}(+) \tag{4.39}$$

and

$$\dot{\hat{x}}(t) = F(t)\hat{x}(t). \tag{4.40}$$

Previous values of the estimates will become the initial conditions for the above equations.

4.4.2 Accommodating Missing Data

It sometimes happens in practice that measurements that had been scheduled to occur over some time interval $(t_{k_1} < t \leq t_{k_2})$ are, in fact, unavailable or unreliable. The estimation accuracy will suffer from the missing information, but the filter can continue to operate without modification. One can continue using the prediction algorithm given in Section 4.4 to continually estimate x_k for $k > k_1$ using the last available estimate $\hat{x}_{k_1}$ until the measurements again become useful (after $k = k_2$).

It is unnecessary to perform the observational update, because there is no information on which to base the conditioning. In practice, the filter is often run with the measurement sensitivity matrix $H = 0$ so that, in effect, the only update performed is the temporal update.

4.5 CORRELATED NOISE SOURCES

4.5.1 Correlation between Plant and Measurement Noise

We want to consider the extensions of the results given in Sections 4.2 and 4.3, allowing correlation between the two noise processes (assumed jointly Gaussian). Let the correlation be given by

$$Ew_{k_1}v_{k_2}^{\mathrm{T}} = C_k\,\Delta(k_2 - k_1) \quad \text{for the discrete-time case,}$$
$$Ew(t_1)v^{\mathrm{T}}(t_2) = C(t)\delta(t_2 - t_1) \quad \text{for the continuous-time case.}$$

For this extension, the discrete-time estimators have the same initial conditions and state estimate extrapolation and error covariance extrapolation equations. However, the measurement update equations in Table 4.3 have been modified as

$$\overline{K}_k = [P_k(-)H_k^{\mathrm{T}} + C_k][H_kP_k(-)H_k^{\mathrm{T}} + R_k + H_kC_k + C_k^{\mathrm{T}}H^{\mathrm{T}}]^{-1},$$
$$P_k(+) = P_k(-) - \overline{K}_k[H_kP_k(-) + C_k^{\mathrm{T}}],$$
$$\hat{x}_k(+) = \hat{x}_k(-) + \overline{K}_k[z_k - H_k\hat{x}_k(-)].$$

Similarly, the continuous-time estimator algorithms can be extended to include the correlation. Equation 4.35 is changed as follows [146, 222]:

$$\overline{K}(t) = [P(t)H^{\mathrm{T}}(t) + C(t)]R^{-1}(t).$$

4.5.2 Time-Correlated Measurements

Correlated measurement noise v_k can be modeled by a shaping filter driven by white Gaussian noise (see Section 3.6). Let the measurement model be given by

$$z_k = H_kx_k + v_k,$$

where

$$v_k = A_{k-1}v_{k-1} + \eta_{k-1} \tag{4.41}$$

and η_k is zero-mean white Gaussian.

Equation 4.1 is augmented by Equation 4.41, and the new state vector $X_k = [x_k \;\; v_k]^{\mathrm{T}}$ satisfies the difference equation:

$$X_k = \begin{bmatrix} x_k \\ \hline v_k \end{bmatrix} = \begin{bmatrix} \Phi_{k-1} & 0 \\ \hline 0 & A_{k-1} \end{bmatrix} \begin{bmatrix} x_{k-1} \\ \hline v_{k-1} \end{bmatrix} + \begin{bmatrix} w_{k-1} \\ \hline \eta_{k-1} \end{bmatrix},$$

$$z_k = [H_k \vdots I]X_k.$$

The measurement noise is zero, $R_k = 0$. The estimator algorithm will work as long as $H_k P_k(-) H_k^{\mathrm{T}} + R_k$ is invertible. Details of numerical difficulties of this problem (when R_k is singular) are given in Chapter 6.

For continuous-time estimators, the augmentation does not work because $\overline{K}(t) = P(t)H^{\mathrm{T}}(t)R^{-1}(t)$ is required. Therefore, $R^{-1}(t)$ must exist. Alternate techniques are required. For detailed information see Gelb et al. [21].

4.6 RELATIONSHIPS BETWEEN KALMAN AND WIENER FILTERS

The Wiener filter is defined for stationary systems in continuous time, and the Kalman filter is defined for either stationary or nonstationary systems in either discrete time or continuous time, but with finite-state dimension. To demonstrate the connections on problems satisfying both sets of constraints, take the continuous-time Kalman–Bucy estimator equations of Section 4.3, letting F, G, and H be constants, the noises be stationary (Q and R constant), and the filter reach steady state (P constant). That is, as $t \to \infty$, then $\dot{P}(t) \to 0$. The Riccati differential equation from Section 4.3 becomes the algebraic Riccati equation

$$0 = FP(\infty) + P(\infty)F^{\mathrm{T}} + GQG^{\mathrm{T}} - P(\infty)H^{\mathrm{T}}R^{-1}HP(\infty)$$

for continuous-time systems. The positive-definite solution of this algebraic equation is the steady-state value of the covariance matrix, $[P(\infty)]$. The Kalman–Bucy filter equation in steady state is then

$$\dot{\hat{x}}(t) = F\hat{x} + \overline{K}(\infty)[z(t) - H\hat{x}(t)].$$

Take the Laplace transform of both sides of this equation, assuming that the initial conditions are equal to zero, to obtain the following transfer function:

$$[sI - F + \overline{K}H]\hat{x}(s) = \overline{K}z(s),$$

where the Laplace transforms $\mathcal{L}\hat{x}(t) = \hat{x}(s)$ and $\mathcal{L}z(t) = z(s)$. This has the solution

$$\hat{x}(s) = [sI - F + \overline{K}H]^{-1}\overline{K}z(s),$$

where the steady-state gain

$$\overline{K} = P(\infty)H^{\mathrm{T}}R^{-1}.$$

This transfer function represents the steady-state Kalman–Bucy filter, which is identical to the Wiener filter [30].

4.7 QUADRATIC LOSS FUNCTIONS

The Kalman filter minimizes *any* quadratic loss function of estimation error. Just the fact that it is *unbiased* is sufficient to prove this property, but saying that the estimate is unbiased is equivalent to saying that $\hat{x} = E\langle x \rangle$. That is, the estimated value is the *mean* of the probability distribution of the state.

4.7.1 Quadratic Loss Functions of Estimation Error

A *loss function* or *penalty function*[6] is a real-valued function of the outcome of a random event. A loss function reflects the *value* of the outcome. Value concepts can be somewhat subjective. In gambling, for example, your perceived loss function for the outcome of a bet may depend upon your personality and current state of winnings, as well as on how much you have riding on the bet.

Loss Functions of Estimates. In estimation theory, the perceived loss is generally a function of *estimation error* (the difference between an estimated function of the outcome and its actual value), and it is generally a monotonically increasing function of the absolute value of the estimation error. In other words, bigger errors are valued less than smaller errors.

Quadratic Loss Functions. If x is a real n-vector (variate) associated with the outcome of an event and $\hat{x}$ is an estimate of x, then a quadratic loss function for the estimation error $\hat{x} - x$ has the form

$$L(\hat{x} - x) = (\hat{x} - x)^{\mathrm{T}} M (\hat{x} - x), \tag{4.42}$$

where M is a *symmetric positive-definite matrix*. One may as well assume that M is symmetric, because the skew-symmetric part of M does not influence the quadratic loss function. The reason for assuming positive definiteness is to assure that the loss is zero only if the error is zero, and loss is a monotonically increasing function of the absolute estimation error.

4.7.2 Expected Value of a Quadratic Loss Function

Loss and Risk. The expected value of loss is sometimes called *risk*. It will be shown that the expected value of a quadratic loss function of the estimation error

[6]These are concepts from decision theory, which includes estimation theory. The theory might have been built just as well on more optimistic concepts, such as "gain functions," "benefit functions," or "reward functions," but the nomenclature seems to have been developed by pessimists. This focus on the negative aspects of the problem is unfortunate, and you should not allow it to dampen your spirit.

$\hat{x} - x$ is a quadratic function of $\hat{x} - E\langle x\rangle$, where $E\langle\hat{x}\rangle = E\langle x\rangle$. This demonstration will depend upon the following identities:

$$\hat{x} - x = (\hat{x} - E\langle x\rangle) - (x - E\langle x\rangle), \tag{4.43}$$

$$\underset{x}{E}\langle x - E\langle x\rangle\rangle = 0, \tag{4.44}$$

$$\begin{aligned} &\underset{x}{E}\langle (x - E\langle x\rangle)^{\mathrm{T}} M (x - E\langle x\rangle)\rangle \\ &\quad = \underset{x}{E}\langle \mathrm{trace}[(x - E\langle x\rangle)^{\mathrm{T}} M (x - E\langle x\rangle)]\rangle && (4.45) \\ &\quad = \underset{x}{E}\langle \mathrm{trace}[M(x - E\langle x\rangle)(x - E\langle x\rangle)^{\mathrm{T}}]\rangle && (4.46) \\ &\quad = \mathrm{trace}[M\underset{x}{E}\langle (x - E\langle x\rangle)(x - E\langle x\rangle)^{\mathrm{T}}] && (4.47) \\ &\quad = \mathrm{trace}[MP], && (4.48) \end{aligned}$$

$$P \stackrel{\mathrm{def}}{=} \underset{x}{E}\langle (x - E\langle x\rangle)(x - E\langle x\rangle)^{\mathrm{T}}\rangle. \tag{4.49}$$

Risk of a Quadratic Loss Function. In the case of the quadratic loss function defined above, the expected loss (risk) will be

$$\begin{aligned} \mathcal{R}(\hat{x}) &= \underset{x}{E}\langle L(\hat{x} - x)\rangle && (4.50) \\ &= \underset{x}{E}\langle (\hat{x} - x)^{\mathrm{T}} M (\hat{x} - x)\rangle && (4.51) \\ &= \underset{x}{E}\langle [(\hat{x} - E\langle x\rangle) - (x - E\langle x\rangle)]^{\mathrm{T}} M [(\hat{x} - E\langle x\rangle) - (x - E\langle x\rangle)]\rangle && (4.52) \\ &= \underset{x}{E}\langle (\hat{x} - E\langle x\rangle)^{\mathrm{T}} M (\hat{x} - E\langle x\rangle) + (x - E\langle x\rangle)^{\mathrm{T}} M (x - E\langle x\rangle)\rangle \\ &\quad - \underset{x}{E}\langle (\hat{x} - E\langle x\rangle)^{\mathrm{T}} M (x - E\langle x\rangle) + (x - E\langle x\rangle)^{\mathrm{T}} M (\hat{x} - E\langle x\rangle)\rangle && (4.53) \\ &= (\hat{x} - E\langle x\rangle)^{\mathrm{T}} M (\hat{x} - E\langle x\rangle) + \underset{x}{E}\langle (x - E\langle x\rangle)^{\mathrm{T}} M (x - E\langle x\rangle)\rangle \\ &\quad - (\hat{x} - E\langle x\rangle)^{\mathrm{T}} M \underset{x}{E}\langle (x - E\langle x\rangle)\rangle - \underset{x}{E}\langle (x - E\langle x\rangle)\rangle^{\mathrm{T}} M (\hat{x} - E\langle x\rangle) && (4.54) \\ &= (\hat{x} - E\langle x\rangle)^{\mathrm{T}} M (\hat{x} - E\langle x\rangle) + \mathrm{trace}[MP], && (4.55) \end{aligned}$$

which is a quadratic function of $\hat{x} - E\langle x\rangle$ with the added nonnegative[7] constant trace$[MP]$.

4.7.3 Unbiased Estimates and Quadratic Loss

The *estimate* $\hat{x} = E\langle x\rangle$ minimizes the expected value of any *positive-definite quadratic loss function*. From the above derivation,

$$\mathcal{R}(\hat{x}) \geq \mathrm{trace}[MP] \tag{4.56}$$

[7]Recall that M and P are symmetric and nonnegative definite, and the matrix trace of any product of symmetric nonnegative definite matrices is nonnegative.

and

$$\mathcal{R}(\hat{x}) = \text{trace}[MP] \tag{4.57}$$

only if

$$\hat{x} = E\langle x \rangle, \tag{4.58}$$

where it has been assumed only that the mean $E\langle x \rangle$ and covariance $E_x\langle (x - E\langle x \rangle)(x - E\langle x \rangle)^{\mathrm{T}} \rangle$ are defined for the probability distribution of x. This demonstrates the utility of quadratic loss functions in estimation theory: They always lead to the mean as the estimate with minimum expected loss (risk).

Unbiased Estimates. An estimate $\hat{x}$ is called *unbiased* if the expected estimation error $E_x\langle \hat{x} - x \rangle = 0$. What has just been shown is that an unbiased estimate minimizes the expected value of any quadratic loss function of estimation error.

4.8 MATRIX RICCATI DIFFERENTIAL EQUATION

The need to solve the Riccati equation is perhaps the greatest single cause of anxiety and agony on the part of people faced with implementing a Kalman filter. This section presents a brief discussion of solution methods for the Riccati *differential* equation for the Kalman–Bucy filter. An analogous treatment of the discrete-time problem for the Kalman filter is presented in the next section. A more thorough treatment of the Riccati equation can be found in the book by Bittanti et al. [54].

4.8.1 Transformation to a Linear Equation

The Riccati differential equation was first studied in the eighteenth century as a nonlinear scalar differential equation, and a method was derived for transforming it to a linear matrix differential equation. That same method works when the dependent variable of the original Riccati differential equation is a matrix. That solution method is derived here for the matrix Riccati differential equation of the Kalman–Bucy filter. An analogous solution method for the discrete-time matrix Riccati equation of the Kalman filter is derived in the next section.

Matrix Fractions. A matrix product of the sort AB^{-1} is called a *matrix fraction*, and a representation of a matrix M in the form

$$M = AB^{-1}$$

will be called a *fraction decomposition* of M. The matrix A is the *numerator* of the fraction, and the matrix B is its *denominator*. It is necessary that the matrix denominator be nonsingular.

Linearization by Fraction Decomposition. The Riccati differential equation is nonlinear. However, a fraction decomposition of the covariance matrix results in a linear differential equation for the numerator and denominator matrices. The numerator and denominator matrices will be functions of time, such that the product $A(t)B^{-1}(t)$ satisfies the matrix Riccati differential equation and its boundary conditions.

Derivation. By taking the derivative of the matrix fraction $A(t)B^{-1}(t)$ with respect to t and using the fact[8] that

$$\frac{d}{dt}B^{-1}(t) = -B^{-1}(t)\dot{B}(t)B^{-1}(t),$$

one can arrive at the following decomposition of the matrix Riccati differential equation, where GQG^{T} has been reduced to an equivalent Q:

$$\dot{A}(t)B^{-1}(t) - A(t)B^{-1}(t)\dot{B}(t)B^{-1}(t)$$

$$= \frac{d}{dt}\{A(t)B^{-1}(t)\} \tag{4.59}$$

$$= \frac{d}{dt}P(t) \tag{4.60}$$

$$= F(t)P(t) + P(t)F^{\mathrm{T}}(t) - P(t)H^{\mathrm{T}}(t)R^{-1}(t)H(t)P(t) + Q(t) \tag{4.61}$$

$$= F(t)A(t)B^{-1}(t) + A(t)B^{-1}(t)F^{\mathrm{T}}(t) - A(t)B^{-1}(t)H^{\mathrm{T}}(t)R^{-1}(t)H(t)A(t)B^{-1}(t) + Q(t), \tag{4.62}$$

$$\dot{A}(t) - A(t)B^{-1}(t)\dot{B}(t) = F(t)A(t) + A(t)B^{-1}(t)F^{\mathrm{T}}(t)B(t) - A(t)B^{-1}(t)H^{\mathrm{T}}(t)R^{-1}(t)H(t)A(t) + Q(t)B(t), \tag{4.63}$$

$$\dot{A}(t) - A(t)B^{-1}(t)\{\dot{B}(t)\} = F(t)A(t) + Q(t)B(t) - A(t)B^{-1}(t) \times \{H^{\mathrm{T}}(t)R^{-1}(t)H(t)A(t) - F^{\mathrm{T}}(t)B(t)\}, \tag{4.64}$$

$$\dot{A}(t) = F(t)A(t) + Q(t)B(t), \tag{4.65}$$

$$\dot{B}(t) = H^{\mathrm{T}}(t)R^{-1}(t)H(t)A(t) - F^{\mathrm{T}}(t)B(t), \tag{4.66}$$

$$\frac{d}{dt}\begin{bmatrix} A(t) \\ B(t) \end{bmatrix} = \begin{bmatrix} F(t) & Q(t) \\ H^{\mathrm{T}}(t)R^{-1}(t)H(t) & -F^{\mathrm{T}}(t) \end{bmatrix}\begin{bmatrix} A(t) \\ B(t) \end{bmatrix}. \tag{4.67}$$

The last equation is a linear first-order matrix differential equation. The dependent variable is a $2n \times n$ matrix, where n is the dimension of the underlying state variable.

[8]This formula is derived in Appendix B, Equation B.10.

Hamiltonian Matrix. This is the name[9] given the matrix

$$\Psi(t) = \begin{bmatrix} F(t) & Q(t) \\ H^{\mathrm{T}}(t)R^{-1}(t)H(t) & -F^{\mathrm{T}}(t) \end{bmatrix} \tag{4.68}$$

of the matrix Riccati differential equation.

Boundary Constraints. The initial values of $A(t)$ and $B(t)$ must also be constrained by the initial value of $P(t)$. This is easily satisfied by taking $A(t_0) = P(t_0)$ and $B(t_0) = I$, the identity matrix.

4.8.2 Time-Invariant Problem

In the time-invariant case, the Hamiltonian matrix Ψ is also time-invariant. As a consequence, the solution for the numerator A and denominator B of the matrix fraction can be represented in matrix form as the product

$$\begin{bmatrix} A(t) \\ B(t) \end{bmatrix} = e^{\Psi t} \begin{bmatrix} P(0) \\ I \end{bmatrix},$$

where $e^{\Psi t}$ is a $2n \times 2n$ matrix.

4.8.3 Scalar Time-Invariant Problem

For this problem, the numerator A and denominator B of the "matrix fraction" AB^{-1} will be scalars, but Ψ will be a 2×2 matrix. We will here show how its exponential can be obtained in closed form. This will illustrate an application of the linearization procedure, and the results will serve to illuminate properties of the solutions—such as their dependence on initial conditions and on the scalar parameters F, H, R, and Q.

Linearizing the Differential Equation. The scalar time-invariant Riccati differential equation and its linearized equivalent are

$$\dot{P}(t) = FP(t) + P(t)F - P(t)HR^{-1}HP(t) + Q,$$

$$\begin{bmatrix} \dot{A}(t) \\ \dot{B}(t) \end{bmatrix} = \begin{bmatrix} F & Q \\ HR^{-1}H & -F \end{bmatrix} \begin{bmatrix} A(t) \\ B(t) \end{bmatrix},$$

respectively, where the symbols F, H, R, and Q represent scalar parameters (constants) of the application, t is a free (independent) variable, and the dependent variable P is constrained as a function of t by the differential equation. One can solve

[9]After the Irish mathematician and physicist William Rowan Hamilton (1805–1865).

this equation for P as a function of the free variable t *and* as a function of the parameters F, H, R, and Q.

Fundamental Solution of Linear Time-Invariant Differential Equation. The linear time-invariant differential equation has the general solution

$$\begin{bmatrix} A(t) \\ B(t) \end{bmatrix} = e^{\Psi t} \begin{bmatrix} P(0) \\ 1 \end{bmatrix},$$

$$\Psi = \begin{bmatrix} F & Q \\ \dfrac{H^2}{R} & -F \end{bmatrix}.$$

This matrix exponential will now be evaluated by using the characteristic vectors of Ψ, which are arranged as the column vectors of the matrix

$$M = \begin{bmatrix} \dfrac{-Q}{F+\phi} & \dfrac{-Q}{F-\phi} \\ 1 & 1 \end{bmatrix}, \quad \phi = \sqrt{F^2 + \frac{H^2 Q}{R}},$$

with inverse

$$M^{-1} = \begin{bmatrix} \dfrac{-H^2}{2\phi R} & \dfrac{H^2 Q}{2H^2 Q + 2F^2 R - 2F\phi R} \\ \dfrac{H^2}{2\phi R} & \dfrac{H^2 Q}{2H^2 Q + 2F^2 R + 2F\phi R} \end{bmatrix},$$

by which it can be diagonalized as

$$M^{-1}\Psi M = \begin{bmatrix} \lambda^2 & 0 \\ 0 & \lambda_1 \end{bmatrix},$$

$$\lambda_2 = -\frac{H^2 Q + F^2 R}{\phi R}, \quad \lambda_1 = \frac{H^2 Q + F^2 R}{\phi R},$$

with the characteristic values of Ψ along its diagonal. The exponential of the diagonalized matrix, multiplied by t, will be

$$e^{M^{-1}\Psi M t} = \begin{bmatrix} e^{\lambda_2 t} & 0 \\ 0 & e^{\lambda_1 t} \end{bmatrix}.$$

Using this, one can write the fundamental solution of the linear homogeneous time-invariant equation as

$$
\begin{aligned}
e^{\Psi t} &= \sum_{k=0}^{\infty} \frac{1}{k!} t^k \Psi^k \\
&= M\left(\sum_{k=0}^{\infty} \frac{1}{k!} [M^{-1}\Psi M]^k\right) M^{-1} \\
&= M e^{M^{-1}\Psi M t} M^{-1} \\
&= M \begin{bmatrix} e^{\lambda_2 t} & 0 \\ 0 & e^{\lambda_1 t} \end{bmatrix} M^{-1} \\
&= \frac{1}{2e^{\phi t}\phi} \begin{bmatrix} \phi(\psi(t)+1) + F(\psi(t)-1) & Q(1-\psi(t)) \\ \dfrac{H^2(\psi(t)-1)}{R} & F(1-\psi(t)) + \phi(1+\psi(t)) \end{bmatrix}, \\
\psi(t) &= e^{2\phi t}
\end{aligned}
$$

and the solution of the linearized system as

$$
\begin{aligned}
\begin{bmatrix} A(t) \\ B(t) \end{bmatrix} &= e^{\Psi t} \begin{bmatrix} P(0) \\ 1 \end{bmatrix} \\
&= \frac{1}{2e^{\phi t}\phi} \begin{bmatrix} P(0)[\phi(\psi(t)+1) + F(\psi(t)-1)] - \dfrac{Q(\psi(t)-1)}{R^2} \\ \dfrac{P(0)H^2(\psi(t)-1)}{R} - \phi(\psi(t)+1) - F(\psi(t)-1) \end{bmatrix}.
\end{aligned}
$$

General Solution of Scalar Time-Invariant Riccati Equation. The general solution formula may now be composed from the previous results as

$$
\begin{aligned}
P(t) &= A(t)/B(t) \\
&= \frac{\mathcal{N}_P(t)}{\mathcal{D}_P(t)}, \qquad (4.69) \\
\mathcal{N}_P(t) &= R[P(0)(\phi + F) + Q] + R[P(0)(\phi - F) - Q]e^{-2\phi t} \\
&= R\left[P(0)\left(\sqrt{F^2 + \frac{H^2 Q}{R}} + F\right) + Q\right] \\
&\quad + R\left[P(0)\left(\sqrt{F^2 + \frac{H^2 Q}{R}} - F\right) - Q\right]e^{-2\phi t}, \qquad (4.70)
\end{aligned}
$$

$$\begin{aligned}\mathcal{D}_P(t) &= [H^2P(0) + R(\phi - F)] - [H^2P(0) - R(F + \phi)]e^{-2\phi t} \\ &= \left[H^2P(0) + R\left(\sqrt{F^2 + \frac{H^2Q}{R}} - F\right)\right] \\ &\quad - \left[H^2P(0) - R\left(\sqrt{F^2 + \frac{H^2Q}{R}} + F\right)\right]e^{-2\phi t}. \end{aligned} \tag{4.71}$$

Singular Values of Denominator. The denominator $\mathcal{D}_P(t)$ can easily be shown to have a zero for t_0 such that

$$e^{-2\phi t_0} = 1 + 2\frac{R}{H^2} \times \frac{H^2[P(0)\phi + Q] + FR(\phi - F)}{H^2P^2(0) - 2FRP(0) - QR}.$$

However, it can also be shown that $t_0 < 0$ if

$$P(0) > -\frac{R}{H^2}(\phi - F),$$

which is a nonpositive lower bound on the initial value. This poses no particular difficulty, however, since $P(0) \geq 0$ anyway. (We will see in the next section what would happen if this condition were violated.)

Boundary values. Given the above formulas for $P(t)$, its numerator $\mathcal{N}(t)$, and its denominator $\mathcal{D}(t)$, one can easily show that they have the following limiting values:

$$\begin{aligned}\lim_{t\to 0} \mathcal{N}_P(t) &= 2P(0)R\sqrt{F^2 + \frac{H^2Q}{R}}, \\ \lim_{t\to 0} \mathcal{D}_P(t) &= 2R\sqrt{F^2 + \frac{H^2Q}{R}}, \\ \lim_{t\to 0} P(t) &= P(0), \\ \lim_{t\to\infty} P(t) &= \frac{R}{H^2}\left(F + \sqrt{F^2 + \frac{H^2Q}{R}}\right). \end{aligned} \tag{4.72}$$

4.8.4 Parametric Dependence of the Scalar Time-Invariant Solution

The previous solution of the scalar time-invariant problem will now be used to illustrate its dependence on the parameters F, H, R, Q, and $P(0)$. There are two fundamental algebraic functions of these parameters that will be useful in characterizing the behavior of the solutions: the asymptotic solution as $t \to \infty$ and the time constant of decay to this steady-state solution.

Decay Time Constant. The only time-dependent terms in the expression for $P(t)$ are those involving $e^{-2\phi t}$. The fundamental decay time constant of the solution is then the algebraic function

$$\tau(F, H, R, Q) = 2\sqrt{F^2 + \frac{H^2 Q}{R}} \tag{4.73}$$

of the problem parameters. Note that this function does not depend upon the initial value of P.

Asymptotic and Steady-State Solutions. The asymptotic solution of the scalar time-invariant Riccati differential equation as $t \to \infty$ is given in Equation 4.72. It should be verified that this is also the solution of the corresponding *steady-state* differential equation

$$\dot{P} = 0,$$
$$P^2(\infty)H^2R^{-1} - 2FP(\infty) - Q = 0,$$

which is also called the *algebraic*[10] Riccati equation. This quadratic equation in $P(\infty)$ has two solutions, expressible as algebraic functions of the problem parameters:

$$P(\infty) = \frac{FR \pm \sqrt{H^2QR + F^2R^2}}{H^2}.$$

The two solutions correspond to the two values for the signum ($\pm$). There is no cause for alarm, however. The solution that agrees with Equation 4.72 is the nonnegative one. The other solution is nonpositive. We are only interested in the nonnegative solution, because the variance P of uncertainty is, by definition, nonnegative.

Dependence on Initial Conditions. For the scalar problem, the initial conditions are parameterized by $P(0)$. The dependence of the solution on its initial value is not continuous everywhere, however. The reason is that there are two solutions to the steady-state equation. The nonnegative solution is *stable* in the sense that initial conditions sufficiently near to it converge to it asymptotically. The nonpositive

[10]So called because it is an algebraic equation, not a differential equation. That is, it is constructed from the operations of algebra, not those of the differential calculus. The term by itself is ambiguous in this usage, however, because there are two entirely different forms of the algebraic Riccati equation. One is derived from the Riccati *differential* equation, and the other is derived from the *discrete-time* Riccati equation. The results are both algebraic equations, but they are significantly different in structure.

solution is *unstable* in the sense that infinitesimal perturbations of the initial condition cause the solution to diverge from the nonpositive steady-state solution and converge, instead, to the nonnegative steady-state solution.

Convergent and Divergent Solutions. The *eventual* convergence of a solution to the nonnegative steady-state value may pass through infinity to get there. That is, the solution may initially *diverge*, depending on the initial values. This type of behavior is shown in Figure 4.3, which is a multiplot of solutions to an example of the Riccati equation with

$$F = 0, \qquad H = 1, \qquad R = 1, \qquad Q = 1,$$

for which the corresponding continuous-time algebraic (quadratic) Riccati equation

$$\begin{aligned} \dot{P}(\infty) &= 0, \\ 2FP(\infty) - \frac{[P(\infty)H]^2}{R} + Q &= 0, \\ 1 - [P(\infty)]^2 &= 0 \end{aligned}$$

has the two solutions $P(\infty) = \pm 1$. The Riccati differential equation has the closed-form solution

$$P(t) = \frac{e^{2t}[1 + P(0)] - [1 - P(0)]}{e^{2t}[1 + P(0)] + [1 - P(0)]}$$

in terms of the initial value $P(0)$. Solutions of the initial-value problem with different initial values are plotted over the time interval $0 \leq t \leq 2$. All solutions except the one with $P(0) = -1$ appear to converge eventually to $P(\infty) = 1$, but those that

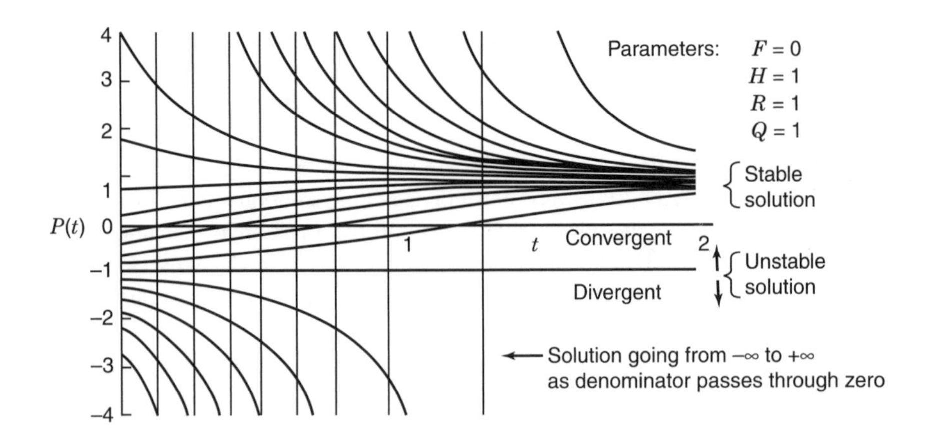

Fig. 4.3 Solutions of the scalar time-invariant Riccati equation.

disappear off the bottom of the graph diverge to $-\infty$, then converge to $P(\infty) = +1$ from the top of the graph. The vertical Lines on the plot are the times at which solutions starting with $P(0) < -1$ pass through $P(t) = \pm\infty$ on their way to $P(\infty) = 1$. This phenomenon occurs at the zeros of the denominator in the expression for $P(t)$, which occur at time

$$t^* = \log_e \sqrt{\frac{P(0) - 1}{P(0) + 1}}$$

for $P(0) < -1$. Solutions with $P(0) > -1$ converge without this discontinuous behavior.

Convergent and Divergent Regions. The line at $P = -1$ in Figure 4.3 separates initial values into two regions, characterized by the stability of solutions to the initial-value problem. Solutions with initial values above that line converge to the positive steady-state solution. Solutions starting below that line diverge.

4.8.5 Convergence Issues

It is usually risky to infer properties of high-order systems from those of lower order. However, the following general trends are apparent in the behavior of the closed-form solution of the scalar time-invariant Riccati differential equation:

1. The solution eventually converges exponentially to the nonnegative steady-state solution. The decay time constant varies as $(F^2 + H^2Q/R)^{1/2}$, which increases with $|F|$, $|H|$, and Q and decreases as R increases (for $R > 0$ and $Q > 0$).
2. Solutions are not uniformly exponentially convergent, however. The initial value does not influence the *asymptotic* decay rate, but it can influence the initial response. In particular, convergence for initial values nearer the unstable steady-state solution is hampered initially.
3. The stable asymptotic solution is

$$P(\infty) = \frac{R}{H^2}\left(F + \sqrt{F^2 + \frac{H^2Q}{R}}\right),$$

which is influenced by both the sign *and* magnitude of F but only by the magnitudes of H, R, and Q.

Stability properties of general (higher order) systems have been proved by Potter [119].

Even Unstable Dynamic Systems Have Convergent Riccati Equations. Note that the corresponding equation for the variance of the *state*

$$\frac{d}{dt}P(t) = FP + PF^{\mathrm{T}} + Q$$

has the general solution

$$P(t) = \frac{(e^{2Ft} - 1)Q}{2F} + e^{2Ft}P(0)$$

in the scalar case. This dynamic system is unstable if $F > 0$, because the solution $P(t) \to +\infty$ as $t \to \infty$. However, the corresponding Riccati equation (with the conditioning term) approaches a finite limit.

4.8.6 Closed-Form Solution of the Algebraic Riccati Equation

We have seen in the previous subsections the difficulty of obtaining a solution of the general Riccati differential equation in "closed form" (i.e., as a formula in the parameters of the model), even for the simplest (scalar) problem. The following example illustrates the difficulty of obtaining closed-form solutions for the *algebraic* Riccati equation, as well, for a simple model.

EXAMPLE 4.2: Solving the Algebraic Riccati Equation in Continuous-Time for the Harmonic Resonator Problem The problem is to characterize the asymptotic uncertainty in estimating the state (position and velocity) of a damped harmonic resonator driven by Gaussian noise, given noisy measurements of position. The system model for this problem has been derived in Examples 2.2, 2.3, 2.6, 2.7, 3.9, 3.10, and 3.11. The resulting algebraic Riccati equation for this problem in continuous-time has the form

$$\begin{aligned}
0 &= FP + PF^{\mathrm{T}} - PH^{\mathrm{T}}R^{-1}HP + Q,\\
F &= \begin{bmatrix} 0 & 1 \\ -\dfrac{1+\omega^2\tau^2}{\tau^2} & \dfrac{-2}{\tau} \end{bmatrix},\\
H &= \begin{bmatrix} 1 & 0 \end{bmatrix},\\
Q &= \begin{bmatrix} 0 & 0 \\ 0 & q \end{bmatrix},
\end{aligned}$$

which is equivalent to the three scalar equations

$$\begin{aligned}
0 &= -p_{11}^2 + 2Rp_{12},\\
0 &= -R(1+\omega^2\tau^2)p_{11} - 2R\tau p_{12} - \tau^2 p_{11}p_{12} + R\tau^2 p_{22},\\
0 &= -\tau^2 p_{12}^2 - 2R(1+\omega^2\tau^2)p_{12} - 4R\tau p_{22} + Rq\tau^2.
\end{aligned}$$

The first and last of these can be solved as linear equations in the variables p_{12} and p_{22}

$$p_{12} = \frac{p_{11}^2}{2R},$$
$$p_{22} = \frac{Rq\tau^2 - \tau^2 p_{12}^2 - 2R(1+\omega^2\tau^2)p_{12}}{4R\tau}$$

in terms of p_{11}. Substitution of these expressions into the middle scalar equation yields the following quartic equation in p_{11}:

$$0 = \tau^3 p_{11}^4 + 8R\tau^2 p_{11}^3 + 20R^2\tau(5+\omega^2\tau^2)p_{11}^2 + 16R^3(1+\omega^2\tau^2)p_{11} - 4R^3 q\tau^3.$$

This may appear to be a relatively simple quartic equation, but its solution is a rather laborious and tedious process. It has four solutions, only one of which yields a nonnegative covariance matrix P:

$$p_{11} = \frac{R(1-b)}{\tau},$$
$$p_{12} = \frac{R(1-b)^2}{2\tau^2},$$
$$p_{22} = \frac{R}{\tau^3}\big(-6 + 2\omega^2\tau^2 - 4a + (4+a)b\big),$$
$$a = \sqrt{(1+\omega^2\tau^2)^2 + \frac{q\tau^4}{R}}, \qquad b = \sqrt{2(1-\omega^2\tau^2+a)}.$$

Because there is no general formula for solving higher order polynomial equations (i.e., beyond quartic), this relatively simple example is at the limit of complexity for finding closed-form solutions to algebraic Riccati equations by purely algebraic means. Beyond this relatively low level of complexity, it is necessary to employ numerical solution methods. Numbers do not always provide us as much insight into the characteristics of the solution as formulas do, but they are all we can get for most problems of practical significance.

4.8.7 Newton–Raphson Solution of the Algebraic Riccati Differential Equation

The *Newton–Raphson solution* of n differentiable functional equations

$$0 = f_1(x_1, x_2, x_3, \ldots, x_n),$$
$$0 = f_2(x_1, x_2, x_3, \ldots x_n),$$
$$0 = f_3(x_1, x_2, x_3, \ldots, x_n),$$
$$\vdots$$
$$0 = f_n(x_1, x_2, x_3, \ldots, x_n)$$

in n unknowns $x_1, x_2, x_3, \ldots, x_n$ is the iterative vector procedure

$$x \leftarrow x - \mathcal{F}^{-1} f(x) \tag{4.74}$$

using the vector and matrix variables

$$\begin{aligned}
x &= [x_1 \quad x_2 \quad x_3 \quad \cdots \quad x_n]^{\mathrm{T}}, \\
f(x) &= [f_1(x) \quad f_2(x) \quad f_3(x) \quad \cdots \quad f_n(x)]^{\mathrm{T}}, \\
\mathcal{F} &= \begin{bmatrix}
\frac{\partial f_1}{\partial x_1} & \frac{\partial f_1}{\partial x_2} & \frac{\partial f_1}{\partial x_3} & \cdots & \frac{\partial f_1}{\partial x_n} \\
\frac{\partial f_2}{\partial x_1} & \frac{\partial f_2}{\partial x_2} & \frac{\partial f_2}{\partial d_3} & \cdots & \frac{\partial f_2}{\partial x_n} \\
\frac{\partial f_3}{\partial x_1} & \frac{\partial f_3}{\partial x_2} & \frac{\partial f_3}{\partial x_3} & \cdots & \frac{\partial f_3}{\partial x_n} \\
\vdots & \vdots & \vdots & \ddots & \vdots \\
\frac{\partial f_n}{\partial x_1} & \frac{\partial f_n}{\partial x_2} & \frac{\partial f_n}{\partial x_3} & \cdots & \frac{\partial f_n}{\partial x_n}
\end{bmatrix}.
\end{aligned}$$

Application of this vector-oriented procedure to matrix equations is generally done by "vectorizing" the matrix of unknowns and using Kronecker products to "matricize" $\mathcal{F}$ from what would otherwise be four-dimensional data structures. However, the general approach does not take advantage of the symmetry constraints in the matrix Riccati differential equation. There are two such constraints: one on the symmetry of the Riccati equation itself and another on the symmetry of the solution, P. Therefore, in solving the steady-state $n \times n$ matrix Riccati differential equation, there are effectively only $n(n+1)/2$ independent scalar equations in $n(n+1)/2$ scalar unknowns. The $n(n+1)/2$ scalar unknowns can be taken as the upper triangular elements of P, and the $n(n+1)/2$ scalar equations can be taken as those equating upper triangular terms of the matrix equation. We will first describe the equations by which the matrix equation and the matrix unknown can be vectorized, then show the form that the variables of the Newton–Raphson solution will take for this vectorization.

Vectorizing Formulas. If one lets the indices i and j stand for the row and column, respectively, of the terms in the matrix Riccati equation, then the respective elements of the *upper triangular* parts of the matrix equation can be vectorized by the single index p, where

$$\begin{aligned} 1 &\le j \le n, \\ 1 &\le i \le j, \\ p &= \tfrac{1}{2} j(j-1) + i, \\ 1 &\le p \le \tfrac{1}{2} n(n+1). \end{aligned}$$

Similarly, the upper triangular part of P can be mapped into a singly subscripted array x with index q, according to the rules

$$\begin{aligned} 1 &\le \ell \le n, \\ 1 &\le k \le \ell, \\ q &= \tfrac{1}{2} \ell(\ell-1) + k, \\ 1 &\le q \le \tfrac{1}{2} n(n+1), \end{aligned}$$

whereby $P_{k\ell}$ is mapped into x_q.

Values of Variables for Newton–Raphson Solution of Steady-State Matrix Riccati Differential Equation. The solution is an implementation of the recursion formula 4.74 with

$$f_p = \mathcal{Z}_{ij}, \tag{4.75}$$

$$\mathcal{Z} = FP + PF^{\mathrm{T}} - PH^{\mathrm{T}}R^{-1}HP + Q, \tag{4.76}$$

$$x_q = P_{k\ell}, \tag{4.77}$$

$$p = \tfrac{1}{2} j(j-1) + i, \tag{4.78}$$

$$q = \tfrac{1}{2} \ell(\ell-1) + k, \tag{4.79}$$

$$\begin{aligned} \mathcal{F}_{pq} &= \frac{\partial f_p}{\partial x_q} \\ &= \frac{\partial \mathcal{Z}_{ij}}{\partial P_{k\ell}} \\ &= \Delta_{j\ell} \mathcal{S}_{ik} + \Delta_{ik} \mathcal{S}_{j\ell}, \end{aligned} \tag{4.80}$$

$$\mathcal{S} = F - PH^{\mathrm{T}}R^{-1}H, \tag{4.81}$$

$$\Delta_{ab} \stackrel{\text{def}}{=} \begin{cases} 1 & \text{if} \quad a = b, \\ 0 & \text{if} \quad a \ne b. \end{cases} \tag{4.82}$$

The least obvious of these is Equation 4.80, which will now be derived.

"Dot" Notation for Row and Column Submatrices. For any matrix M, let the notation $M_{\cdot j}$ [with a dot $(\cdot)$ where the row index should be] stand for the jth column of M. When this notation is applied to the identity matrix I, $I_{\cdot j}$ will equal a column vector with 1 in the jth row and zeros elsewhere. As a vector, it has the property that

$$MI_{\cdot j} = M_{\cdot j}$$

for any conformable matrix M.

Matrix Partial Derivatives. With this notation, one can write matrix partial derivatives as follows:

$$\frac{\partial P}{\partial P_{k\ell}} = I_{\cdot k} I_{\cdot \ell}^{\mathrm{T}}, \tag{4.83}$$

$$\frac{\partial \mathcal{Z}}{\partial P_{k\ell}} = F\frac{\partial P}{\partial P_{k\ell}} + \frac{\partial P}{\partial P_{k\ell}} F^{\mathrm{T}} - \frac{\partial P}{\partial P_{k\ell}} H^{\mathrm{T}} R^{-1} H P - P H^{\mathrm{T}} R^{-1} H \frac{\partial P}{\partial P_{k\ell}} \tag{4.84}$$

$$= F I_{\cdot k} I_{\cdot \ell}^{\mathrm{T}} + I_{\cdot k} I_{\cdot \ell}^{\mathrm{T}} F^{\mathrm{T}} - I{\cdot}k I_{\cdot \ell}^{\mathrm{T}} H^{\mathrm{T}} R^{-1} H P - P H^{\mathrm{T}} R^{-1} H \frac{\partial \partial}{\partial P_{k\ell}} \tag{4.85}$$

$$= F_{\cdot k} I_{\cdot \ell}^{\mathrm{T}} + I_{\cdot k} F_{\cdot \ell}^{\mathrm{T}} - I_{\cdot k} M_{\cdot \ell}^{\mathrm{T}} - M_{\cdot k} I_{\cdot \ell}^{\mathrm{T}} \tag{4.86}$$

$$= (F - M)_{\cdot k} I_{\cdot \ell}^{\mathrm{T}} + I_{\cdot k} (F - M)_{\cdot \ell}^{\mathrm{T}} \tag{4.87}$$

$$= \mathcal{S}_{\cdot k} I_{\cdot \ell}^{\mathrm{T}} + I_{\cdot k} \mathcal{S}_{\cdot \ell}^{\mathrm{T}}, \tag{4.88}$$

$$\mathcal{S} = F - M, \tag{4.89}$$

$$M = P H^{\mathrm{T}} R^{-1} H. \tag{4.90}$$

Note that, on the right-hand side of Equation 4.88, the first term $(\mathcal{S}_{\cdot k} I_{\cdot \ell}^{\mathrm{T}})$ has only one nonzero column—the ℓth column. Similarly, the other term $(I_{\cdot k} \mathcal{S}_{\cdot \ell}^{\mathrm{T}})$ has only one nonzero row—its kth row. Consequently, the element in the ith row and jth column of this matrix will be the expression given in Equation 4.80. This completes its derivation.

Computational Complexity. The number of floating-point operations per iteration for this solution method is dominated by the inversion of the $n(n+1)/2 \times n(n+1)/2$ matrix $\mathcal{F}$, which requires somewhat more than $n^6/8$ flops.

4.8.8 MacFarlane–Potter–Fath Eigenstructure Method

Steady-State Solution of Time-Invariant Matrix Riccati Differential Equation. It was discovered independently by MacFarlane [197], Potter [209], and Fath [158] that the solution $P(\infty)$ of the continuous-time form of the steady-state matrix Riccati differential equation can be expressed in the form

$$P(\infty) = AB^{-1},$$
$$\begin{bmatrix} A \\ B \end{bmatrix} = [e_{i_1} \quad e_{i_2} \quad e_{i_3} \quad \cdots \quad e_{i_n}],$$

where the matrices A and B are $n \times n$ and the $2n$-vectors e_{i_k} are characteristic vectors of the continuous-time system Hamiltonian matrix

$$\Psi_c = \begin{bmatrix} F & Q \\ H^{\mathrm{T}}R^{-1}H & -F^{\mathrm{T}} \end{bmatrix}.$$

This can be formalized in somewhat greater generality as a lemma:

LEMMA 1 If A and B are $n \times n$ matrices such that B is nonsingular and

$$\Psi_c \begin{bmatrix} A \\ B \end{bmatrix} = \begin{bmatrix} A \\ B \end{bmatrix} D \tag{4.91}$$

for an $n \times n$ matrix D, then $P = AB^{-1}$ satisfies the steady-state matrix Riccati differential equation

$$0 = FP + PF^{\mathrm{T}} - PH^{\mathrm{T}}R^{-1}HP + Q.$$

Proof: Equation 4.91 can be written as two equations,

$$AD = FA + QB, \qquad BD = H^{\mathrm{T}}R^{-1}HA - F^{\mathrm{T}}B.$$

If one multiplies both of these on the right by B^{-1} and the last of these on the left by AB^{-1}, one obtains the equivalent equations

$$ADB^{-1} = FAB^{-1} + Q,$$
$$ADB^{-1} = AB^{-1}H^{\mathrm{T}}R^{-1}HAB^{-1} - AB^{-1}F^{\mathrm{T}},$$

or, taking the differences of the left-hand sides and substituting P for AB^{-1},

$$0 = FP + PF^{\mathrm{T}} - PH^{\mathrm{T}}R^{-1}HP + Q,$$

which was to be proved.

In the case that A and B are formed in this way from n characteristic vectors of Ψ_c, the matrix D will be a diagonal matrix of the corresponding characteristic values. (Check it out for yourself.) Therefore, to obtain the steady-state solution of the matrix Riccati differential equation by this method, it suffices to find n characteristic vectors of Ψ_c such that the corresponding B-matrix is nonsingular. (As will be shown in the next section, the same trick works for the discrete-time matrix Riccati equation.)

4.9 MATRIX RICCATI EQUATION IN DISCRETE TIME

4.9.1 Linear Equations for Matrix Fraction Propagation

The representation of the covariance matrix as a matrix fraction is also sufficient to transform the nonlinear discrete-time Riccati equation for the estimation uncertainty into a linear form. The discrete-time problem differs from the continuous-time problem in two important aspects:

1. The numerator and denominator matrices will be propagated by a $2n \times 2n$ transition matrix and not by differential equations. The approach is otherwise similar to that for the continuous-time Riccati equation, but the resulting $2n \times 2n$ state transition matrix for the recursive updates of the numerator and denominator matrices is a bit more complicated than the coefficient matrix for the linear form of the continuous-time matrix Riccati equation.
2. There are two distinct values of the discrete-time covariance matrix at any discrete-time step—the a priori value and the a posteriori value. The a priori value is of interest in computing Kalman gains, and the a posteriori value is of interest in the analysis of estimation uncertainty.

The linear equations for matrix fraction propagation of the a priori covariance matrix are derived below. The method is then applied to obtain a closed-form solution for the scalar time-invariant Riccati equation in discrete time and to a method for exponential speedup of convergence to the asymptotic solution.

4.9.2 Matrix Fraction Propagation of the a priori Covariance

LEMMA 2 If the state transition matrices Φ_k are nonsingular and

$$P_k(-) = A_k B_k^{-1} \tag{4.93}$$

is a nonsingular matrix solution of the discrete-time Riccati equation at time t_k, then

$$P_{k+1}(-) = A_{k+1} B_{k+1}^{-1} \tag{4.93}$$

is a solution at time t_{k+1}, where

$$\begin{bmatrix} A_{k+1} \\ B_{k+1} \end{bmatrix} = \begin{bmatrix} Q_k & I \\ I & 0 \end{bmatrix} \begin{bmatrix} \Phi_k^{-\mathrm{T}} & 0 \\ 0 & \Phi_k \end{bmatrix} \begin{bmatrix} H_k^{\mathrm{T}} R_k^{-1} H_k & I \\ I & 0 \end{bmatrix} \begin{bmatrix} A_k \\ B_k \end{bmatrix} \tag{4.94}$$

$$= \begin{bmatrix} \Phi_k + Q_k \Phi_k^{-\mathrm{T}} H_k^{\mathrm{T}} R_k^{-1} H_k & Q_k \Phi_k^{-\mathrm{T}} \\ \Phi_k^{-\mathrm{T}} H_k^{\mathrm{T}} R_k^{-1} H_k & \Phi_k^{-\mathrm{T}} \end{bmatrix} \begin{bmatrix} A_k \\ B_k \end{bmatrix}. \tag{4.95}$$

Proof: The following annotated sequence of equalities starts with the product $A_{k+1} B_{k+1}^{-1}$ as defined, and proves that it equals P_{k+1}:

$$\begin{aligned}
A_{k+1} B_{k+1}^{-1} &= \{[\Phi_k + Q_k \Phi_k^{-\mathrm{T}} H_k^{-\mathrm{T}} R_k^{-1} H_k] A_k + Q_k \Phi_k^{-\mathrm{T}} B_k\} \\
&\quad \times \{\Phi_k^{-\mathrm{T}} [H_k^{\mathrm{T}} R_k^{-1} H_k A_k B_k^{-1} + I] B_k\}^{-1} && \text{(definition)} \\
&= \{[\Phi_k + Q_k \Phi_k^{-\mathrm{T}} H_k^{\mathrm{T}} R_k^{-1} H_k] A_k + Q_k \Phi_k^{-\mathrm{T}} B_k\} \\
&\quad \times B_k^{-1} \{H_k^{\mathrm{T}} R_k^{-1} H_k A_k B_k^{-1} + I\}^{-1} \Phi_k^{\mathrm{T}} && \text{(factor } B_k\text{)} \\
&= \{[\Phi_k + Q_k \Phi_k^{-\mathrm{T}} H_k^{\mathrm{T}} R_k^{-\mathrm{T}} H_k] A_k B_k^{-1} + Q_k \Phi_k^{-\mathrm{T}}\} \\
&\quad \times \{H_k^{\mathrm{T}} R_k^{-1} H_k A_k B_k^{-1} + I\}^{-1} \Phi_k^{\mathrm{T}} && \text{(distribute } B_k\text{)} \\
&= \{[\Phi_k + Q_k \Phi_k^{-\mathrm{T}} H_k^{\mathrm{T}} R_k^{-1} H_k] P_k(-) + Q_k \Phi_k^{-\mathrm{T}}\} \\
&\quad \times \{H_k^{\mathrm{T}} R_k^{-1} H_k P_k(-) + I\}^{-1} \Phi_k^{\mathrm{T}} && \text{(definition)} \\
&= \{\Phi_k P_k(-) + Q_k \Phi_k^{-\mathrm{T}} [H_k^{\mathrm{T}} R_k^{-1} H_k P_k(-) + I]\} \\
&\quad \times \{H_k^{\mathrm{T}} R_k^{-\mathrm{T}} H_k P_k(-) + I\}^{-1} \Phi_k^{\mathrm{T}} && \text{(regroup)} \\
&= \Phi_k P_k(-) \{H_k^{\mathrm{T}} R_k^{-\mathrm{T}} H_k P_k(-) + I\}^{-1} \Phi_k^{\mathrm{T}} + Q_k \Phi_k^{-\mathrm{T}} \Phi_k^{\mathrm{T}} && \text{(distribute)} \\
&= \Phi_k \{H_k^{\mathrm{T}} R_k^{-1} H_k + P_k^{-1}\}^{-1} \Phi_k^{\mathrm{T}} + Q_k \\
&= \Phi_k \{P_k(-) - P_k(-) H_k^{\mathrm{T}} [H_k P(-) H_k^{\mathrm{T}} + R_k]^{-1} \\
&\quad \times H_k P_k(-)\} \Phi_k^{\mathrm{T}} + Q_k && \text{(Hemes)} \\
&= P_{k+1}(-), && \text{(Riccati)},
\end{aligned}$$

where the "Hemes inversion formula" is given in Appendix B. This completes the proof.

This lemma is used below to derive a closed-form solution for the steady-state Riccati equation in the scalar time-invariant case and in Chapter 7 to derive a fast iterative solution method for the matrix time-invariant case.

4.9.3 Closed-Form Solution of the Scalar Time-Invariant Case

Because this case can be solved in closed form, it serves to illustrate the application of the linearization method derived above.

Characteristic Values and Vectors. The linearization will yield the following 2×2 transition matrix for the numerator and denominator matrices representing the covariance matrix as a matrix fraction:

$$\Psi = \begin{bmatrix} Q_k & I \\ I & 0 \end{bmatrix} \begin{bmatrix} \Phi_k^{-T} & 0 \\ 0 & \Phi_k \end{bmatrix} \begin{bmatrix} H_k^T R_k^{-1} H_k & I \\ I & 0 \end{bmatrix}$$
$$= \begin{bmatrix} \Phi + \dfrac{H^2 Q}{\Phi R} & \dfrac{Q}{\Phi} \\ \dfrac{H^2}{\Phi R} & \dfrac{1}{\Phi} \end{bmatrix}.$$

This matrix has characteristic values

$$\lambda_1 = \frac{H^2 Q + R(\Phi^2 + 1) + \sigma}{2\Phi R}, \qquad \lambda_2 = \frac{H^2 Q + R(\Phi^2 + 1) - \sigma}{2\Phi R},$$
$$\sigma = \sigma_1 \sigma_2,$$
$$\sigma_1 = \sqrt{H^2 Q + R(\Phi + 1)^2}, \qquad \sigma_2 = \sqrt{H^2 Q + R(\Phi - 1)^2},$$

with ratio

$$\rho = \frac{\lambda_2}{\lambda_1}$$
$$= \frac{\psi - [H^2 Q + R(\Phi^2 + 1)]\sigma}{2\Phi^2 R^2}$$
$$\leq 1,$$
$$\psi = [H^2 Q + R(\Phi^2 + 1)]^2 - 2R^2\Phi^2$$
$$= H^4 Q^2 + 2H^2 QR + 2H^2 \Phi^2 QR + R^2 + \Phi^4 R^2.$$

The corresponding characteristic vectors are the column vectors of the matrix

$$M = \begin{bmatrix} \dfrac{-2QR}{H^2 QR(\Phi^2 - 1) + \sigma} & \dfrac{-2QR}{H^2 QR(\Phi^2 - 1) - \sigma} \\ 1 & 1 \end{bmatrix},$$

the inverse of which is

$$M^{-1} = \begin{bmatrix} -\dfrac{H^2}{\sigma_2\sigma_1} & \dfrac{H^2Q - R + \Phi^2 R + \sigma_2\sigma_1}{2\sigma_2\sigma_1} \\ \dfrac{H^2}{\sigma_2\sigma_1} & \dfrac{-(H^2Q) + R - \Phi^2 R + \sigma_2\sigma_1}{2\sigma_2\sigma_1} \end{bmatrix}$$

$$= \frac{1}{4QR\sigma_1\sigma_2}\begin{bmatrix} \tau_1\tau_2 & 2QR\tau_1 \\ -\tau_1\tau_2 & -2QR\tau_2 \end{bmatrix},$$

$$\tau_1 = H^2Q + R(\Phi^2 - 1) + \sigma, \quad \tau_2 = H^2Q + R(\Phi^2 - 1) - \sigma.$$

Closed-Form Solution. This will have the form

$$P_k = A_k B_K^{-\mathrm{T}}$$

for

$$\begin{bmatrix} A_k \\ B_k \end{bmatrix} = \Psi^k \begin{bmatrix} P_0 \\ 1 \end{bmatrix}$$

$$= M \begin{bmatrix} \lambda_1^k & 0 \\ 0 & \lambda_2^k \end{bmatrix} M^{-1} \begin{bmatrix} P_0 \\ 1 \end{bmatrix}.$$

This can be expressed in the form

$$P_k = \frac{(P_0\tau_2 + 2QR) - (P_0\tau_1 + 2QR)\rho^k}{(2H^2P_0 - \tau_1) - (2H^2P_0 - \tau_2)\rho^k},$$

which is similar in structure to the closed-form solution for the scalar time-invariant Riccati differential equation. In both cases, the solution is a ratio of linear functions of an exponential time function. In the discrete-time case, the discrete-time power ρ^k serves essentially the same function as the exponential function $e^{-2\phi t}$ in the closed-form solution of the differential equation. Unlike the continuous-time solution, however, this discrete-time solution can "skip over" zeros of the denominator.

4.9.4 MacFarlane–Potter–Fath Eigenstructure Method

Steady-State Solution of Time-Invariant Discrete-Time Matrix Riccati Equation. The method presented in Section 4.8.8 for the steady-state solution of the time-invariant matrix Riccati differential equation (i.e., in continuous time) also applies to the Riccati equation in discrete time. As before, it is formalized as a lemma:

LEMMA 3 If A and B are $n \times n$ matrices such that B is nonsingular and

$$\Psi_d \begin{bmatrix} A \\ B \end{bmatrix} = \begin{bmatrix} A \\ B \end{bmatrix} D \tag{4.96}$$

for an $n \times n$ nonsingular matrix D, then $P_\infty = AB^{-1}$ satisfies the steady-state discrete-time matrix Riccati equation

$$P_\infty = \Phi\{P_\infty - P_\infty H^T[HP_\infty H^T + R]^{-1}HP_\infty\}\Phi^T + Q.$$

Proof: If $P_k = AB^{-1}$, then it was shown in Lemma 2 that $P_{k+1} = \acute{A}\acute{B}^{-1}$, where

$$\begin{aligned}
\begin{bmatrix} \acute{A} \\ \acute{B} \end{bmatrix} &= \begin{bmatrix} (\Phi_k + Q_k\Phi_k^{-T}H_k^T R_k^{-1}H_k) & Q_k\Phi_k^{-T} \\ \Phi_k^{-T}H_k^T R_k^{-1}H_k & \Phi_k^{-T} \end{bmatrix} \begin{bmatrix} A \\ B \end{bmatrix} \\
&= \Psi_d \begin{bmatrix} A \\ B \end{bmatrix} \\
&= \begin{bmatrix} A \\ B \end{bmatrix} D \\
&= \begin{bmatrix} AB \\ BD \end{bmatrix}.
\end{aligned}$$

Consequently,

$$\begin{aligned}
P_{k+1} &= \acute{A}\acute{B}^{-1} \\
&= (AD)(BD)^{-1} \\
&= ADD^{-1}B^{-1} \\
&= AB^{-1} \\
&= P_k.
\end{aligned}$$

That is, AB^{-1} is a steady-state solution, which was to be proved.

In practice, A and B are formed from n characteristic vectors of Ψ_d. The matrix D will be a diagonal matrix of the corresponding nonzero characteristic values.

4.10 RELATIONSHIPS BETWEEN CONTINUOUS AND DISCRETE RICCATI EQUATIONS

4.10.1 Relationship between $Q(t)$ and Q_k

Some of the mathematical relationships between the covariance matrices of the continuous-time and discrete-time process noise models were examined in Chapter 3. They will now be reexamined from a slightly different perspective.

The process noise covariance matrices appearing in the continuous and discrete Riccati equations have the same symbol (Q) but different physical units. They are not identical but they are related. The relationship can be derived from the propagation equation for the estimation error, $\tilde{x} = \hat{x} - x$. Between discrete observations, it is propagated according to the equations

$$\frac{d}{dt}\tilde{x}(t) = F(t)\tilde{x}(t) + w(t),$$

$$\tilde{x}_{k+1} = \Phi_k \tilde{x}_k + \int_{t_k}^{t_{k+1}} \Phi(t_{k+1}, s) w(s)\, ds,$$

$$\begin{aligned}
\Phi_k P_k \Phi_k^{\mathrm{T}} + Q_k &= P_{k+1} \\
&= E\langle \tilde{x}_{k+1} \tilde{x}_{k+1}^{\mathrm{T}} \rangle \\
&= \Phi_k E\langle \tilde{x}_k \tilde{x}_k^{\mathrm{T}} \rangle \Phi_k^{\mathrm{T}} + \int_{t_k}^{t_{k+1}} \int_{t_k}^{t_{k+1}} \Phi(t_{k+1}, s) E\langle w(s) w^{\mathrm{T}}(t) \rangle \Phi^{\mathrm{T}}(t_{k+1}, t)\, ds\, dt \\
&= \Phi_k P_k \Phi_k^{\mathrm{T}} + \int_{t_k}^{t_{k+1}} \Phi(t_{k+1}, \tau) Q(\tau) \Phi^{\mathrm{T}}(t_{k+1}, \tau)\, d\tau,
\end{aligned}$$

from which

$$Q_k = \int_{t_k}^{t_{k+1}} \Phi(t_{k+1}, \tau) Q(\tau) \Phi^{\mathrm{T}}(t_{k+1}, \tau)\, d\tau.$$

Here, the symbol Q_k on the left of the equal sign is the one for the discrete-time Riccati equation and the function $Q(\tau)$ on the right is for the Riccati differential equation.

This relationship has special forms in special cases:

- In problems with constant states

  $$F = 0 \quad \text{and} \quad \Phi = I,$$

  the solution is

  $$Q_k = (t_{k+1} - t_k)\bar{Q},$$

  where $\bar{Q}$ is the time-averaged value of $Q(\tau)$ on the interval $t_k \leq \tau < t_{k+1}$.

- In the general time-invariant case,

$$Q_{\Delta t} = e^{F\Delta t}\left\{\int_0^{\Delta t} e^{-F\tau} Q_\tau e^{-F^{\mathrm{T}}\tau}\,d\tau\right\} e^{F^{\mathrm{T}}\Delta t},$$

where Q_τ is the constant covariance of process noise for continuous time, and $Q_{\Delta t}$ is its constant counterpart for discrete-time intervals equal to Δt.

4.10.2 Relationship between *R*(*t*) and R_k

This depends upon the way that the discrete-time sensor processes the noise. If it is an integrating sensor, then

$$v_k = \int_{t_{k-1}}^{t_k} v(t)\,dt, \tag{4.97}$$

$$R_k = \bar{R} \tag{4.98}$$

$$= \frac{1}{t_k - t_{k-1}} \int_{t_{k-1}}^{t_k} R(t)\,dt, \tag{4.99}$$

where $\bar{R}$ is the time-averaged value of $R(t)$ over the interval $t_{k-1} < t \le t_k$.

4.11 MODEL EQUATIONS FOR TRANSFORMED STATE VARIABLES

The question to be addressed here is what happens to the Kalman filter model equations when the state variables and measurement variables are redefined by linear transformations? The answer to this question can be derived as a set of formulas, giving the new model equations in terms of the parameters of "old" model equations and the linear transformations relating the two sets of variables. In Chapter 7, these formulas will be used to simplify the model equations.

4.11.1 Linear Transformations of State Variables

These are changes of variables by which the "new" state and measurement variables are linear combinations of the respective old state and measurement variables. Such transformations can be expressed in the form

$$\acute{x}_k = A_k x_k, \tag{4.100}$$

$$\acute{z}_k = B_k H_k, \tag{4.101}$$

where x and z are the old variables and $\acute{x}$ and $\acute{z}$ are the new state vector and measurement, respectively.

Matrix Constraints. One must further assume that for each discrete-time index k, A_k is a *nonsingular* $n \times n$ matrix. The requirements on B_k are less stringent. One need only assume that it is conformable for the product $B_k H_k$, that is, that B_k is a matrix with ℓ columns. The dimension of $\acute{z}_k$ is arbitrary, and can depend on k.

4.11.2 New Model Equations

With the above assumptions, the corresponding state, measurement, and state uncertainty covariance equations of the Kalman filter model are transformed to the following forms:

$$\acute{x}_{k+1} = \acute{\Phi}_k \acute{x}_k + \acute{w}_k, \tag{4.102}$$

$$\acute{z} = \acute{H}_k \acute{x}_k + \acute{v}_k, \tag{4.103}$$

$$\acute{P}_k(+) = \acute{P}_k(-) - \acute{P}_k(-)\acute{H}_k[\acute{H}_k \acute{P}_k(-)\acute{H}_k^{\mathrm{T}} + \acute{R}_k]\acute{H}_k \acute{P}_k(-), \tag{4.104}$$

$$\acute{P}_{k+1}(-) = \acute{\Phi}_k \acute{P}_k(+)\acute{\Phi}_k^{\mathrm{T}} + \acute{Q}_k, \tag{4.105}$$

where the new model parameters are

$$\acute{\Phi}_k = A_k \Phi_k A_k^{-1}, \tag{4.106}$$

$$\acute{H}_k = B_k H_k A_k^{-1}, \tag{4.107}$$

$$\acute{Q}_k = E\langle \acute{w}_k \acute{w}_k^{\mathrm{T}} \rangle \tag{4.108}$$

$$= A_k Q_k A_k^{\mathrm{T}}, \tag{4.109}$$

$$\acute{R} = E\langle \acute{v}_k \acute{v}_k^{\mathrm{T}} \rangle \tag{4.110}$$

$$= B_k R_k B_k^{\mathrm{T}}, \tag{4.111}$$

and the new state estimation uncertainty covariance matrices are

$$\acute{P}_k(\pm) = A_k P_k(\pm) A_k^{\mathrm{T}}. \tag{4.112}$$

4.12 APPLICATION OF KALMAN FILTERS

The Kalman filter has been applied to inertial navigation [15, 45, 167], sensor calibration [168], radar tracking [18], manufacturing [47], economics [30], signal processing [47], and freeway traffic modeling [166]—to cite a few examples. This section shows some applications of the programs provided on the companion floppy diskette. A simple example of a second-order underdamped oscillator is given here to illustrate the application of the equations in Table 4.3. This harmonic oscillator is an approximation of a longitudinal dynamics of an aircraft short period [8].

EXAMPLE 4.3 Consider a linear, underdamped, second-order system with displacement $x_1(t)$, rate $x_2(t)$, damping ratio ζ and (undamped) natural frequency of 5 rad/sec, and constant driving term of 12.0 with additive white noise $w(t)$ normally distributed. The second-order continuous-time dynamic equation

$$\ddot{x}_1(t) + 2\zeta w \dot{x}_1(t) + \omega^2 x_1(t) = 12 + w(t)$$

can be written in state space form via state-space techniques of Chapter 2:

$$\begin{bmatrix} \dot{x}_1(t) \\ \dot{x}_2(t) \end{bmatrix} = \begin{bmatrix} 0 & 1 \\ -\omega^2 & -2\zeta\omega \end{bmatrix} \begin{bmatrix} x_1(t) \\ x_2(t) \end{bmatrix} + \begin{bmatrix} 0 \\ 1 \end{bmatrix} w(t) + \begin{bmatrix} 0 \\ 12 \end{bmatrix}.$$

The observation equation is

$$z(t) = x_1(t) + v(t).$$

Generate 100 data points with plant noise and measurement noise equal to zero. Then estimate $\hat{x}_1(t)$ and $\hat{x}_2(t)$ with the following initial condition and parameter values:

$$\begin{bmatrix} x_1(0) \\ x_2(0) \end{bmatrix} = \begin{bmatrix} 0 & \text{ft} \\ 0 & \text{ft/s} \end{bmatrix},$$

$$P(0) = \begin{bmatrix} 2 & 0 \\ 0 & 2 \end{bmatrix},$$

$$Q = 4.47(\text{ft/s})^2, \quad R = 0.01\ (\text{ft})^2,$$

$$\zeta = 0.2, \quad \omega = 5\ \text{rad/s}.$$

Equations 4.21, 4.24, 4.25, and 4.26 were programmed in MATLAB on a PC (see Appendix A). Figure 4.4 shows the resulting estimates of position and velocity using the noise-free data generated from simulating the above second-order equation. Figure 4.5 shows the corresponding RMS uncertainties in position and velocity (top plot), correlation coefficient between position and velocity (middle plot), and Kalman gains (bottom plot). These results were generated from the accompanying MATLAB program exam 43.m described in Appendix A with sample time = 1 sec.

EXAMPLE 4.4 This example is that of a pulsed *radar tracking system*. In this system, radar pulses are sent out and return signals are processed by the Kalman filter in order to determine the position of maneuvering airborne objects [137]. This example's equations are drawn from IEEE papers [219, 200].

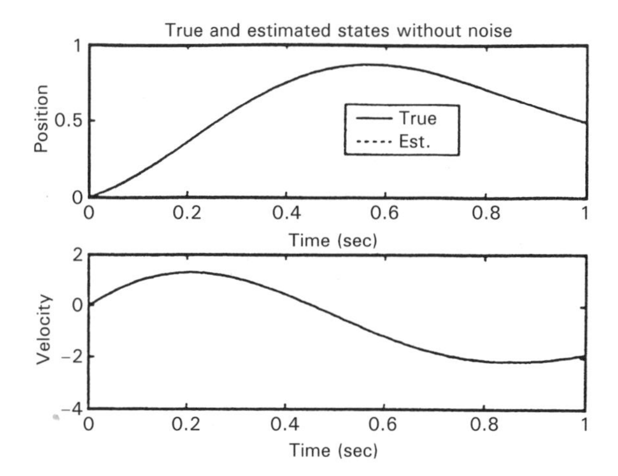

Fig. 4.4 *Estimated position (ft) and velocity (ft/s) versus time (s).*

Difference equations of dynamics equations in state-space formulation are

$$x_k = \begin{bmatrix} 1 & T & 0 & 0 & 0 & 0 \\ 0 & 1 & 1 & 0 & 0 & 0 \\ 0 & 0 & \rho & 0 & 0 & 0 \\ 0 & 0 & 0 & 1 & T & 0 \\ 0 & 0 & 0 & 0 & 1 & 1 \\ 0 & 0 & 0 & 0 & 0 & \rho \end{bmatrix} x_{k-1} + \begin{bmatrix} 0 \\ 0 \\ w_{k-1}^1 \\ 0 \\ 0 \\ w_{k-1}^2 \end{bmatrix}.$$

The discrete-time observation equation is given by

$$z_k = \begin{bmatrix} 1 & 0 & 0 & 0 & 0 & 0 \\ 0 & 0 & 0 & 1 & 0 & 0 \end{bmatrix} x_k + \begin{bmatrix} v_k^1 \\ v_k^2 \end{bmatrix},$$

where

$$\begin{aligned} x_k^{\mathrm{T}} &= [r_k \quad \dot{r}_k \quad U_k^1 \quad \theta_k \quad \dot{\theta}_k \quad U_k^2] \\ r_k &= \text{range of the vehicle at time } k \\ \dot{r}_k &= \text{range rate of the vehicle at time } k \\ U_k^1 &= \text{maneuvering correlated state noise} \\ \theta_k &= \text{bearing of the vehicle at time } k \end{aligned}$$

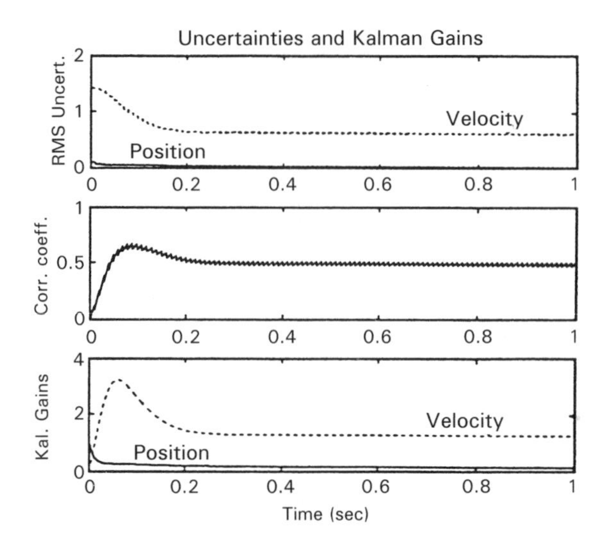

Fig. 4.5 *RMS uncertainties, position and velocity, correlation coefficient, and Kalman gain.*

$\dot{\theta}_k$ = bearing rate of the vehicle at time k

U_k^2 = maneuvering-correlated state noise

T = sampling period in seconds.

$w_k^{\mathrm{T}} = [w_k^1 \quad w_k^2]$ zero-mean white-noise sequences and covariance of σ_1^2 and σ_2^2, respectively

$v_k^{\mathrm{T}} = [v_k^1 \quad v_k^2]$ sensor zero mean white noise sequence and covariance of σ_r^2 and σ_θ^2, respectively,

and w_k and v_k are uncorrelated :

$$\rho = \text{correlation coefficient} = \frac{E[U_k U_{k-1}]}{\sigma_m^2} = \begin{cases} 1 - \lambda T & T \leq \dfrac{1}{\lambda} \\ 0 & T > \dfrac{1}{\lambda} \end{cases},$$

where σ_m^2 is the maneuver variance and λ the inverse of average maneuver duration.

The shaping filter for whitening the maneuver noise is given by

$$U_k^1 = \rho U_{k-1}^1 + w_{k-1}^1,$$

which drives the range rate ($\dot{r}_k$) state of the vehicle, and

$$U_k^2 = \rho U_{k-1}^2 + w_{k-1}^2,$$

which drives the bearing rate (θ_k) state of the vehicle. The derivation of the discrete-time shaping filter is given in Section 3.6 with examples. The range, range rate, bearing, and bearing rate equations have been augmented by the shaping filter equations. The dimension of the state vector is increased from 4×1 to 6×1.

Covariance and gain plots for this system are produced using the Kalman filter program of Appendix A. The following initial covariance (P_0), plant noise (Q), and measurement noise (R) are used to generate the covariance results:

$$P_0 = \begin{bmatrix} \sigma_r^2 & \dfrac{\sigma_r^2}{T} & 0 & 0 & 0 & 0 \\ \dfrac{\sigma_r^2}{T} & \dfrac{2\sigma_r^2}{T^2} + \sigma_1^2 & 0 & 0 & 0 & 0 \\ 0 & 0 & \sigma_1^2 & 0 & 0 & 0 \\ 0 & 0 & 0 & \sigma_\theta^2 & \dfrac{\sigma_\theta^2}{T} & 0 \\ 0 & 0 & 0 & \dfrac{\sigma_\theta^2}{T} & \dfrac{2\sigma_\theta^2}{T^2} + \sigma_2^2 & 0 \\ 0 & 0 & 0 & 0 & 0 & \sigma_2^2 \end{bmatrix},$$

$$Q = \begin{bmatrix} 0 & 0 & 0 & 0 & 0 & 0 \\ 0 & 0 & 0 & 0 & 0 & 0 \\ 0 & 0 & \sigma_1^2 & 0 & 0 & 0 \\ 0 & 0 & 0 & 0 & 0 & 0 \\ 0 & 0 & 0 & 0 & 0 & 0 \\ 0 & 0 & 0 & 0 & 0 & \sigma_2^2 \end{bmatrix},$$

$$R = \begin{bmatrix} \sigma_r^2 & 0 \\ 0 & \sigma_\theta^2 \end{bmatrix},$$

Here $\rho = 0.5$ and $T = 5$, 10, 15 s, respectively. Also,

$$\sigma_r^2 = (1000\,\text{m})^2, \qquad \sigma_\theta^2 = (0.017\,\text{rad})^2,$$
$$\sigma_1^2 = (103/3)^2, \qquad \sigma_2^2 = 1.3 \times 10^{-8}.$$

Some parts of this example are discussed in [100]. Results of covariances and Kalman gain plots are shown in Figures 4.6–4.8. Convergence of the diagonal elements of the covariance matrix is shown in these figures for intervals (5, 10, 15 s). Selected Kalman gain values are shown in the following figures for various values of sampling times. These results were generated using the accompanying MATLAB program exam 44.m described in Appendix A.

4.13 SMOOTHERS

A *smoother* estimates the state of a system at time t using measurements made before *and after* time t. The accuracy of a smoother is generally superior to that of a filter, because it uses more measurements for its estimate. Smoothers are usually divided into three types:

1. *Fixed-interval* smoothers use all the measurements over a fixed interval to estimate the system state at all times in the same interval. This type of smoother is most often used for off-line processing to get the very best estimate of the system state over the entire time interval.
2. *Fixed-point* smoothers estimate the system state at a fixed time in the past, given the measurements up to the current time. This type of smoother is used when the state estimate is needed at only one time in the interval, such as for estimating the miss distance between two objects that are being tracked by radar.

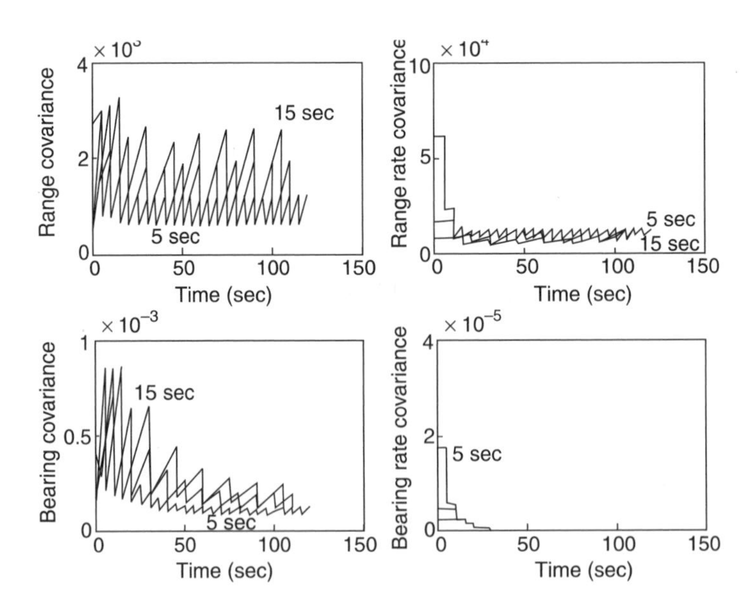

Fig. 4.6 Covariances.

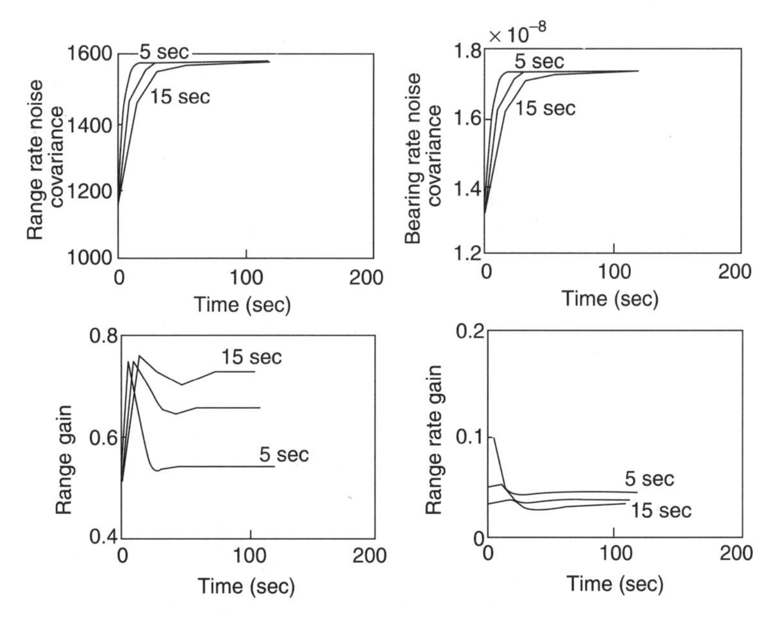

Fig. 4.7 *Covariances and Kalman gains.*

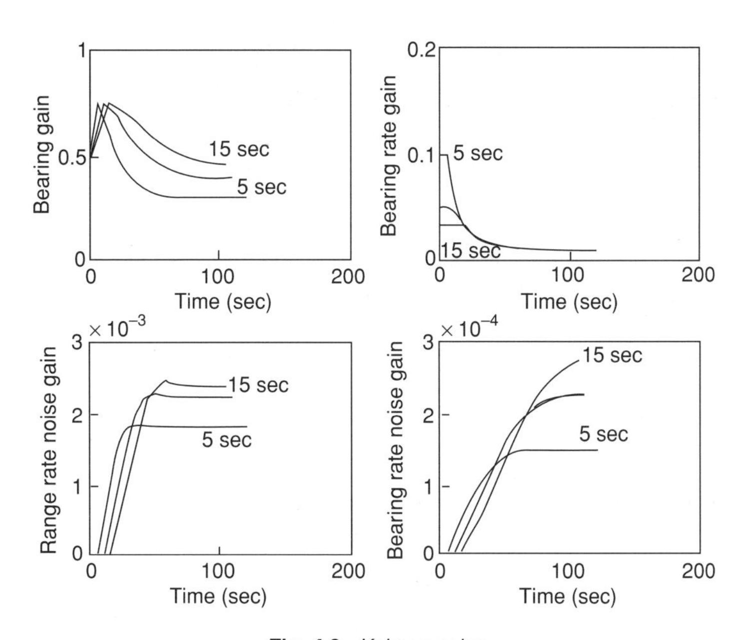

Fig. 4.8 *Kalman gains.*

3. *Fixed-lag* smoothers estimate the system state at a fixed time interval lagging the time of the current measurement. This type of smoother trades off estimate latency for more accuracy.

These can all be derived from the Kalman filter model. The general derivation methodology uses the Kalman filter for measurements up to each time t that the state is to be estimated, combined with another algorithm derived from the Kalman filter for the measurements beyond that time. This second algorithm can be derived by running the Kalman filter *backward* from the last measurement to the measurement just past t, then optimally combining the two independent estimates (forward and backward) of the state at time t based on the two independent sets of measurements. The resulting formulas generally need to be modified for more efficient and robust implementation.

Smoothers derived in this way appeared in the technical literature soon after the introduction of the Kalman filter. We present here a smoother implementation of each type. These are not necessarily in the best forms for implementation. Derivations of these and more numerically stable implementations (including "square-root" and "information" forms) can be found in many textbooks on the general subject of estimation (e.g., Anderson and Moore [1], Bierman [7], Gelb et al. [21]).

Both the fixed-lag smoother and the fixed-point smoother can be implemented in real time, as the measurements are made. The fixed-interval smoother can be implemented by a forward (filtering) pass through all the measurements, followed by a backward (smoothing) pass.

4.13.1 Rauch–Tung–Striebel Two-Pass Smoother

This fixed-interval smoother implementation was derived by H. Rauch, K. Tung, and C. Striebel and published in 1965 [21]. The first (forward) pass uses a Kalman filter but saves the intermediate results $\hat{x}_k(-)$, $\hat{x}_k(+)$, $P_k(-)$, and $P_k(+)$ at each measurement time t_k. The second pass runs backward in time in a sequence from the time t_N of the last measurement, computing the smoothed state estimate from the intermediate results stored on the forward pass. The smoothed estimate (designated by the subscript $[s]$) is initialized with the value

$$\frac{\hat{x}}{\hat{x}[s]N} = \hat{x}_n(+), \tag{4.113}$$

then computed recursively by the formulas

$$\hat{x}_{[s]k} = \hat{x}_k(+) + A_k(\hat{x}_{[s]k+1} - \hat{x}_{k+1}(-)), \tag{4.114}$$

$$A_k = P_k(+)\Phi_k^{\mathrm{T}} P_{k+1}^{-1}(-). \tag{4.115}$$

The covariance of uncertainty of the smoothed estimate can also be computed on the second pass:

$$P_{[s]k} = P_k(+) + A_k(P_{[s]k+1} - P_{k+1}(-))A_k^{\mathrm{T}}, \tag{4.116}$$

although this is not a necessary part of the smoother implementation. It should be computed only if it is of sufficient interest.

The MATLAB m-file RTSvsKF.m, described in Appendix A, demonstrates the relative performance of this smoother and the Kalman filter.

4.13.2 A Fixed-Point Smoother

This type of smoother includes a Kalman filter to estimate the state at the current time t_k using the measurements up to time t_k, then adds the following equations to obtain a smoothed estimate of the state at a fixed time $t_i < t_k$:

$$\hat{x}_{[s]i|k} = \hat{x}_{[s]i|k-1} + B_k\overline{K}_k(z_k - H\hat{x}_k(-)), \tag{4.117}$$

$$B_k = B_{k-1}P_{k-1}(+)\Phi_{k-1}^{\mathrm{T}}P_k^{-1}(-), \tag{4.118}$$

where the subscript notation $[s]i|k$ refers to the smoothed estimate of the state at time t_i, given the measurements up to time t_k. (A derivation and application of this technique to the analysis of inertial navigation system test data may be found in [169].) The values of $\hat{x}(-)$, $\overline{K}_k$, z_k, H_kP, and P are computed in the Kalman filter and the initial value $B_i = I$, the identity matrix. The covariance of uncertainty of the smoothed estimate can also be computed by the formula

$$P_{[s]i|k} = P_{[s]i|k-1} + B_k(P_k(+) - P_k(-))B_k^{\mathrm{T}}, \tag{4.119}$$

although this is not a necessary part of the smoother implementation.

4.13.3 A Fixed-Lag Smoother

The fixed-lag smoother estimates the system state at time $t_{k-\ell}$, given the measurements up to time t_k (usually, the current time). The fixed positive integer ℓ is the fixed lag, equal to the number of discrete time steps between the time at which the state is to be estimated and the time of the last measurement used in estimating it. The memory requirements for fixed-lag smoothers increase with ℓ, because the intermediate Kalman filter values for $\hat{x}_i(+)$, $P_i(+)$, Φ_i, and Q_i must be saved for $k - \ell \leq i \leq k$. [For time-invariant systems, only $\hat{x}_i(+)$ and $P_i(+)$ need to be saved, and the steady state-value of $P_i(+)$ may suffice.] In addition to a Kalman filter implementation for the state estimate at time t_k, the following equations must be implemented to obtain the smoothed estimate of the state at time $t_{k-\ell}$:

$$\begin{aligned}\hat{x}_{[s]k+1-\ell} &= \Phi_{k-\ell}\hat{x}_{[s]k-\ell} + Q_{k-\ell}\Phi_{k-\ell}^{\mathrm{T}}P_{k-\ell}(+)(\hat{x}_{[s]k-\ell} - \hat{x}_{k-\ell}(+)) \\ &\quad + B_{k+1}\overline{K}_{k+1}(z_{k+1} - H_{k+1}\Phi_k\hat{x}_k(+)), \qquad (4.120)\\ B_{k+1} &= B_k P_k(+)\Phi_k^{\mathrm{T}}P_{k+1}^{-1}(-). \qquad (4.121)\end{aligned}$$

The first ℓ steps of a fixed-lag smoother must be implemented as a fixed-point smoother, with the fixed point at the initial time. This procedure also initializes B_k. For time-invariant systems, the steady-state values of the gainlike expressions $Q_{k-\ell}\Phi_{k-\ell}^{\mathrm{T}}P_{k-\ell}(+)$ and $B_{k+1}\overline{K}_{k+1}$ can be used with the stored values of $\hat{x}_k(+)$.

4.14 SUMMARY

4.14.1 Points to Remember

The optimal linear estimator is equivalent to the general (nonlinear) optimal estimator if the random processes x and z are jointly normal. Therefore, the equations for the discrete-time and continuous-time linear optimal estimators can be derived by using the orthogonality principle of Chapter 3. The discrete-time estimator (Kalman filter) has been described, including its implementation equations and block diagram description. The continuous-time estimator (Kalman–Bucy filter) is also described.

Prediction is equivalent to filtering when measurements (system outputs) are not available. Implementation equations for continuous-time and discrete-time predictors have been given, and the problem of missing data has been discussed in detail. The estimator equations for the case that there is correlation between plant and measurement noise sources and correlated measurement errors were discussed. Relationships between stationary continuous-time and Kalman filter and Wiener filters were covered.

Methods for solving matrix Riccati differential equations have been included. Examples discussed include the applications of the Kalman filter to (1) estimating the state (phase and amplitude) of a harmonic oscillator and (2) a discrete-time Kalman filter implementation of a five-dimensional radar tracking problem.

An estimator for the state of a dynamic system at time t, using measurements made *after* time t, is called a *smoother.*

4.14.2 Important Equations to Remember

Kalman Filter. The *discrete-time model* for a linear stochastic system has the form

$$\begin{aligned} x_k &= \Phi_{k-1}x_{k-1} + G_{k-1}w_{k-1}, \\ z_k &= H_k x_k + v_k, \end{aligned}$$

where the zero-mean uncorrelated Gaussian random processes $\{w_k\}$ and $\{v_k\}$ have covariances Q_k and R_k, respectively, at time t_k. The corresponding *Kalman filter equations* have the form

$$\begin{aligned}
\hat{x}_k(-) &= \Phi_{k-1}\hat{x}_{k-1}(+),\\
P_k(-) &= \Phi_{k-1}P_{k-1}(+)\Phi_{k-1}^{\mathrm{T}} + G_{k-1}Q_{k-1}G_{k-1}^{\mathrm{T}},\\
\hat{x}_k(+) &= \hat{x}_k(-) + \overline{K}_k(z_k - H_k\hat{x}_k(-)),\\
\overline{K}_k &= P_k(-)H_k^{\mathrm{T}}(H_kP_k(-)H_k^{\mathrm{T}} + R)^{-1},\\
P_k(+) &= P_k(-) - \bar{K}_kH_kP_k(-),
\end{aligned}$$

where the $(-)$ indicates the a priori values of the variables (before the information in the measurement is used) and the $(+)$ indicates the a posteriori values of the variables (after the information in the measurement is used). The variable $\overline{K}$ is the Kalman gain.

Kalman–Bucy Filter. The *continuous-time model* for a linear stochastic system has the form

$$\begin{aligned}
\frac{d}{dt}x(t) &= F(t)x(t) + G(t)w(t),\\
z(t) &= H(t)x(t) + v(t),
\end{aligned}$$

where the zero-mean uncorrelated Gaussian random processes $\{w(t)\}$ and $\{v(t)\}$ have covariances $Q(t)$ and $R(t)$, respectively, at time t. The corresponding *Kalman–Bucy filter* equations for the estimate $\hat{x}$ of the state variable x, given the output signal z, has the form

$$\begin{aligned}
\frac{d}{dt}\hat{x}(t) &= F(t)\hat{x}(t) + \overline{K}(t)[z(t) - H(t)\hat{x}(t)],\\
\overline{K}(t) &= P(t)H^{\mathrm{T}}(t)R^{-1}(t),\\
\frac{d}{dt}P(t) &= F(t)P(t) + P(t)F^{\mathrm{T}}(t) - \overline{K}(t)R(t)\overline{K}^{\mathrm{T}}(t) + G(t)Q(t)G^{\mathrm{T}}(t).
\end{aligned}$$

PROBLEMS

4.1 A scalar discrete-time random sequence x_k is given by

$$\begin{aligned}
x_{k+1} &= 0.5x_k + w_k,\\
Ex_0 = 0, \quad Ex_0^2 &= 1, \quad Ew_k^2 = 1, \quad Ew_k = 0,
\end{aligned}$$

where w_k is white noise. The observation equation is given by

$$z_k = x_k + v_k$$

$Ev_k = 0$, $Ev_k^2 = 1$, and v_k is also white noise. The terms x_0, w_k, and v_k are all Gaussian. Derive a (nonrecursive) expression for

$$E[x_1 | z_0, z_1, z_2]$$

4.2 For the system given in Problem 4.1:

(a) Write the discrete-time Kalman filter equations.

(b) Provide the correction necessary if z_2 was not received.

(c) Derive the loss in terms of the estimate $\hat{x}_3$ due to missing z_2.

(d) Derive the filter for $k \to \infty$ (steady state).

(e) Repeat (d) when every other observation is missed.

4.3 In a single-dimension example of a radar tracking an object by means of track-while-scan, measurements of the continuous-time target trajectory at some discrete times are made. The process and measurement models are given by

$$\dot{x}(t) = -0.5x(t) + w(t), \quad z_{kT} = x_{kT} + v_{kT},$$

where T is the intersampling interval (assume 1 s for simplicity):

$$\begin{aligned} Ev_k &= Ew(t) = 0, \\ Ew(t_1)w(t_2) &= 1\delta(t_2 - t_1), \\ Ev_{k_1\mathrm{T}}v_{k_2\mathrm{T}} &= 1\Delta(k_2 - k_1), \\ E\langle v_k w^{\mathrm{T}} \rangle &= 0. \end{aligned}$$

Derive the minimum mean-squared-filter of $x(t)$ for all t.

4.4 In Problem 4.3, the measurements are received at discrete times and each measurement is spread over some nonzero time interval (radar beam width nonzero). The measurement equation of Problem 4.3 can be modified to

$$z_{kT+\eta} = x_{kT+\eta} + v_{kT+\eta},$$

where

$$k = 0, 1, 2, \ldots, 0 \le \eta \le \eta_0.$$

Let $T = 1$ s, $\eta_0 = 0.1$ (radar beam width) and $v(t)$ be a zero-mean white Gaussian process with covariance equal to 1. Derive the minimum mean-squared filter of $x(t)$ for all t.

4.5 Prove the condition in the discussion following Equation 4.9 that $Ew_k z_i^{\mathrm{T}} = 0$ for $i = 1, \ldots k$ when w_k and v_k are uncorrelated and white.

4.6 In Example 4.4, use white noise as a driving input to range rate ($\dot{r}_k$) and bearing rate ($\dot{\theta}_k$) equations instead of colored noise. This reduces the dimension of the state vector from 6×1 to 4×1. Formulate the new observation equation. Generate the covariance and Kalman gain plots for the same values of P_0, Q, R, σ_r^2, σ_θ^2, σ_1^2, and σ_2^2.

4.7 For the same problem as Problem 4.6, obtain values of the plant covariance Q for the four-state model such the associated mean-squared estimation uncertainties for range, range rate, bearing, and bearing rate are within 5–10 % of those for the six-state model. (*Hint:* This should be possible because the plant noise is used to model the effects of linearization errors, discretization errors, and other unmodeled effects or approximations. This type of suboptimal filtering will be discussed further in Chapter 7.)

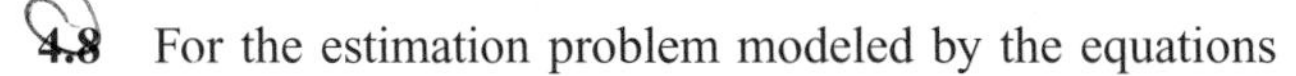

4.8 For the estimation problem modeled by the equations

$$
\begin{aligned}
x_k &= x_{k-1} + w_{k-1}, \\
w_k &\sim \mathcal{N}(0, 30) \text{ and white,} \\
z_k &= x_k + v_k, \\
v_k &\sim \mathcal{N}(0, 20) \text{ and white,} \\
P_0 &= 150,
\end{aligned}
$$

calculate the values of $P_k(+)$, $P_k(-)$, and $\overline{K}_k$ for $k = 1, 2, 3, 4$ and $P_\infty(+)$ (the steady-state value).

4.9 *Parameter estimation problem.* Let x be a zero-mean Gaussian random variable with covariance P_0, and let $z_k = x + v_k$ be an observation of x with noise $v_k \sim \mathcal{N}(0, R)$.

(a) Write the recursion equations for $P_k(+)$, $P_k(-)$, $\overline{K}_k$, and $\hat{x}_k$.

(b) What is the value of x_1 if $R = 0$?

(c) What is the value of x_1 if $R = +\infty$?

(d) Explain the results of (b) and (c) in terms of measurement uncertainty.

4.10 Assume a stochastic system in continuous time modeled by the equations

$$
\begin{aligned}
\dot{x}(t) &= -x(t) + w(t), \\
w(t) &\sim \mathcal{N}(0, 30), \\
z(t) &= x(t) + v(t), \\
v(t) &\sim \mathcal{N}(0, 20).
\end{aligned}
$$

(a) Derive the values of the mean-squared estimation error $P(t)$ and Kalman gain $\overline{K}(t)$ for time $t = 1, 2, 3, 4$.

(b) Solve for the steady state value of P.

4.11 Show that the matrices P_k and $P(t)$ of Equations 4.23 and 4.37 are *symmetric*. That is, $P_k^{\mathrm{T}} = P_k$ and $P^{\mathrm{T}}(t) = P(t)$.

4.12 Derive the observability matrices for Example 4.4 and Problem 4.6 and determine whether these systems are observable.

4.13 A vector discrete-time random sequence x_k is given by

$$x_k = \begin{bmatrix} 1 & 1 \\ 0 & 1 \end{bmatrix} x_{k-1} + \begin{bmatrix} 0 \\ 1 \end{bmatrix} w_{k-1},$$
$$w_k \sim \mathcal{N}(0, 1) \text{ and white.}$$

The observation equation is given by

$$z_k = [1 \quad 0] x_k + v_k,$$
$$v_k \sim \mathcal{N}[0, 2 + (-1)^k] \text{ and white.}$$

Calculate the values of $P_k(+)$, $P_k(-)$ and $\overline{K}_k$ for $k = 1, \ldots, 10$ and $P_\infty(+)$ (the steady-state value) with

$$P_0 = \begin{bmatrix} 10 & 0 \\ 0 & 10 \end{bmatrix}.$$

5

Nonlinear Applications

The principal uses of linear filtering theory are for solving nonlinear problems.
Harold W. Sorenson, in a private conversation

5.1 CHAPTER FOCUS

5.1.1 Nonlinear Estimation Problems

Linear estimators for discrete and continuous systems were derived in Chapter 4. The combination of functional linearity, quadratic performance criteria, and Gaussian statistics is essential to this development. The resulting optimal estimators are simple in form and powerful in effect.

Many dynamic systems and sensors are not absolutely linear, but they are not far from it. Following the considerable success enjoyed by linear estimation methods on linear problems, extensions of these methods were applied to such nonlinear problems. In this chapter, we investigate the model extensions and approximation methods used for applying the methodology of Kalman filtering to these "slightly nonlinear" problems. More formal derivations of these nonlinear filters and predictors can be found in references [1, 21, 23, 30, 36, 75, 112].

5.1.2 Main Points to Be Covered

- Many estimation problems that are of practical interest are nonlinear but "smooth." That is, the functional dependences of the measurement or state dynamics on the system state are nonlinear, but approximately linear for small perturbations in the values of the state variables.
- Methods of linear estimation theory can be applied to such nonlinear problems by linear approximation of the effects of small perturbations in the state of the nonlinear system from a "nominal" value.

- For some problems, the nominal values of the state variables are fairly well known beforehand. These include guidance and control applications for which operational performance depends on staying close to an optimal trajectory. For these applications, the estimation problem can often be effectively linearized about the nominal trajectory and the Kalman gains can be precomputed to relieve the real-time computational burden.
- The nominal trajectory can also be defined "on the fly" as the current best estimate of the actual trajectory. This approach is called *extended Kalman filtering*. It has the advantage that the perturbations include only the state estimation errors, which are generally smaller than the perturbations from any predefined nominal trajectory and therefore better conditioned for linear approximation. The major disadvantage of extended Kalman filtering is the added real-time computational cost of linearization about an unpredictable trajectory, for which the Kalman gains cannot be computed beforehand.
- Extensions of the linear model to include quadratic terms yield optimal filters of greater applicability but increased computational complexity.

5.2 PROBLEM STATEMENT

Suppose that a continuous or discrete stochastic system can be represented by nonlinear plant and measurement models as shown in Table 5.1, with dimensions of the vector and matrix quantities as shown in Table 5.2 and where the symbols $\Delta(k - \ell)$ stand for the *Kronecker delta function* and the symbols $\delta(t - s)$ stand for the *Dirac delta function* (actually, a *generalized function*).

The function f is a continuously differentiable function of the state vector x, and the function h is a continuously differentiable function of the state vector.

Whereas affine (i.e., linear and additive) transformations of Gaussian RVs have Gaussian distributions, the same is not always true in the nonlinear case. Consequently, it is not necessary that w and v be Gaussian. They may be included as arguments of the nonlinear functions f and h, respectively. However, the initial value

TABLE 5.1 Nonlinear Plant and Measurement Models

Model	Continuous Time	Discrete Time
Plant	$\dot{x} = f(x, t) + w(t)$	$x_k = f(x_{k-1}, k-1) + w_{k-1}$
Measurement	$z(t) = h(x(t), t) + v(t)$	$z_k = h(x_k, k) + v_k$
Plant noise	$E\langle w(t)\rangle = 0$ $E\langle w(t)w^T(s)\rangle = \delta(t-s)Q(t)$	$E\langle w_k\rangle = 0$ $E\langle w_k w_i^T\rangle = \Delta(k-i)Q_k$
Measurement noise	$E\langle v(t)\rangle = 0$ $E\langle v(t)v^T(s)\rangle = \delta(t-s)R(t)$	$E\langle v_k\rangle = 0$ $E\langle v_k v_i^T\rangle = \Delta(k-i)R_k$

TABLE 5.2 Dimensions of Vectors and Matrices in Nonlinear Model

Symbol	Dimensions	Symbol	Dimensions
x, f, w	$n \times 1$	z, h, v	$\ell \times 1$
Q	$n \times n$	R	$\ell \times \ell$
Δ, δ	Scalars		

x_0 may be assumed to be a Gaussian random variate with known mean and known $n \times n$ covariance matrix P_0.

The objective is to estimate x_k or $x(t)$ to satisfy a specified performance criterion as given in Chapter 4.

5.3 LINEARIZATION METHODS

Applying linearization techniques to get simple approximate solutions to nonlinear estimation problems requires that f and h be twice-continuously differentiable [112, 133].

5.4 LINEARIZATION ABOUT A NOMINAL TRAJECTORY

5.4.1 Nominal Trajectory

A *trajectory* is a particular solution of a stochastic system, that is, with a particular instantiation of the random variates involved. The trajectory is a vector-valued sequence $\{x_k | k = 0, 1, 2, 3, \ldots\}$ for discrete-time systems and a vector-valued function $x(t)$, $0 \leq t$, for continuous-time systems.

The term "nominal" in this case refers to that trajectory obtained when the random variates assume their expected values. For example, the sequence $\{x_k^{\text{nom}}\}$ obtained as a solution of the equation

$$x_k^{\text{nom}} = f(x_{k-1}^{\text{nom}}, k - 1) \tag{5.1}$$

with zero process noise and with the mean x_0^{nom} as the initial condition would be a nominal trajectory for a discrete-time system.

5.4.2 Perturbations about a Nominal Trajectory

The word "perturbation" has been used by astronomers to describe a minor change in the trajectory of a planet (or any free-falling body) due to secondary forces, such as those produced by other gravitational bodies. Astronomers learned long ago that the actual trajectory can be accurately modeled as the sum of the solution of the two-body problem (which is available in closed form) and a linear dynamic model for the perturbations due to the secondary forces. This technique also works well for many other nonlinear problems, including the problem at hand. In this case, the perturbations are due to the presence of random process noise and errors in the assumed initial conditions.

If the function f in the previous example is continuous, then the state vector x_k at any instant on the trajectory will vary smoothly with small perturbations of the state vector x_{k-1} at the previous instant. These perturbations are the result of "off-nominal" (i.e., off-mean) values of the random variates involved. These random variates include the initial value of the state vector (x_0), the process noise (w_k), and (in the case of the estimated trajectory) the measurement noise (v_k).

If f is continuously differentiable infinitely often, then the influence of the perturbations on the trajectory can be represented by a Taylor series expansion about the nominal trajectory. The likely magnitudes of the perturbations are determined by the variances of the variates involved. If these perturbations are sufficiently small relative to the higher order coefficients of the expansion, then one can obtain a good approximation by ignoring terms beyond some order. (However, one must usually evaluate the magnitudes of the higher order coefficients before making such an assumption.)

Let the symbol δ denote *perturbations* from the nominal,

$$\delta x_k = x_k - x_k^{\text{nom}},$$
$$\delta z_k = z_k - h(x_k^{\text{nom}}, k),$$

so that the Taylor series expansion of $f(x, k-1)$ with respect to x at $x = x_{k-1}^{\text{nom}}$ is

$$x_k = f(x_{k-1}, k-1) \tag{5.2}$$
$$= f(x_{k-1}^{\text{nom}}, k-1) + \left.\frac{\partial f(x, k-1)}{\partial x}\right|_{x=x_{k-1}^{\text{nom}}} \delta x_{k-1}$$
$$+ \text{higher order terms} \tag{5.3}$$
$$= x_k^{\text{nom}} + \left.\frac{\partial f(x, k-1)}{\partial x}\right|_{x=x_{k-1}^{\text{nom}}} \delta x_{k-1}$$
$$+ \text{higher order terms,} \tag{5.4}$$

or

$$\delta x_k = x_k - x_k^{\text{nom}} \tag{5.5}$$
$$= \left.\frac{\partial f(x, k-1)}{\partial x}\right|_{x=x_{k-1}^{\text{nom}}} \delta x_{k-1}$$
$$+ \text{higher order terms.} \tag{5.6}$$

If the higher order terms in δx can be neglected, then

$$\delta x_k \approx \Phi_{k-1}^{[1]} \delta x_{k-1} + w_{k-1}, \tag{5.7}$$

where the first-order approximation coefficients are given by

$$\Phi_{k-1}^{[1]} = \left.\frac{\partial f(x, k-1)}{\partial x}\right|_{x=x_{k-1}^{\text{nom}}} \tag{5.8}$$

$$= \left.\begin{bmatrix} \frac{\partial f_1}{\partial x_1} & \frac{\partial f_1}{\partial x_2} & \frac{\partial f_1}{\partial x_3} & \cdots & \frac{\partial f_1}{\partial x_n} \\ \frac{\partial f_2}{\partial x_1} & \frac{\partial f_2}{\partial x_2} & \frac{\partial f_2}{\partial x_3} & \cdots & \frac{\partial f_2}{\partial x_n} \\ \frac{\partial f_3}{\partial x_1} & \frac{\partial f_3}{\partial x_2} & \frac{\partial f_3}{\partial x_3} & \cdots & \frac{\partial f_3}{\partial x_n} \\ \vdots & \vdots & \vdots & \ddots & \vdots \\ \frac{\partial f_n}{\partial x_1} & \frac{\partial f_n}{\partial x_2} & \frac{\partial f_n}{\partial x_3} & \cdots & \frac{\partial f_n}{\partial x_n} \end{bmatrix}\right|_{x=x_{k-1}^{\text{nom}}}, \tag{5.9}$$

an $n \times n$ constant matrix.

5.4.3 Linearization of h about a Nominal Trajectory

If h is sufficiently differentiable, then the measurement can be represented by a Taylor series:

$$h(x_k, k) = h(x_k^{\text{nom}}, k) + \left.\frac{\partial h(x, k)}{\partial x}\right|_{x=x_k^{\text{nom}}} \delta x_k + \text{higher order terms}, \tag{5.10}$$

or

$$\delta z_k = \left.\frac{\partial h(x, k)}{\partial x}\right|_{x=x_k^{\text{nom}}} \delta x_k + \text{ higher order terms}. \tag{5.11}$$

If the higher-order terms in this expansion can be ignored, then one can represent the perturbation in z_k as

$$\delta z_k = H_k^{[1]} \delta x_k, \tag{5.12}$$

where the first-order variational term is

$$H_k^{[1]} = \left.\frac{\partial h(x,k)}{\partial x}\right|_{x=x_k^{\text{nom}}} \tag{5.13}$$

$$= \left.\begin{bmatrix} \frac{\partial h_1}{\partial x_1} & \frac{\partial h_1}{\partial x_2} & \frac{\partial h_1}{\partial x_3} & \cdots & \frac{\partial h_1}{\partial x_n} \\ \frac{\partial h_2}{\partial x_1} & \frac{\partial h_2}{\partial x_2} & \frac{\partial h_2}{\partial x_3} & \cdots & \frac{\partial h_2}{\partial x_n} \\ \frac{\partial h_3}{\partial x_1} & \frac{\partial h_3}{\partial x_2} & \frac{\partial h_3}{\partial x_3} & \cdots & \frac{\partial h_3}{\partial x_n} \\ \vdots & \vdots & \vdots & & \vdots \\ \frac{\partial h_\ell}{\partial x_1} & \frac{\partial h_\ell}{\partial x_2} & \frac{\partial h_\ell}{\partial x_3} & \cdots & \frac{\partial h_\ell}{\partial x_n} \end{bmatrix}\right|_{x=x_k^{\text{nom}}}, \tag{5.14}$$

which is an $\ell \times n$ constant matrix.

5.4.4 Summary of Perturbation Equations in the Discrete Case

From Equations 5.7 and 5.12, the linearized equations about nominal values are

$$\delta x_k = \Phi_{k-1}^{[1]} \delta x_{k-1} + w_{k-1}, \tag{5.15}$$

$$\delta z_k = H_k^{[1]} \delta x_k + v_k. \tag{5.16}$$

If the problem is such that the actual trajectory x_k is sufficiently close to the nominal trajectory x_k^{nom} so that the higher order terms in the expansion can be ignored, then this method transforms the problem to a linear problem.

5.4.5 Continuous Case

In the continuous case, the corresponding nonlinear differential equations for plant and observation are

$$\dot{x}(t) = f(x(t), t) + G(t)w(t), \tag{5.17}$$

$$z(t) = h(x(t), t) + v(t), \tag{5.18}$$

with the dimensions of the vector quantities the same as in the discrete case.

Similar to the case of the discrete system, the linearized differential equations can be derived as

$$\delta\dot{x}(t) = \left(\frac{\partial f(x(t), t)}{\partial x(t)} \bigg|_{x(t)=x^{\text{nom}}} \right) \delta x(t) + G(t)w(t) \tag{5.19}$$

$$= F^{[1]}\delta x(t) + G(t)w(t), \tag{5.20}$$

$$\delta z(t) = \left(\frac{\partial h(x(t), t)}{\partial x(t)} \bigg|_{x(t)=x^{\text{nom}}} \right) \delta x(t) + v(t) \tag{5.21}$$

$$= H^{[1]}\delta x(t) + v(t). \tag{5.22}$$

Equations 5.20 and 5.22 represent linearized continuous model equations. The variables $\delta x(t)$ and $\delta z(t)$ are the perturbations about the nominal values as in discrete case.

5.5 LINEARIZATION ABOUT THE ESTIMATED TRAJECTORY

The problem with linearization about the nominal trajectory is that the deviation of the actual trajectory from the nominal trajectory tends to increase with time. As the deviation increases, the significance of the higher order terms in the Taylor series expansion of the trajectory also increases.

A simple but effective remedy for the deviation problem is to replace the nominal trajectory with the estimated trajectory, that is, to evaluate the Taylor series expansion about the estimated trajectory. If the problem is sufficiently observable (as evidenced by the covariance of estimation uncertainty), then the deviations between the estimated trajectory (along which the expansion is made) and the actual trajectory will remain sufficiently small that the linearization assumption is valid [112, 113].

The principal drawback to this approach is that it tends to increase the real-time computational burden. Whereas Φ, H, and $\overline{K}$ for linearization about a nominal trajectory may have been precomputed off-line, they must be computed in real time as functions of the estimate for linearization about the estimated trajectory.

5.5.1 Matrix Evaluations for Discrete Systems

The only modification required is to replace x_{k-1}^{nom} by $\hat{x}_{k-1}$ and x_k^{nom} by $\hat{x}_k$ in the evaluations of partial derivatives. Now the matrices of partial derivatives become

$$\Phi^{[1]}(\hat{x}, k) = \frac{\partial f(x, k)}{\partial x} \bigg|_{x=\hat{x}_k(-)} \tag{5.23}$$

and

$$H^{[1]}(\hat{x}, k) = \left.\frac{\partial h(x, k)}{\partial x}\right|_{x=\hat{x}_k(-)}. \tag{5.24}$$

5.5.2 Matrix Evaluations for Continuous Systems

The matrices have the same general form as for linearization about a nominal trajectory, except for the evaluations of the partial derivatives:

$$F^{[1]}(t) = \left.\frac{\partial f(x(t), t)}{\partial x(t)}\right|_{x=\hat{x}(t)} \tag{5.25}$$

and

$$H^{[1]}(t) = \left.\frac{\partial h(x(t), t)}{\partial x(t)}\right|_{x=\hat{x}(t)}. \tag{5.26}$$

5.5.3 Summary of Implementation Equations

For discrete systems linearized about the estimated state,

$$\delta x_k = \Phi_{k-1}^{[1]} \, \delta x_{k-1} + w_{k-1}, \tag{5.27}$$

$$\delta z_k = H_k^{[1]} \, \delta x_k + v_k. \tag{5.28}$$

For continuous systems linearized about the estimated state,

$$\dot{\delta}_x(t) = F^{[1]}(t) \, \delta x(t) + G(t)w(t), \tag{5.29}$$

$$\delta z(t) = H^{[1]} \, \delta x(t) + v(t). \tag{5.30}$$

5.6 DISCRETE LINEARIZED AND EXTENDED FILTERING

These two approaches to Kalman filter approximations for nonlinear problems yield decidedly different implementation equations. The linearized filtering approach generally has a more efficient real-time implementation, but it is less robust against nonlinear approximation errors than the extended filtering approach.

The real-time implementation of the linearized version can be made more efficient by precomputing the measurement sensitivities, state transition matrices, and Kalman gains. This off-line computation is not possible for the extended Kalman filter, because these implementation parameters will be functions of the real-time state estimates.

Nonlinear Approximation Errors. The extended Kalman filter generally has better robustness because it uses linear approximation over smaller ranges of state space. The linearized implementation assumes linearity over the range of the trajectory perturbations plus state estimation errors, whereas the extended Kalman filter assumes linearity only over the range of state estimation errors. The expected squared magnitudes of these two ranges can be analyzed by comparing the solutions of the two equations

$$P_{k+1} = \Phi_k^{[1]} P_k \Phi_k^{[1]\mathrm{T}} + Q_k,$$
$$P_{k+1} = \Phi_k^{[1]} \{P_k - P_k H_k^{\mathrm{T}} [H_k P_k H_k^{\mathrm{T}} + R_k]^{-1} H_k P_k\} \Phi_k^{[1]\mathrm{T}} + Q_k.$$

The first of these is the equation for the covariance of trajectory perturbations, and the second is the equation for the a priori covariance of state estimation errors. The solution of the second equation provides an idea of the ranges over which the extended Kalman filter uses linear approximation. The sum of the solutions of the two equations provides an idea of the ranges over which the linearized filter assumes linearity. The nonlinear approximation errors can be computed as functions of perturbations (for linearized filtering) or estimation errors (for extended filtering) δx by the formulas

$$\varepsilon_x = f(x + \delta x) - f(x) - \frac{\partial f}{\partial x} \delta x,$$
$$\varepsilon_z = \overline{K}\left(h(x + \delta x) - h(x) - \frac{\partial h}{\partial x} \delta x \right),$$

where ε_x is the error in the temporal update of the estimated state variable due to nonlinearity of the dynamics and ε_z is the error in the observational update of the estimated state variable due to nonlinearity of the measurement. As a rule of thumb for practical purposes, the magnitudes of these errors should be dominated by the RMS estimation uncertainties. That is, $|\varepsilon|^2 \ll \operatorname{trace} P$ for the ranges of δx expected in implementation.

5.6.1 Linearized Kalman Filter

The block diagram of Figure 5.1 shows the data flow of the estimator linearized about a nominal trajectory of the state dynamics. Note that the operations within the dashed box have no inputs. These are the computations for the nominal trajectory. Because they have no inputs from the rest of the estimator, they can be precomputed off-line.

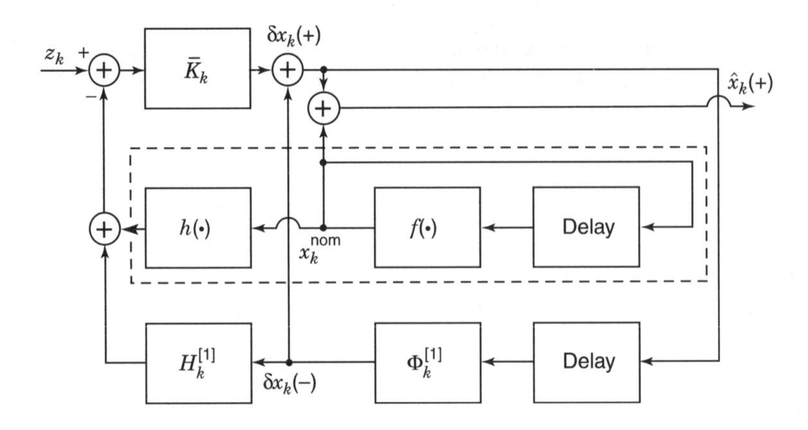

Fig. 5.1 *Estimator linearized about a "nominal" state.*

The models and implementation equations for the linearized discrete Kalman filter that were derived in Section 5.4 are summarized in Table 5.3. Note that the last three equations in this table are identical to those of the "standard" Kalman filter.

5.7 DISCRETE EXTENDED KALMAN FILTER

The essential idea of the extended Kalman filter was proposed by Stanley F. Schmidt, and it has been called the "Kalman–Schmidt" filter [122, 123, 136].

The models and implementation equations of the extended Kalman filter that were derived in Section 5.5 are summarized in Table 5.4. The last three equations in this table are the same as those for the "standard" Kalman filter, but the other equations are noticeably different from those of the linearized Kalman filter in Table 5.3.

EXAMPLE 5.1 Consider the discrete-time system

$$\begin{aligned}
x_k &= x_{k-1}^2 + w_{k-1}, \\
z_k &= x_k^3 + v_k, \\
Ev_k &= Ew_k = 0, \\
Ev_{k_1}v_{k_2} &= 2\Delta(k_2 - k_1), \\
Ew_{k_1}w_{k_2} &= \Delta(k_2 - k_1), \\
Ex(0) &= \hat{x}_0 = 2, \\
x_k^{\text{nom}} &= 2, \\
P_0(+) &= 1,
\end{aligned}$$

TABLE 5.3 Discrete Linearized Kalman Filter Equations

Nonlinear nominal trajectory model:

$$x_k^{\text{nom}} = f_{k-1}(x_{k-1}^{\text{nom}})$$

Linearized perturbed trajectory model:

$$\delta x \stackrel{\text{def}}{=} x - x^{\text{nom}}$$

$$\delta x_k \approx \left.\frac{\partial f_{k-1}}{\partial x}\right|_{x=x_{k-1}^{\text{nom}}} \delta x_{k-1} + w_{k-1}$$

$$w_k \sim \mathcal{N}(0, Q_k)$$

Nonlinear measurement model:

$$z_k = h_k(x_k) + v_k, \qquad v_k \sim \mathcal{N}(0, R_k)$$

Linearized approximation equations:

Linear perturbation prediction:

$$\widehat{\delta x}_k(-) = \Phi_{k-1}^{[1]}\, \widehat{\delta x}_{k-1}(+), \qquad \Phi_{k-1}^{[1]} \approx \left.\frac{\partial f_{k-1}}{\partial x}\right|_{x=x_{k-1}^{\text{nom}}}$$

Conditioning the predicted perturbation on the measurement:

$$\widehat{\delta x}_k(+) = \widehat{\delta x}_k(-) + \overline{K}_k[z_k - h_k(x_k^{\text{nom}}) - H_k^{[1]}\widehat{\delta x}_k(-)]$$

$$H_k^{[1]} \approx \left.\frac{\partial h_k}{\partial x}\right|_{x=x_k^{\text{nom}}}$$

Computing the a priori covariance matrix:

$$P_k(-) = \Phi_{k-1}^{[1]} P_{k-1}(+) \Phi_{k-1}^{[1]\text{T}} + Q_{k-1}$$

Computing the Kalman gain:

$$\overline{K}_k = P_k(-)H_k^{[1]\text{T}}[H_k^{[1]}P_k(-)H_k^{[1]\text{T}} + R_k]^{-1}$$

Computing the a posteriori covariance matrix:

$$P_k(+) = \{I - \overline{K}_k H_k^{[1]}\}P_k(-)$$

TABLE 5.4 Discrete Extended Kalman Filter Equations

Nonlinear dynamic model:

$$x_k = f_{k-1}(x_{k-1}) + w_{k-1}, \qquad w_k \sim \mathcal{N}(0, Q_k)$$

Nonlinear measurement model:

$$z_k = h_k(x_k) + v_k, \qquad v_k \sim \mathcal{N}(0, R_k)$$

Nonlinear implementation equations:

Computing the predicted state estimate:

$$\hat{x}_k(-) = f_{k-1}(\hat{x}_{k-1}(+))$$

Computing the predicted measurement:

$$\hat{z}_k = h_k(\hat{x}_k(-))$$

Linear approximation equations:

$$\Phi_{k-1}^{[1]} \approx \left.\frac{\partial f_k}{\partial x}\right|_{x=\hat{x}_{k-1}(-)}$$

Conditioning the predicted estimate on the measurement:

$$\hat{x}_k(+) = \hat{x}_k(-) + \overline{K}_k(z_k - \hat{z}_k), \qquad H_k^{[1]} \approx \left.\frac{\partial h_k}{\partial x}\right|_{x=\hat{x}_k(-)}$$

Computing the a priori covariance matrix:

$$P_k(-) = \Phi_{k-1}^{[1]} P_{k-1}(+) \Phi_{k-1}^{[1]\mathrm{T}} + Q_{k-1}$$

Computing the Kalman gain:

$$\overline{K}_k = P_k(-) H_k^{[1]\mathrm{T}} [H_k^{[1]} P_k(-) H_k^{[1]\mathrm{T}} + R_k]^{-1}$$

Computing the a posteriori covariance matrix

$$P_k(+) = \{I - \overline{K}_k H_k^{[1]}\} P_k(-)$$

for which one can use the "nominal" solution equations from Table 5.3,

$$
\begin{aligned}
\Phi^{[1]}(x_k^{\text{nom}}) &= \frac{\partial}{\partial x}[x^2]\bigg|_{x=x^{\text{nom}}} \\
&= 4, \\
H^{[1]}(x_k^{\text{nom}}) &= \frac{\partial}{\partial x}(x^3)\bigg|_{x=x^{\text{nom}}} \\
&= 12,
\end{aligned}
$$

to obtain the discrete linearized filter equations

$$
\begin{aligned}
\hat{x}_k(+) &= \widehat{\delta x}_k(+) + 2, \\
\widehat{\delta x}_k(+) &= 4\widehat{\delta x}_{k-1}(+) + \overline{K}_k[z_k - 8 - 48\widehat{\delta x}_{k-1}(+)], \\
P_k(-) &= 16P_{k-1}(+) + 1, \\
P_k(+) &= [1 - 12\overline{K}_k]P_k(-), \\
\overline{K}_k &= \frac{12P_k(-)}{144P_k(-) + 2)}.
\end{aligned}
$$

Given the measurements z_k, $k = 1, 2, 3$, the values for $P_k(-)$, $\overline{K}_k$, $P_k(+)$, and $\hat{x}(+)$, can be computed. If z_k are not given, then $P_k(-)$, $\overline{K}_k$, and $P_k(+)$ can be computed for covariance analysis results. For large k with very small Q and R, the difference $\hat{x}_k - x_k^{\text{nom}}$ will not stay small, and the results become meaningless.

This situation can be improved by using the extended Kalman filter as discussed in Section 5.7:

$$
\begin{aligned}
\hat{x}_k(+) &= \hat{x}_{k-1}^2(+) + \overline{K}_k\{z_k - [\hat{x}_k(-)]^3\}, \\
P_k(-) &= 4[\hat{x}_{k-1}(-)]^2 P_{k-1}(+) + 1, \\
\overline{K}_k &= \frac{3P_k(-)[\hat{x}_k(-)]^2}{9[\hat{x}_k(-)]^4 P_k(-) + 2}, \\
P_k(+) &= \{1 - 3\overline{K}_k[\hat{x}_k(-)]^2\}P_k(-).
\end{aligned}
$$

These equations are now more complex but should work, provided Q and R are small.

5.8 CONTINUOUS LINEARIZED AND EXTENDED FILTERS

The essential equations defining the continuous form of the extended Kalman filter are summarized in Table 5.5. The linearized Kalman filter equations will have x^{nom} in place of $\hat{x}$ as the argument in the evaluations of nonlinear functions and their derivatives.

TABLE 5.5 Continuous Extended Kalman Filter Equations

Nonlinear dynamic model:

$$\dot{x}(t) = f(x(t), t) + w(t) \qquad w(t) \sim \mathcal{N}(0, Q(t))$$

Nonlinear measurement model:

$$z(t) = h(x(t), t) + v(t) \qquad v(t) \sim \mathcal{N}(0, R(t))$$

Implementation equations:

Differential equation of the state estimate:

$$\dot{\hat{x}}(t) = f(\hat{x}(t), t) + \overline{K}(t)[z(t) - \hat{z}(t)]$$

Predicted measurement:

$$\hat{z}(t) = h(\hat{x}(t), t)$$

Linear approximation equations:

$$F^{[1]}(t) \approx \left.\frac{\partial f(x, t)}{\partial x}\right|_{x=\hat{x}(t)}$$

$$H^{[1]}(t) \approx \left.\frac{\partial h(x, t)}{\partial x}\right|_{x=\hat{x}(t)}$$

Kalman gain equations:

$$\dot{P}(t) = F^{[1]}(t)P(t) + P(t)F^{[1]\mathrm{T}}(t) + G(t)Q(t)G^{\mathrm{T}}(t) - \overline{K}(t)R(t)\overline{K}^{\mathrm{T}}(t)$$

$$\overline{K}(t) = P(t)H^{[1]\mathrm{T}}(t)R^{-1}(t)$$

5.8.1 Higher Order Estimators

The linearized and extended Kalman filter equations result from truncating a Taylor series expansion of $f(x, t)$ and $h(x, t)$ after the linear terms. Improved model fidelity may be achieved at the expense of an increased computational burden by keeping the second-order terms as well [21, 31, 75].

5.9 BIASED ERRORS IN QUADRATIC MEASUREMENTS

Quadratic dependence of a measurement on the state variables introduces an approximation error ε when the *expected value* of the measurement is approximated by the formula $\hat{z} = h(\hat{x}) + \varepsilon \approx h(\hat{x})$. It will be shown that the approximation is biased (i.e., $E\langle\varepsilon\rangle \neq 0$) and how the expected error $E\langle\varepsilon\rangle$ can be calculated and compensated in the Kalman filter implementation.

Quadratic Measurement Model. For the sake of simplicity, we consider the case of a scalar measurement. (The resulting correction can be applied to each component of a measurement vector, however.) Suppose that its dependence on the state vector can be represented in the form

$$z = h(x) \tag{5.31}$$

$$= H_1 x + x^{\mathrm{T}} H_2 x + v, \tag{5.32}$$

where H_1 represents the linear (first-order) dependence of the measurement component on the state vector and H_2 represents the quadratic (second-order) dependence. The matrix H_1 will then be a $1 \times n$-dimensioned array and H_2 will be an $n \times n$-dimensioned array, where n is the dimension of the state vector.

Quadratic Error Model. If one defines the estimation error as $\tilde{x} = \hat{x} - x$, then the expected measurement

$$\hat{z} = E\langle h(x)\rangle \tag{5.33}$$

$$= E\langle H_1 x + x^{\mathrm{T}} H_2 x\rangle \tag{5.34}$$

$$= E\langle H_1(\hat{x} - \tilde{x}) + (\hat{x} - \tilde{x})^{\mathrm{T}} H_2(\hat{x} - \tilde{x})\rangle \tag{5.35}$$

$$= H_1\hat{x} + \hat{x}^{\mathrm{T}} H_2 \hat{x} + E\langle \tilde{x}^{\mathrm{T}} H_2 \tilde{x}\rangle \tag{5.36}$$

$$= h(\hat{x}) + E\langle \text{trace } [\tilde{x}^{\mathrm{T}} H_2 \tilde{x}]\rangle \tag{5.37}$$

$$= h(\hat{x}) + E\langle \text{trace } [H_2(\tilde{x}\tilde{x}^{\mathrm{T}})]\rangle \tag{5.38}$$

$$= h(\hat{x}) + \text{trace } [H_2 P(-)] \tag{5.39}$$

$$= h(\hat{x}) + \varepsilon, \tag{5.40}$$

$$\varepsilon = \text{trace } [H_2 P(-)], \tag{5.41}$$

$$P(-) = E\langle \tilde{x}\tilde{x}^{\mathrm{T}}\rangle, \tag{5.42}$$

where $P(-)$ is the covariance matrix of a priori estimation uncertainty. The quadratic error correction should be added in the extended Kalman filter implementation.

EXAMPLE 5.2 Quadratic measurement functions commonly occur in the calibration of linear sensors for which the *scale factor* s (the ratio between variations of its output z and variations of its input y) and *bias* b (the value of the output when the input is zero) are also part of the system state vector, along with the input y itself:

$$x = [s \quad b \quad y], \tag{5.43}$$

$$z = h(x) + v \tag{5.44}$$

$$= sy + b + v. \tag{5.45}$$

The measurement is proportional to the product of the two states $x_1 = s$ and $x_3 = y$. The quadratic form of the second-order measurement model in this example is:

$$H_2 = \frac{1}{2}\frac{\partial^2}{\partial x^2} h(x)\bigg|_{x=0} \tag{5.46}$$

$$= \frac{1}{2}\begin{bmatrix} \frac{\partial^2 h}{\partial s^2} & \frac{\partial^2 h}{\partial s \partial b} & \frac{\partial^2 h}{\partial s \partial y} \\ \frac{\partial^2 h}{\partial s \partial b} & \frac{\partial^2 h}{\partial b^2} & \frac{\partial^2 h}{\partial b \partial y} \\ \frac{\partial^2 h}{\partial s \partial y} & \frac{\partial^2 h}{\partial b \partial y} & \frac{\partial^2 h}{\partial y^2} \end{bmatrix}_{s=b=y=0} \tag{5.47}$$

$$= \begin{bmatrix} 0 & 0 & \frac{1}{2} \\ 0 & 0 & 0 \\ \frac{1}{2} & 0 & 0 \end{bmatrix}, \tag{5.48}$$

and the correct form for the expected measurement is

$$\hat{z} = h(\hat{x}) + \text{trace}\left\{ \begin{bmatrix} 0 & 0 & \frac{1}{2} \\ 0 & 0 & 0 \\ \frac{1}{2} & 0 & 0 \end{bmatrix} P(-) \right\} \tag{5.49}$$

$$= \hat{s}\hat{y} + \hat{b} + p_{13}(-), \tag{5.50}$$

where $p_{13}(-)$ is the a priori covariance between the scale factor uncertainty and the input uncertainty.

5.10 APPLICATION OF NONLINEAR FILTERS

EXAMPLE 5.3: Damping Parameter Estimation This example uses Example 4.3 from Chapter 4. Assume that ζ (damping coefficient) is unknown and is a constant. Therefore, the damping coefficient can be modeled as a state vector and its value is estimated via linearized and extended Kalman estimators.

The conversion from a parameter estimation problem to a state estimation problem shown in Example 4.3 results in a nonlinear problem.

Let

$$x_3(t) = \zeta \tag{5.51}$$

and

$$\dot{x}_3(t) = 0 \tag{5.52}$$

Then the plant equation of Example 4.3 becomes

$$\begin{bmatrix} \dot{x}_1(t) \\ \dot{x}_2(t) \\ \dot{x}_3(t) \end{bmatrix} = \begin{bmatrix} x_2 \\ -\omega^2 x_1 - 2x_2 x_3 \omega \\ 0 \end{bmatrix} + \begin{bmatrix} 0 \\ 1 \\ 0 \end{bmatrix} w(t) + \begin{bmatrix} 0 \\ 12 \\ 0 \end{bmatrix}. \tag{5.53}$$

The observation equation is

$$z(t) = x_1(t) + v(t) \tag{5.54}$$

One hundred data points were generated with plant noise and measurement noise set equal to zero, $\zeta = 0.1$, $w = 5$ rad/s, and initial conditions

$$\begin{bmatrix} x_1(0) \\ x_2(0) \\ x_3(0) \end{bmatrix} = \begin{bmatrix} 0\ \text{ft} \\ 0\ \text{ft/s} \\ 0 \end{bmatrix},$$

$$P(0) = \begin{bmatrix} 2 & 0 & 0 \\ 0 & 2 & 0 \\ 0 & 0 & 2 \end{bmatrix},$$

$$Q = 4.47(\text{ft/s})^2,$$

$$R = 0.01\text{ft}^2.$$

The discrete nonlinear plant and linear observation equations for this model are

$$x_1^k = x_1^{k-1} + T x_2^{k-1}, \tag{5.55}$$

$$x_2^k = -25T x_1^{k-1} + (1 - 10T x_3^{k-1}) x_2^{k-1} + 12T + T w_{k-1}, \tag{5.56}$$

$$x_3^k = x_3^{k-1}, \tag{5.57}$$

$$z_k = x_1^k + v_k. \tag{5.58}$$

The relevant equations from Table 5.3 (discrete linearized Kalman filter equations) and Table 5.4 (discrete extended Kalman filter equations) have been programmed in MATLAB as exam53.m on the accompanying diskette. T is sampling interval.

Figure 5.2 shows the estimated position, velocity, and damping factor states (note non-convergence due to vanishing gain) using the noise-free data generated from simulating the second-order equation (the same data as in Example 4.1). Figure 5.3 shows the corresponding RMS uncertainties from the extended Kalman filter. (See Appendix A for descriptions of exam53.m and modified versions.)

For the noise-free data, the linearized and extended Kalman filter (EKF) results are very close. But for noisy data, convergence for the discrete linearized results is

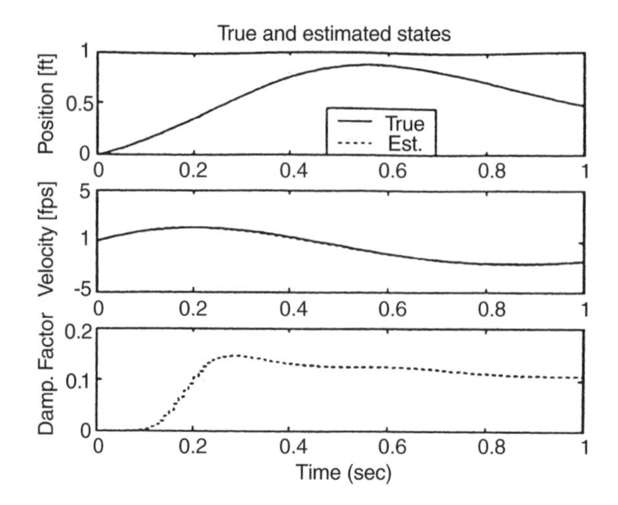

Fig. 5.2 State variables estimated by extended Kalman filter.

not as fast, compared to convergence for the extended filter. Results are a little better with the EKF [122, 196, 200, 202, 211].

EXAMPLE 5.4 Inertial reference systems maintain a *computational reference frame,* which is a set of orthogonal reference axes defined with respect to the inertial sensors (gyroscopes and accelerometers). The attitude error of an inertial reference system is a set of rotations about these axes, representing the rotations between where the system *thinks* these axes are and where they *really* are.

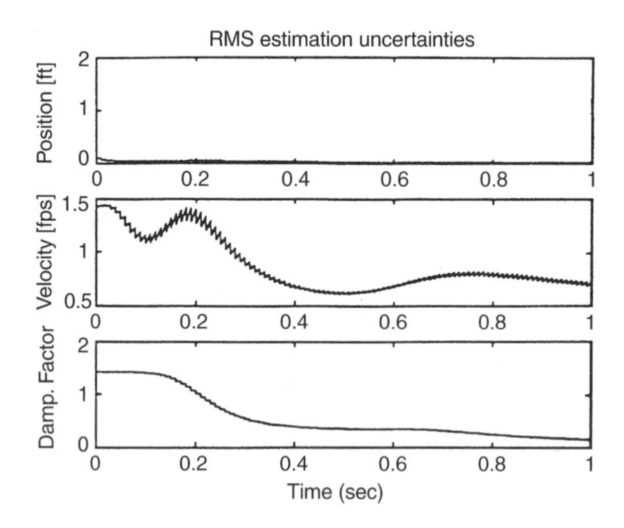

Fig. 5.3 RMS uncertainties in estimates.

Gyroscope Filter. This error can be represented by a model of the form

$$\dot{\Psi} = \Psi \otimes \omega + \varepsilon, \tag{5.59}$$

where

$\Psi = 3 \times 1$ vector containing the attitude alignment errors between the sensor axes frame and the computational reference frame

$\otimes =$ vector cross-product operator

$\omega = 3 \times 1$ vector of platform inertial angular rate from the trajectory generator

$\varepsilon = 3 \times 1$ vector of composite gyroscope drift rates (algebraic sum of all error sources)

This attitude error model can be augmented by a 48-state model of the gyroscope parameters (see the beginning paragraphs of Chapter 3) as random walks and random ramps. The first-order vector differential equation in a state-space form for the augmented 51×1 state vector is

$$\dot{x}^g(t) = F(t)x^g(t) + w_1(t), \tag{5.60}$$

where the 51-component state vector $x^g(t)$ is composed of the nonredundant components of the following arrays:

Symbol	Ψ	b_g	h_g^1	U_g^1	K_g^1	T_g	b_{gt}	U_{gt}^1
Dimension	3×1	3×1	3×3	3×3	3×6	3×1	3×1	3×6
Subvector	3×1	3×1	9×1	9×1	15×1	3×1	3×1	6×1
$x-$Indices	$1-3$	$4-6$	$7-15$	$16-24$	$25-39$	$40-42$	$43-45$	$46-51$

(5.61)

The symbol at the top is the array name, with its dimensions shown below it, and the bottom dimension refers to the dimension of the corresponding subvector of its non-redundant terms in the system state vector, shown at the bottom. The matrices h_g, U_g, K_g, and U_{gt} are defined as follows:

h_g is a 3×3 matrix containing unknown scale factor (S_{gi}) and linear axes alignment errors (Δ_{ij}) as components $(i, j = 1, 2, 3)$:

$$\begin{bmatrix} S_{g1} & \Delta_{12} & \Delta_{13} \\ \Delta_{21} & S_{g2} & \Delta_{23} \\ \Delta_{31} & \Delta_{32} & S_{g3} \end{bmatrix} \tag{5.62}$$

U_g is a 3×3 matrix of unknown gyroscope mass unbalance parameters $d_{k,j}$:

$$\begin{bmatrix} d_{I1} & d_{01} & d_{S1} \\ d_{S2} & d_{I2} & d_{02} \\ d_{03} & d_{S3} & d_{I3} \end{bmatrix} \tag{5.63}$$

K_g is a 3×6 matrix of unknown gyroscope compliance (g-squared) errors (k_{kji}):

$$\begin{bmatrix} k_{II1} & k_{001} & k_{SS1} & k_{I01} & k_{IS1} & k_{S01} \\ k_{SS2} & k_{II2} & k_{002} & k_{IS2} & k_{S02} & k_{I02} \\ k_{003} & k_{SS3} & k_{II3} & k_{S03} & k_{I03} & k_{IS3} \end{bmatrix} \tag{5.64}$$

U_{gt} is a 3×6 matrix of unknown gyroscope mass unbalance trend parameters.
h_g, U_g, k_g, U_{gt} have been redimensioned rowwise to form column vectors h_g^1, U_g^1, K_g^1, and U_{gt}^1.
b_g is a 3×1 vector of unknown gyroscope fixed-drift rate parameters.
T_g is a 3×1 vector of unknown nonlinear gyroscope torquer scale factor errors, with elements δS_{gi}.
b_{gt} is a 3×1 vector of unknown gyroscope fixed-drift trend parameters.

In expanded form,

$$\begin{bmatrix} \dot{\Psi} \\ \dot{b}_g \\ \dot{h}_g^1 \\ \dot{U}_g^1 \\ \dot{K}_g^1 \\ \dot{T}_g \\ \dot{b}_{gt} \\ \dot{U}_{gt}^1 \end{bmatrix} = \begin{bmatrix} F_{11} & F_{12} & F_{13} & F_{14} & F_{15} & F_{16} & F_{17} & F_{18} \\ 0 & 0 & 0 & 0 & 0 & 0 & F_{27} & 0 \\ 0 & 0 & 0 & 0 & 0 & 0 & 0 & 0 \\ 0 & 0 & 0 & 0 & 0 & 0 & 0 & F_{48} \\ 0 & 0 & 0 & 0 & 0 & 0 & 0 & 0 \\ 0 & 0 & 0 & 0 & 0 & 0 & 0 & 0 \\ 0 & 0 & 0 & 0 & 0 & 0 & 0 & 0 \\ 0 & 0 & 0 & 0 & 0 & 0 & 0 & 0 \end{bmatrix} \begin{bmatrix} \Psi \\ b_g \\ h_g^1 \\ U_g^1 \\ K_g^1 \\ T_g \\ b_{gt} \\ U_{gt}^1 \end{bmatrix} + \begin{bmatrix} w_{\Psi}(t) \\ w_{bg}(t) \\ w_{hg}(t) \\ w_{ug}(t) \\ w_{kg}(t) \\ w_{tg}(t) \\ w_{bgt}(t) \\ w_{ugt}(t) \end{bmatrix}, \tag{5.65}$$

where

$$w_1^{\mathrm{T}}(t) = [w_{\Psi}^{\mathrm{T}}(t) \quad w_{bg}^{\mathrm{T}}(t) \quad w_{hg}^{\mathrm{T}}(t) \quad w_{ug}^{\mathrm{T}}(t) \quad w_{kg}^{\mathrm{T}}(t) \quad w_{tg}^{\mathrm{T}}(t) \quad w_{bgt}^{\mathrm{T}}(t) \quad w_{ugt}^{\mathrm{T}}(t)] \tag{5.66}$$

is a noise vector of unmodeled effects and

$$F_{11} = \begin{bmatrix} 0 & \omega_3 & -\omega_2 \\ -\omega_3 & 0 & \omega_1 \\ \omega_2 & -\omega_1 & 0 \end{bmatrix}, \qquad F_{12} = \begin{bmatrix} 1 & 0 & 0 \\ 0 & 1 & 0 \\ 0 & 0 & 1 \end{bmatrix}, \tag{5.67}$$

$$F_{13} = \left[\begin{array}{ccc|ccc|ccc} \omega_1 & \omega_2 & \omega_3 & 0 & 0 & 0 & 0 & 0 & 0 \\ 0 & 0 & 0 & \omega_1 & \omega_2 & \omega_3 & 0 & 0 & 0 \\ 0 & 0 & 0 & 0 & 0 & 0 & \omega_1 & \omega_2 & \omega_3 \end{array}\right], \tag{5.68}$$

$$F_{14} = \left[\begin{array}{ccc|ccc|ccc} \beta_1 & \beta_2 & \beta_3 & 0 & 0 & 0 & 0 & 0 & 0 \\ 0 & 0 & 0 & \beta_1 & \beta_2 & \beta_3 & 0 & 0 & 0 \\ 0 & 0 & 0 & 0 & 0 & 0 & \beta_1 & \beta_2 & \beta_3 \end{array}\right], \tag{5.69}$$

$$F_{15} = \left[\begin{array}{ccccc|ccccc|ccccc} \beta_{11} & \beta_{33} & \beta_{12} & \beta_{13} & \beta_{23} & 0 & 0 & 0 & 0 & 0 & 0 & 0 & 0 & 0 & 0 \\ 0 & 0 & 0 & 0 & 0 & \beta_{11} & \beta_{22} & \beta_{12} & \beta_{13} & \beta_{23} & 0 & 0 & 0 & 0 & 0 \\ 0 & 0 & 0 & 0 & 0 & 0 & 0 & 0 & 0 & 0 & \beta_{22} & \beta_{33} & \beta_{12} & \beta_{13} & \beta_{23} \end{array}\right], \tag{5.70}$$

$$\beta_{ij} \stackrel{\text{def}}{=} \beta_i \beta_j, \tag{5.71}$$

$$F_{16} = \begin{bmatrix} |\omega_1| & 0 & 0 \\ 0 & |\omega_2| & 0 \\ 0 & 0 & |\omega_3| \end{bmatrix}, \tag{5.72}$$

$$F_{17} = \begin{bmatrix} 1 & 0 & 0 \\ 0 & 1 & 0 \\ 0 & 0 & 1 \end{bmatrix} t, \tag{5.73}$$

$$F_{18} = \begin{bmatrix} \beta_1 & \beta_3 & 0 & 0 & 0 & 0 \\ 0 & 0 & \beta_1 & \beta_2 & 0 & 0 \\ 0 & 0 & 0 & 0 & \beta_2 & \beta_3 \end{bmatrix} t, \tag{5.74}$$

$$F_{27} = \begin{bmatrix} 1 & 0 & 0 \\ 0 & 1 & 0 \\ 0 & 0 & 1 \end{bmatrix}, \tag{5.75}$$

$$F_{48} = \begin{bmatrix} 1 & 0 & 0 & 0 & 0 & 0 \\ 0 & 0 & 0 & 0 & 0 & 0 \\ 0 & 1 & 0 & 0 & 0 & 0 \\ 0 & 0 & 1 & 0 & 0 & 0 \\ 0 & 0 & 0 & 1 & 0 & 0 \\ 0 & 0 & 0 & 0 & 0 & 0 \\ 0 & 0 & 0 & 0 & 0 & 0 \\ 0 & 0 & 0 & 0 & 1 & 0 \\ 0 & 0 & 0 & 0 & 0 & 1 \end{bmatrix}, \tag{5.76}$$

β is a 3×1 vector of vertical direction cosines (normalized gravity)

$$\beta = (\beta_1, \beta_2, \beta_3)^{\mathrm{T}}, \tag{5.77}$$

β^1 is a (6×1) with products of β_i as components

$$(\beta_1^2, \beta_2^2, \beta_3^2, \beta_1\beta_2, \beta_1\beta_3, \beta_2\beta_3), \tag{5.78}$$

and ω_i and β_i are time dependent. Thus a different system description matrix is computed in each filter cycle.

The corresponding difference equation of continuous Equation 5.60,

$$\dot{x}^g(t) = F(t)x^g(t) + w_1(t), \tag{5.79}$$

is

$$x_j^g = \Phi_{j,j-1}^g x_{j-1}^g + w_{j-1}^g, \tag{5.80}$$

where the gyroscope state transition matrix $\Phi_{j,j-1}^g$ is approximated by the first two terms of the power series expansion of the exponential function,

$$\Phi_{j,j-1}^g = I + Ft + F^2 \frac{t^2}{2} + \text{higher order terms} \tag{5.81}$$

and w_j^g is normally distributed white noise with zero mean and covariance Q and accounts for gyroscope modeling and truncation errors,

$$w_j^g \sim \mathcal{N}(0, Q). \tag{5.82}$$

The scalar t is the filter cycle time.

The gyro observation equation is

$$\overset{2\times 1}{z_j^g} = \overset{51\times 1}{H^g x_j^g} + v_j^g, \tag{5.83}$$

where

$$\underset{2\times 51}{H^g} = \underset{2\times 3}{\begin{bmatrix} \alpha^{\mathrm{T}} \\ \gamma^{\mathrm{T}} \end{bmatrix}} \underset{3\times 51}{[\Phi_{11}\Phi_{12}\Phi_{13}\Phi_{14}\Phi_{15}\Phi_{16}\Phi_{17}\Phi_{18}]}$$

α is north direction cosine vector (3×1), γ is west direction cosine vector (3×1), and $\Phi_{1m}(m = 1, 2, 3, 4, 5, 6, 7, 8)$ are the appropriate submatrices of the gyroscope state transition matrix $\Phi^g_{j,j-1}$, and $v^g_j \sim \mathcal{N}(0, R)$, which includes noise and errors from sensors.

Accelerometer filter. The difference equation for the accelerometers is

$$X^A_j = \Phi^A_{j,j-1} X^A_{j-1} + w^A_{j-1}, \tag{5.84}$$

where

$$\Phi^A_{j,j-1} = I. \tag{5.85}$$

The 12×1 accelerometer state vector X^A_j is composed of

$$(X^A)^{\mathrm{T}} = [b_A\, S_1\, \delta_{12}\, S_2\, \delta_{13}\, \delta_{23}\, S_3\, (FX1^{\mathrm{T}} - FI1^{\mathrm{T}})] \tag{5.86}$$

where b_A, $FX1$, and $FI1$ are 3×1 vectors as defined:

b_A is a 3×1 vector of unknown accelerometer biases, normalized to the magnitude of gravity;

$FI1$ is a 3×1 unknown acceleration-squared nonlinearity for acceleration along the accelerometer input axis; and

$FX1$ is a 3×1 unknown acceleration-squared nonlinearity for acceleration normal to the accelerometer input axis.

Here, $S_1, \delta_{12}, S_2, \delta_{13}, \delta_{23}, S_3$ are elements of matrix h_A.

Twelve unknown parameters are modeled as random walk and $w^A_j \sim \mathcal{N}(0, Q)$ and white noise includes accelerometer modeling and truncation errors:

$$h_A = \begin{bmatrix} S_1 & \delta_{12} & \delta_{13} \\ 0 & S_2 & \delta_{23} \\ 0 & 0 & S_3 \end{bmatrix}, \tag{5.87}$$

where

S_i = unknown accelerometer scale factor errors $(i = 1, 2, 3)$

δ_{ij} = unknown accelerometer axes nonorthogonalities (misalignments) $(1_i \cdot 1_j)$

Here, β_m is a 3-vector $(\beta_1, \beta_2, \beta_3)^{\mathrm{T}}$ of midpoint components of acceleration in platform coordinates,

$$\beta_m^2 = \begin{bmatrix} \beta_1^2 & 0 & 0 \\ 0 & \beta_2^2 & 0 \\ 0 & 0 & \beta_3^2 \end{bmatrix}. \tag{5.88}$$

The accelerometer observation equation is

$$Z_j^A = H^A X_j^A + V_j^A, \tag{5.89}$$

where

$$H^A = [\beta_1, \beta_3, \beta_1^2, \beta_1, \beta_2, \beta_1, \beta_3, \beta_2^2, \beta_2, \beta_3, \beta_3^2, (1-\beta_1^2)\beta_1, (1-\beta_2^2)\beta_2, (1-\beta_3^2)\beta_3] \tag{5.90}$$

and V_j^A is $\sim \mathcal{N}(0, R)$ and white, including sensor noise.

The dimension of the observation vector

$$Z = (Z^g, Z^A)^{\mathrm{T}} \tag{5.91}$$

is 3×1. The EKF equations of Table 5.4 are applied to Equations 5.80 and 5.83 to obtain gyroscope estimates. The EKF equations are applied to Equations 5.84 and 5.89 to obtain accelerometer estimates.

Typical plots of gyroscope fixed-drift, accelerometer fixed-drift, and scale factor estimates are shown in Figures 5.4, 5.5, and 5.6, respectively. Innovation sequence plots for accelerometer and gyroscope are shown in Figures 5.7 and 5.8, respectively. The results are completely described in reference [170].

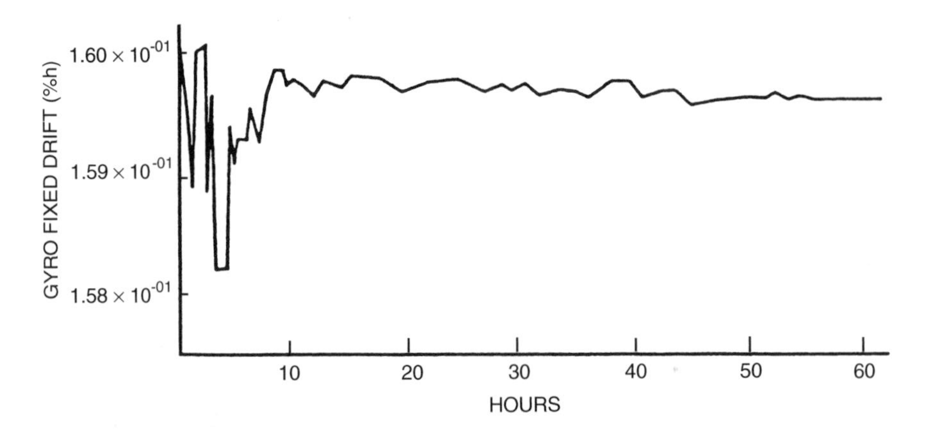

Fig. 5.4 *Gyro fixed-drift rate estimates.*

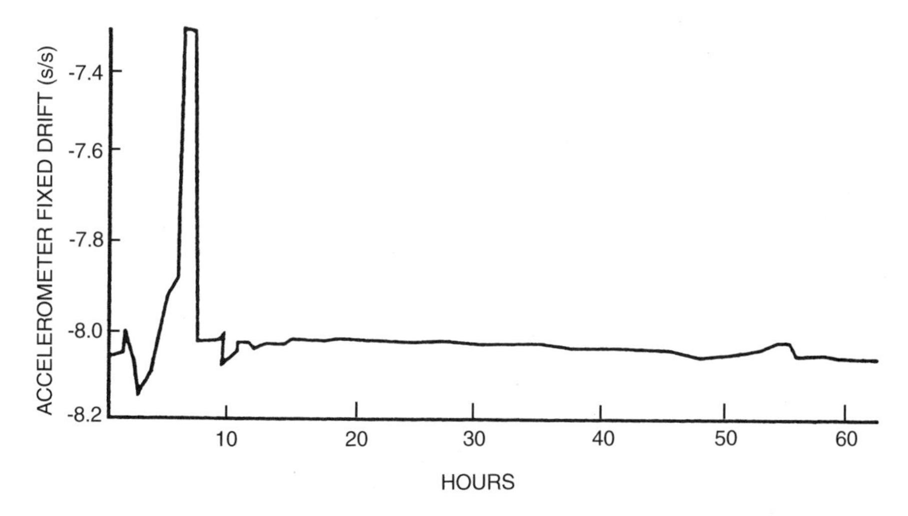

Fig. 5.5 Accelerometer fixed drift estimates.

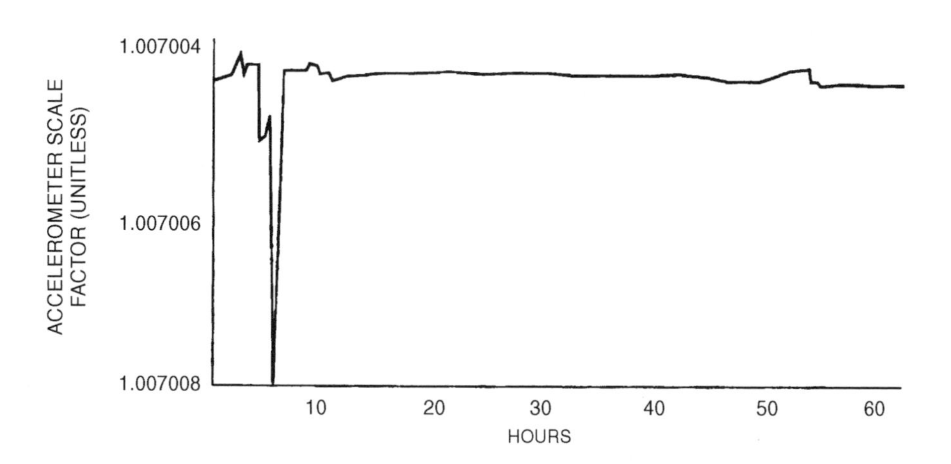

Fig. 5.6 Accelerometer scale factor estimates.

EXAMPLE 5.5 This is an application of a discrete-time, extended Kalman filter to estimate the parameters in a macroscopic freeway traffic model [166].

A dynamic equation is given by

$$u_j^{n+1} = u_j^n + \Delta t \left\{ -u_j^n \frac{u_j^n - u_{j-1}^n}{\Delta x} - \frac{1}{T}\left[u_j^n - a - b\rho_j^n + \frac{v(\rho_{j+1}^n - \rho_j^n)}{\rho_j^n \,\Delta x} \right] \right\} + w_j^n$$

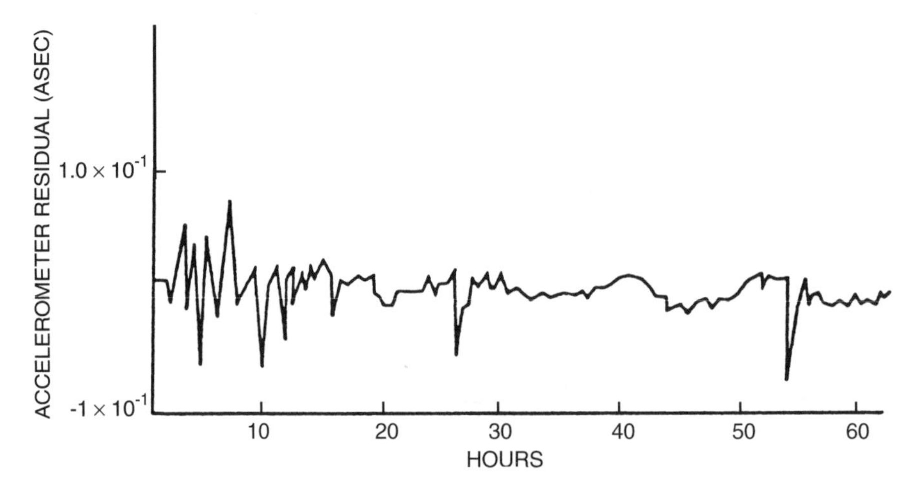

Fig. 5.7 Accelerometer residuals.

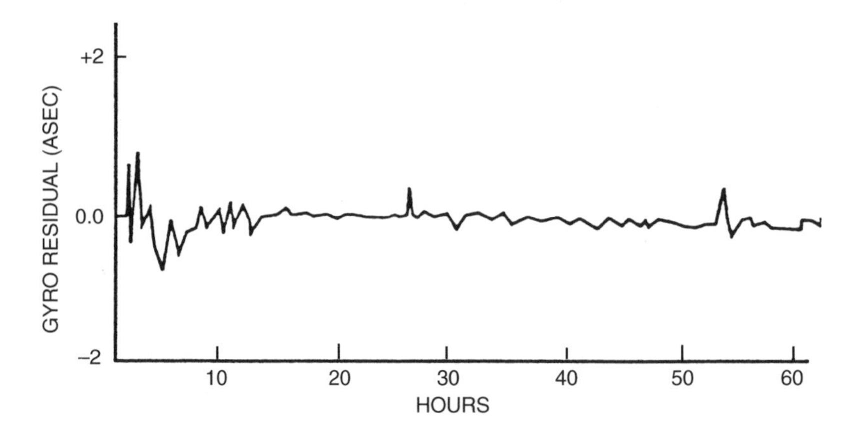

Fig. 5.8 Gyroscope residuals.

Define

$\Delta x_j =$ freeway section length (miles)

$\rho_j^n \equiv \rho(x_j, x_{j+1}; n\,\Delta t), j = 0, \ldots, N-1$ (vehicles per unit length in the section between x_j and x_{j+1} at time $n\,\Delta t$)

$u_j^n \equiv u(x_j, x_{j+1}; n\,\Delta t), j = 0, \cdots, N-1$ (average of speeds of the vehicles in the section between x_j and x_{j+1} at time $n\,\Delta t$)

$T =$ reaction time

$\Delta t =$ discrete time interval

$a =$ parameter
$b =$ parameter
$u_e(\rho_j^n) = a + b\rho_j$
$v =$ sensitivity factor
$q_j^n \equiv q[x_j; (n-1)\,\Delta t, n\,\Delta t], j = 0, 1, \ldots, N$ (vehicles per unit time passing the location x_j between $(n-1)\,\Delta t$ and $n\,\Delta t$)

The observation equation

$$z_j^n = u_j^n + v_j^n$$

The objective is to estimate the parameters $1/T$, v, b, a, where a and b are estimated by least-squares curve filtering to speed and density data, as shown in Figure 5.9. Then, the remaining unknown parameters are $1/T$ and v. The technique of state augmentation is applied. The parameters $1/T$ and v/T are modeled by constants as shown. Let

$$x_{1j}^n = \frac{1}{T}, \qquad x_{2j}^n = \frac{v}{T}.$$

Then

$$x_{1j}^{n+1} = x_{1j}^n, \qquad x_{2j}^{n+1} = x_{2j}^n.$$

The nonlinear dynamic equation becomes

$$\begin{bmatrix} u_j^{n+1} \\ x_{1j}^{n+1} \\ x_{2j}^{n+1} \end{bmatrix} = \begin{bmatrix} u_j^n - \dfrac{\Delta t}{\Delta x}(u_j^n)^2 + \dfrac{\Delta t}{\Delta x} u_j^n u_{j-1}^n - \Delta t\, x_{1j}^n(u_j^n - a - b\rho_j^n) - \dfrac{\Delta t}{\Delta x} x_{2j}^n \dfrac{\rho_{j+1}^n}{\rho_j^n} - 1 \\ x_{1j}^n \\ x_{2j}^n \end{bmatrix} + \begin{bmatrix} 1 \\ 0 \\ 0 \end{bmatrix} w_j^n.$$

The observation equation is

$$z_j^n = [1 \quad 0 \quad 0] \begin{bmatrix} u_j^n \\ x_{1j}^n \\ x_{2j}^n \end{bmatrix} + v_j^n,$$

where w_j^n, v_j^n are zero-mean white Gaussian with covariance Q and R, respectively.

A four-lane, 6000-ft-long section of freeway with no on or off ramps and no accidents was simulated on an IBM 360-44 computer. Using a digital simulation of a microscopic model, eight files of data sets (high flow cases), each containing 20 min of data, section mean speed, and density at 1.5-s intervals were generated.

To demonstrate the application and performance of the methodology of identifying parameters, results from a digital simulation are shown. It has been observed that speed bears a fairly consistent relationship to density. The equilibrium speed–density relationship $u_e(\rho)$ is taken as a least-squares fit by a straight line to the above data and is shown in Figure 5.9. Two parameters, a and b, are estimated by this procedure.

The extended Kalman filter algorithm is applied to estimate $1/T$ and v/T from the above data. For purposes of numerical computation, it is convenient to define dimensionless variables through the use of nominal values. The nominal values used in the parameter identification algorithm are as follows:

$$\text{Nominal section mean speed } (u) = 40 \text{ mph},$$
$$\text{Nominal value of reaction time } (T) = 30 \text{ s},$$
$$\text{Nominal value of sensitivity factor } (v) = 4.0\text{mi}^2/\text{h}.$$

The time step size $\Delta t = 4.5$ s and the space step size $\Delta x_j = 0.5$ miles are chosen from the stability considerations. Figures 5.10 and 5.11 show the typical normalized values of $1/T$ and v/T for the middle (jth) section of a piece of freeway. Actual values of the parameters can be computed by using the above nominal values.

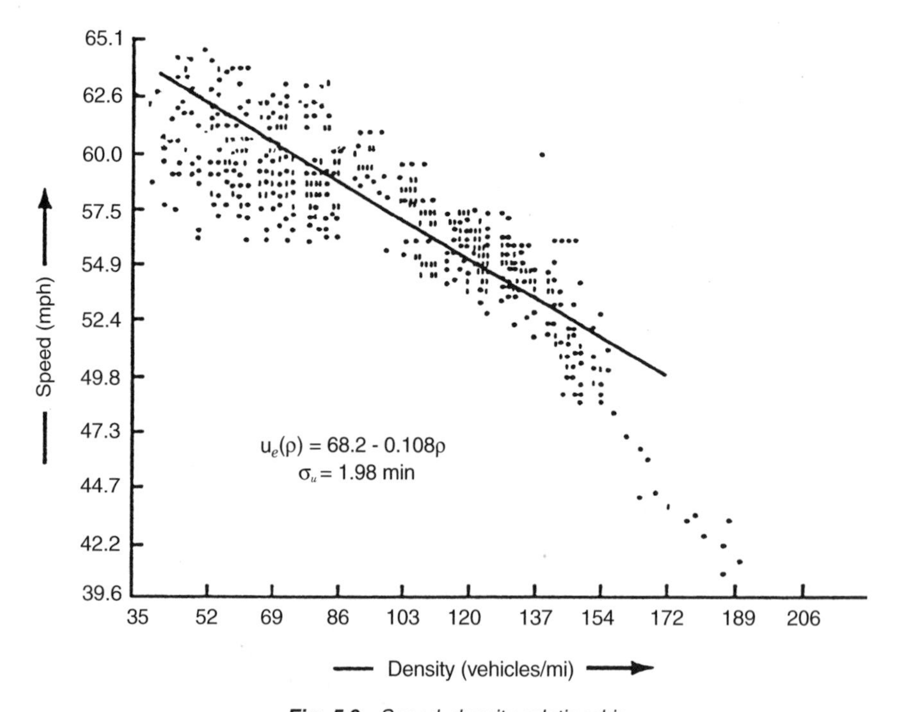

Fig. 5.9 *Speed–density relationship.*

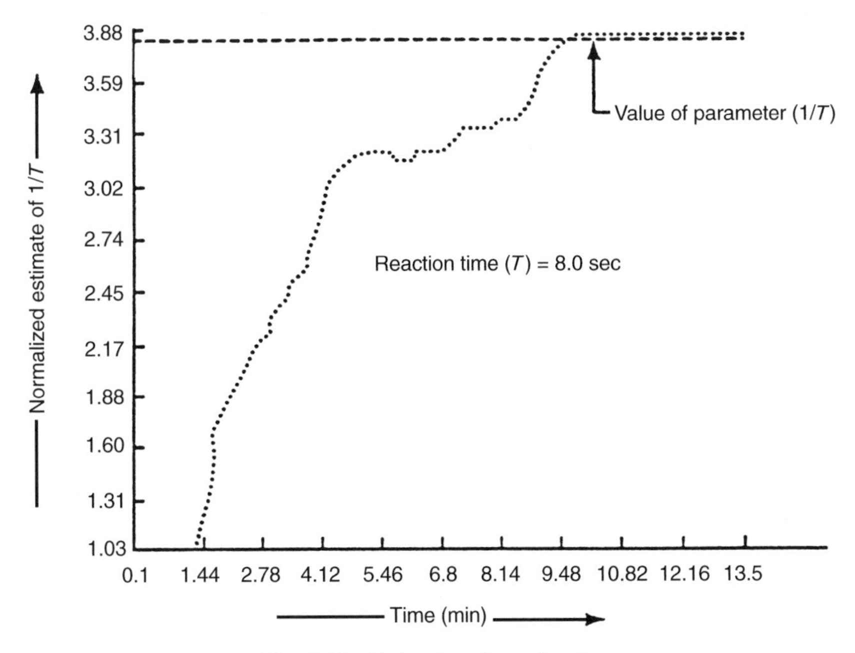

Fig. 5.10 *Estimation of reaction time.*

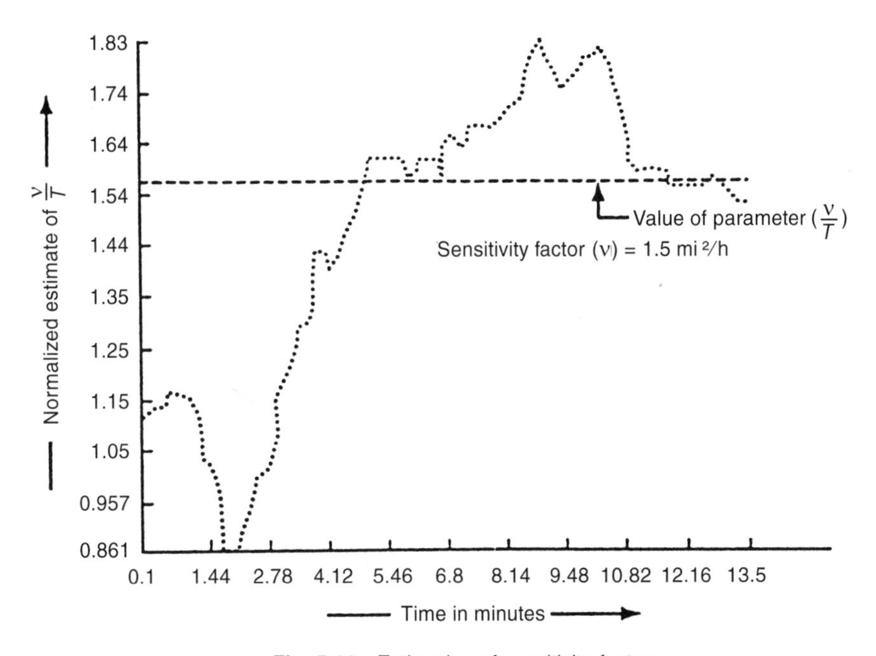

Fig. 5.11 *Estimation of sensitivity factor.*

To test the resultant model as a predictor of future traffic conditions, the estimated values of $1/T, v/T, a,$ and b are used in the model equation. Section density ρ_j^n and output flow q_{j+1}^{n+1} are computed from the flow. The final model is used to predict density and speed of the middle section by using the available data from the adjoining sections, that is $u_{j-1}^n, \rho_{j-1}^n, \rho_{j+1}^n,$ and q_j^{n+1}. This model is particularly effective in predicting speed of traffic flow and density in one section of the freeway over 15-min intervals. This time interval is adequate for traffic responsive control. The single-section density prediction results from the model and actual density are shown in Figure 5.12. The single-section speed prediction results from the model and actual section speed are shown in Figure 5.13. These results show that the final model with the parameter values estimated by the above procedures predicts the traffic conditions (density and speed) satisfactorily.

5.11 SUMMARY

Discrete-time and continuous-time estimators derived in Chapter 4 can be applied to nonlinear problems using the following approaches:

1. Linearization of nonlinear plant and observation models about a fixed *nominal trajectory*. See Table 5.3 for the discrete-time linearized Kalman filter using this approach.

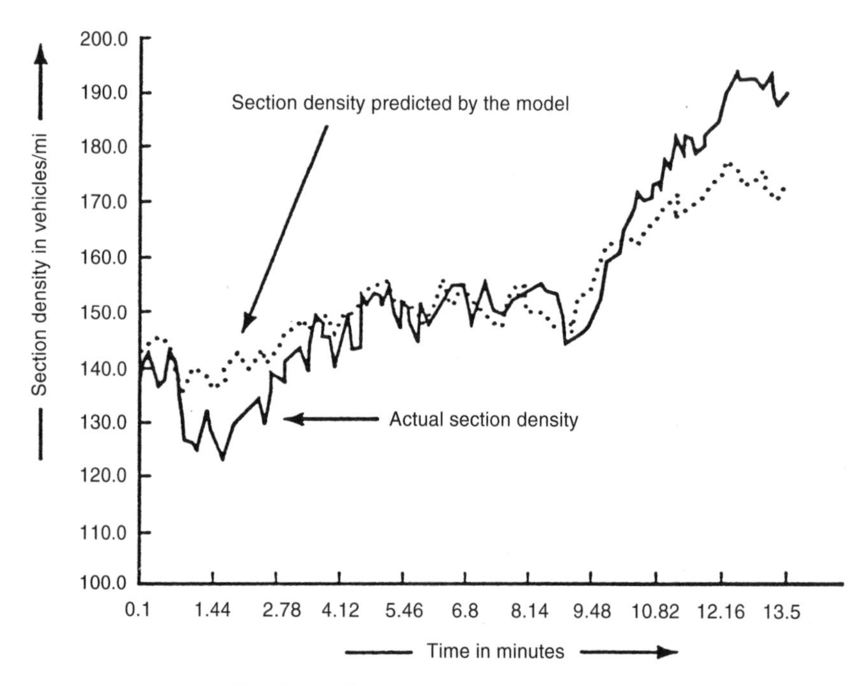

Fig. 5.12 *Single-section density prediction.*

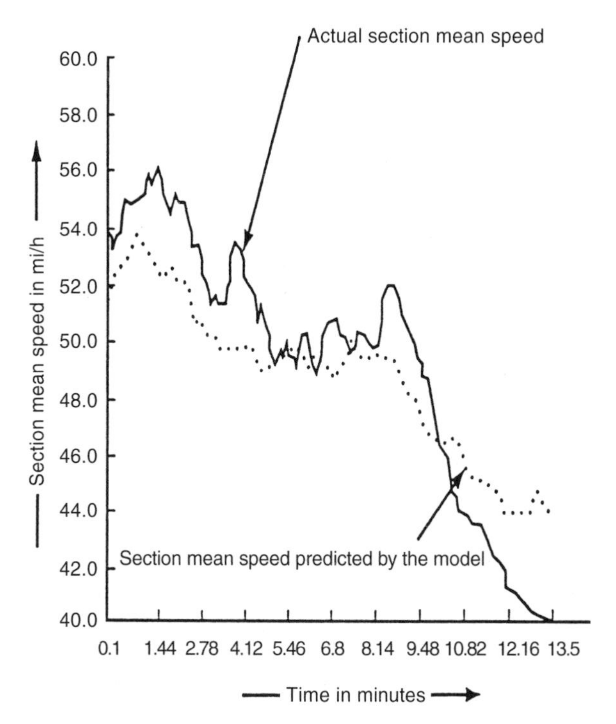

Fig. 5.13 *Single-section speed prediction.*

2. Extended Kalman filtering, which requires linearization of plant and observation equations about the *estimated trajectory* at every time step $[\hat{x}_k(-)]$). See Tables 5.4 and 5.5 for the filter equations.
3. Higher order estimators, which are outside the scope of the treatment in this chapter, although references are provided.

The *parameter identification problem* is a common nonlinear estimation problem. It is the problem of estimating a model parameter that occurs as a coefficient of a dynamic system state variable—either as a dynamic coefficient or as a measurement sensitivity. When this estimation problem is solved simultaneously with the state estimation problem (via state vector augmentation), the linear model becomes nonlinear. As an example, the state estimation problem for a harmonic oscillator (Example 4.3) is extended to a problem in which the parameter ζ (oscillator damping coefficient) is also unknown. The system model equations are augmented with the damping coefficient as an additional state variable. Further large-dimensional examples include the problems of calibration and alignment of an inertial measuring unit (IMU) and the identification of the parameters of a freeway traffic model.

PROBLEMS

5.1 A scalar stochastic sequence x_k is given by

$$x_k = -0.1x_{k-1} + \cos x_{k-1} + w_{k-1}, \qquad z_k = x_k^2 + v_k,$$
$$Ew_k = 0 = Ev_k, \qquad \text{cov } w_k = \Delta(k_2 - k_1), \qquad \text{cov } v_k = 0.5\,\Delta(k_2 - k_1),$$
$$Ex_0 = 0, \qquad P_0 = 1, \qquad x_k^{\text{nom}} = 1.$$

Determine the linearized and extended Kalman estimator equations.

5.2 A scalar stochastic process $x(t)$ is given by

$$\dot{x}(t) = -0.5x^2(t) + w(t),$$
$$z(t) = x^3(t) + v(t),$$
$$Ew(t) = Ev(t) = 0$$
$$\text{cov } w_t = \delta(t_1 - t_2), \qquad \text{cov } v(t) = 0.5\delta(t_1 - t_2),$$
$$Ex_0 = 0, \qquad P_0 = 1, \qquad x^{\text{nom}} = 1.$$

Determine the linearized and extended Kalman estimator equations.

5.3 **(a)** Verify the results of Example 5.3 (noise-free simulation data).
(b) Estimate the states from a noisy data.
(c) Compare the results of linearized and extended Kalman filters.

Assume that the plant noise is normally distributed with mean zero and covariance 0.2 and measurement noise is normally distributed with mean zero and covariance 0.001.

5.4 Derive the linearized and EKF equation for the following equations:

$$x_k = f(x_{k-1}, k-1) + Gw_{k-1}, \qquad z_k = h(x_k, k) + v_k.$$

5.5 Given the following plant and measurement model for a scalar dynamic system:

$$\dot{x}(t) = ax(t) + w(t), \qquad z(t) = x(t) + v(t),$$
$$w(t) \sim \mathcal{N}(0, 1), \qquad v(t) \sim \mathcal{N}(0, 2)$$
$$Ex(0) = 1$$
$$Ew(t)v(t) = 0$$
$$P(0) = 2,$$

Assume an unknown constant parameter a and derive an estimator for a, given $z(t)$.

5.6 Let $\mathbf{r}$ represent the position vector to a magnet with dipole moment vector $\mathbf{m}$. The magnetic field vector $\mathbf{H}$ at the origin of the coordinate system in which $\mathbf{r}$ is measured is given by the formula

$$\mathbf{B} = \frac{\mu_0}{4\pi|\mathbf{r}|^5}\{3\mathbf{r}\mathbf{r}^{\mathrm{T}} - |\mathbf{r}|^2\mathbf{I}\}\mathbf{m} \tag{5.92}$$

in SI units.

(a) Derive the measurement sensitivity matrix for **H** as the measurement and **m** as the state vector.

(b) Derive the sensitivity matrix for **r** as the state vector.

(c) If **r** is known but **m** is to be estimated from measurements of **B**, is the estimation problem linear?

5.7 Generate the error covariance results for the plant and measurement models given in Example 5.3 with the appropriate values of process and measurement noise covariance, and initial state estimation error covariance.

5.8 Generate the error covariance results for the plant and measurement models given in Example 5.4 with the appropriate values of process and measurement noise covariance and initial state estimation error covariance.

5.9 This problem is taken from reference [46]. The equations of motion for the space vehicle are given below:

$$\ddot{r} - r\dot{\theta}^2 + \frac{k}{r^2} = w_r(t), \qquad r\ddot{\theta} + 2\dot{r}\dot{\theta} = w_\theta(t),$$

where r is range, θ is bearing angle, k is a constant, and $w_r(t)$ and $w_\theta(t)$ are small random forcing functions in the r and θ directions.

The observation equation is given by

$$z(t) = \begin{bmatrix} \sin^{-1}\dfrac{R_e}{r} \\ \alpha_0 - \theta \end{bmatrix},$$

where R_e = earth radius and α_0 is a constant.

Linearize these equations about $r_{\text{nom}} = R_0$ and $\theta_{\text{nom}} = w_0 t$.

6

Implementation Methods

There is a great difference between theory and practice.
Giacomo Antonelli (1806–1876)[1]

6.1 CHAPTER FOCUS

Up to this point, we have discussed what Kalman filters are and how they are supposed to behave. Their theoretical performance has been shown to be characterized by the covariance matrix of estimation uncertainty, which is computed as the solution of a matrix Riccati differential equation or difference equation.

However, soon after the Kalman filter was first implemented on computers, it was discovered that the observed mean-squared estimation errors were often much larger than the values predicted by the covariance matrix, even with simulated data. The variances of the filter estimation errors were observed to diverge from their theoretical values, and the solutions obtained for the Riccati equation were observed to have negative variances, an embarrassing example of a theoretical impossibility. The problem was eventually determined to be caused by computer roundoff, and alternative implementation methods were developed for dealing with it.

This chapter is primarily concerned with

1. how computer roundoff can degrade Kalman filter performance,
2. alternative implementation methods that are more robust against roundoff errors, and
3. the relative computational costs of these alternative implementations.

[1]In a letter to the Austrian Ambassador, as quoted by Lytton Strachey in *Eminent Victorians* [101]. Cardinal Antonelli was addressing the issue of papal infallibility, but the same might be said about the infallibility of numerical processing systems.

6.1.1 Main Points to Be Covered

The main points to be covered in this chapter are the following:

1. Computer roundoff errors can and do seriously degrade the performance of Kalman filters.
2. Solution of the matrix Riccati equation is a major cause of numerical difficulties in the conventional Kalman filter implementation, from the standpoint of computational load as well as from the standpoint of computational errors.
3. Unchecked error propagation in the solution of the Riccati equation is a major cause of degradation in filter performance.
4. Asymmetry of the covariance matrix of state estimation uncertainty is a symptom of numerical degradation and a cause of numerical instability, and measures to symmetrize the result can be beneficial.
5. Numerical solution of the Riccati equation tends to be more robust against roundoff errors if *Cholesky factors* or *modified Cholesky factors* of the covariance matrix are used as the dependent variables.
6. Numerical methods for solving the Riccati equation in terms of Cholesky factors are called *factorization methods*, and the resulting Kalman filter implementations are collectively called *square-root filtering*.
7. Information filtering is an alternative state vector implementation that improves numerical stability properties. It is especially useful for problems with very large initial estimation uncertainty.

6.1.2 Topics Not Covered

1. **Parametric Sensitivity Analysis.** The focus here is on numerically stable implementation methods for the Kalman filter. Numerical analysis of *all* errors that influence the performance of the Kalman filter would include the effects of errors in the assumed values of all model parameters, such as Q, R, H, and Φ. These errors also include truncation effects due to finite precision. The sensitivities of performance to these types of modeling errors can be modeled mathematically, but this is not done here.

2. **Smoothing Implementations.** There have been significant improvements in smoother implementation methods beyond those presented in Chapter 4. The interested reader is referred to the surveys by Meditch [201] (methods up to 1973) and McReynolds [199] (up to 1990) and to earlier results by Bierman [140] and by Watanabe and Tzafestas [234].

3. **Parallel Computer Architectures for Kalman Filtering.** The operation of the Kalman filter can be speeded up, if necessary, by performing some operations in parallel. The algorithm listings in this chapter indicate those loops that can be performed in parallel, but no serious attempt is made to define specialized algorithms to exploit concurrent processing capabilities. An overview of theoretical approaches to this problem is presented by Jover and Kailath [175].

6.2 COMPUTER ROUNDOFF

Roundoff errors are a side effect of computer arithmetic using fixed- or floating-point data words with a fixed number of bits. Computer roundoff is a fact of life for most computing environments.

EXAMPLE 6.1: Roundoff Errors In binary representation, the rational numbers are transformed into sums of powers of 2, as follows:

$$\begin{aligned}1 &= 2^0\\ 3 &= 2^0 + 2^1\\ \frac{1}{3} &= \frac{1}{4} + \frac{1}{16} + \frac{1}{64} + \frac{1}{256} + \cdots\\ &= 0_b 0101010101010101010101010\ldots,\end{aligned}$$

where the subscript "b" represents the "binary point" in binary representation (so as not to be confused with the "decimal point" in decimal representation). When 1 is divided by 3 in an IEEE/ANSI standard [107] single-precision floating-point arithmetic, the 1 and the 3 can be represented precisely, but their ratio cannot. The binary representation is limited to 24 bits of mantissa.[2] The above result is then rounded to the 24-bit approximation (starting with the leading "1"):

$$\begin{aligned}\frac{1}{3} &\approx 0_b 0101010101010101010101011\\ &= \frac{11184811}{33554432}\\ &= \frac{1}{3} - \frac{1}{100663296},\end{aligned}$$

giving an approximation error magnitude of about 10^{-8} and a relative approximation error of about 3×10^{-8}. The difference between the true value of the result and the value approximated by the processor is called *roundoff error.*

[2] The mantissa is the part of the binary representation starting with the leading nonzero bit. Because the leading significant bit is always a "1," it can be omitted and replaced by the sign bit. Even including the sign bit, there are effectively 24 bits available for representing the magnitude of the mantissa.

6.2.1 Unit Roundoff Error

Computer roundoff for floating-point arithmetic is often characterized by a single parameter $\varepsilon_{\text{roundoff}}$, called the *unit roundoff error*, and defined in different sources as the largest number such that either

$$1 + \varepsilon_{\text{roundoff}} \equiv 1 \text{ in machine precision} \tag{6.1}$$

or

$$1 + \varepsilon_{\text{roundoff}}/2 \equiv 1 \text{ in machine precision.} \tag{6.2}$$

The name "eps" in MATLAB is the parameter satisfying the second of these equations. Its value may be found by typing "eps⟨RETURN⟩" (i.e., typing "eps" without a following semicolon, followed by hitting the RETURN or ENTER key) in the MATLAB command window. Entering "-log2(eps)" should return the number of bits in the mantissa of the standard data word.

6.2.2 Effects of Roundoff on Kalman Filter Performance

Many of the roundoff problems discovered in the earlier years of Kalman filter implementation occurred on computers with much shorter wordlengths than those available in most MATLAB implementations and less accurate implementations of bit-level arithmetic than the current ANSI standards.

However, the next example (from [156]) demonstrates that roundoff can still be a problem in Kalman filter implementations in MATLAB environments and how a problem that is well-conditioned, as posed, can be made ill-conditioned by the filter implementation.

EXAMPLE 6.2 Let I_n denote the $n \times n$ identity matrix. Consider the filtering problem with measurement sensitivity matrix

$$H = \begin{bmatrix} 1 & 1 & 1 \\ 1 & 1 & 1+\delta \end{bmatrix}$$

and covariance matrices

$$P_0 = I_3 \quad \text{and} \quad R = \delta^2 I_2$$

where $\delta^2 < \varepsilon_{\text{roundoff}}$ but $\delta > \varepsilon_{\text{roundoff}}$. In this case, although H clearly has rank = 2 in machine precision, the product HP_0H^{T} *with roundoff* will equal

$$\begin{bmatrix} 3 & 3+\delta \\ 3+\delta & 3+2\delta \end{bmatrix},$$

which is singular. The result is unchanged when R is added to HP_0H^T. In this case, then, the filter observational update fails because the matrix $HP_0H^T + R$ is not invertible.

Sneak Preview of Alternative Implementations. Figure 6.1 illustrates how the standard Kalman filter and some of the alternative implementation methods perform on the variably ill-conditioned problem of Example 6.2 (implemented as MATLAB m-file shootout.m on the accompanying diskette) as the conditioning parameter $\delta \to 0$. All solution methods were implemented in the same precision (64-bit floating point) in MATLAB. The labels on the curves in this plot correspond to the names of the corresponding m-file implementations on the accompanying diskette. These are also the names of the authors of the corresponding methods, the details of which will be presented further on.

For this particular example, the accuracies of the methods labeled "Carlson" and "Bierman" appear to degrade more gracefully than the others as $\delta \to \varepsilon$, the machine precision limit. The Carlson and Bierman solutions still maintain about 9 digits ($\approx$ 30 bits) of accuracy at $\delta \approx \sqrt{\varepsilon}$, when the other methods have essentially no bits of accuracy in the computed solution.

This one example, by itself, does not prove the general superiority of the Carlson and Bierman solutions for the observational updates of the Riccati equation. The full implementation will require a compatible method for performing the temporal update, as well. (However, the observational update had been the principal source of difficulty with the conventional implementation.)

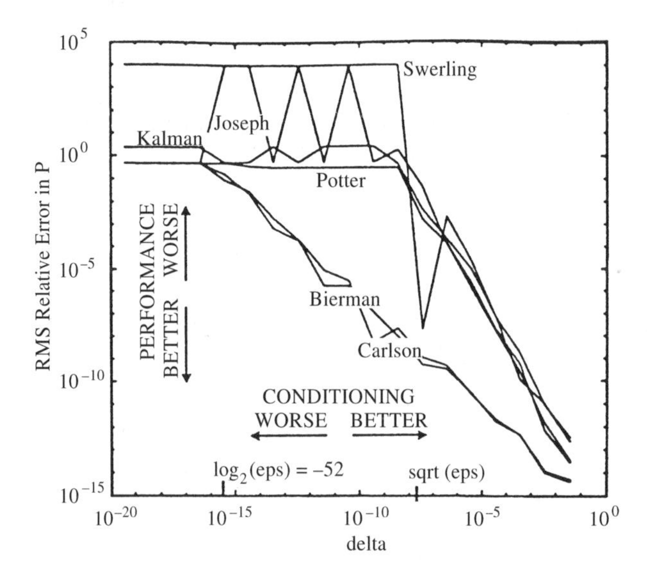

Fig. 6.1 Degradation of Riccati equation observational updates with problem conditioning.

6.2.3 Terminology of Numerical Error Analysis

We first need to define some general terms used in characterizing the influence of roundoff errors on the accuracy of the numerical solution to a given computation problem.

Robustness and Numerical Stability. These terms are used to describe qualitative properties of arithmetic problem-solving methods. *Robustness* refers to the relative insensitivity of the solution to errors of some sort. *Numerical stability* refers to robustness against roundoff errors.

Precision versus Numerical Stability. Relative roundoff errors can be reduced by using more precision (i.e., more bits in the mantissa of the data format), but the accuracy of the result is also influenced by the accuracy of the initial parameters used and the procedural details of the implementation method. Mathematically equivalent implementation methods can have very different numerical stabilities at the same precision.

Numerical Stability Comparisons. Numerical stability comparisons can be slippery. Robustness and stability of solution methods are matters of degree, but implementation methods cannot always be totally ordered according to these attributes. Some methods are considered more robust than others, but their relative robustness can also depend upon intrinsic properties of the problem being solved.

Ill-Conditioned and Well-Conditioned Problems. In the analysis of numerical problem-solving methods, the qualitative term "conditioning" is used to describe the sensitivity of the error in the output (solution) to variations in the input data (problem). This sensitivity generally depends on the input data and the solution method.

A problem is called *well-conditioned* if the solution is not "badly" sensitive to the input data and *ill-conditioned* if the sensitivity is "bad." The definition of what is bad generally depends on the uncertainties of the input data and the numerical precision being used in the implementation. One might, for example, describe a matrix A as being "ill-conditioned with respect to inversion" if A is "close" to being singular. The definition of "close" in this example could mean within the uncertainties in the values of the elements of A or within machine precision.

EXAMPLE 6.3: Condition Number of a Matrix The sensitivity of the solution x of the linear problem $Ax = b$ to uncertainties in the input data (A and b) and roundoff errors is characterized by the *condition number* of A, which can be defined as the ratio

$$\mathrm{cond}(A) = \frac{\max_x \|Ax\|/\|x\|}{\min_x \|Ax\|/\|x\|} \tag{6.3}$$

if A is nonsingular and as ∞ if A is singular. It also equals the ratio of the largest and smallest characteristic values of A. Note that the condition number will always be ≥ 1 because max $\geq$ min. As a general rule in matrix inversion, condition numbers *close to 1* are a *good omen,* and increasingly larger values are cause for increasing concern over the validity of the results.

The *relative error* in the computed solution $\hat{x}$ of the equation $Ax = b$ is defined as the ratio $\|\hat{x} - x\| / \|x\|$ of the magnitude of the error to the magnitude of x.

As a rule of thumb, the maximum relative error in the computed solution is bounded above by $c_A \varepsilon_{\text{roundoff}} \text{cond}(A)$, where $\varepsilon_{\text{roundoff}}$ is the unit roundoff error in computer arithmetic (defined in Section 6.2.1) and the positive constant c_A depends on the dimension of A. The problem of computing x, given A and b, is considered ill-conditioned if adding 1 to the condition number of A in computer arithmetic has no effect. That is, the logical expression $1 + \text{cond}(A) = \text{cond}(A)$ evaluates to true.

Consider an example with the coefficient matrix

$$A = \begin{bmatrix} 1 & L & 0 \\ 0 & 1 & L \\ 0 & 0 & 1 \end{bmatrix},$$

where

$$\begin{aligned} L &= 2^{64} \\ &= 18{,}446{,}744{,}073{,}709{,}551{,}616, \end{aligned}$$

which is such that computing L^2 would cause overflow in ANSI standard single-precision arithmetic.

The condition number of A will then be

$$\text{cond}(A) \approx 3.40282 \times 10^{38}.$$

This is about 31 orders of magnitude beyond where the rule-of-thumb test for ill-conditioning would fail in this precision ($\approx 2 \times 10^7$). One would then consider A extremely ill-conditioned for inversion (which it is) even though its determinant equals 1.

Programming note: For the general linear equation problem $Ax = b$, it is not necessary to invert A explicitly in the process of solving for x, and numerical stability is generally improved if matrix inversion is avoided. The MATLAB matrix divide (using x = A\b) does this.

6.2.4 Ill-Conditioned Kalman Filtering Problems

For Kalman filtering problems, the solution of the associated Riccati equation should equal the covariance matrix of actual estimation uncertainty, which should be optimal with respect to all quadratic loss functions. The computation of the Kalman (optimal) gain depends on it. If this does not happen, the problem is considered ill-conditioned. Factors that contribute to such ill-conditioning include the following:

1. Large uncertainties in the values of the matrix parameters Φ, Q, H, or R. Such modeling errors are not accounted for in the derivation of the Kalman filter.
2. Large ranges of the actual values of these matrix parameters, the measurements, or the state variables—all of which can result from poor choices of scaling or dimensional units.
3. Ill-conditioning of the intermediate result $R^* = HPH^{\mathrm{T}} + R$ for inversion in the Kalman gain formula.
4. Ill-conditioned theoretical solutions of the matrix Riccati equation—without considering numerical solution errors. With numerical errors, the solution may become indefinite, which can destabilize the filter estimation error.
5. Large matrix dimensions. The number of arithmetic operations grows as the square or cube of matrix dimensions, and each operation can introduce roundoff errors.
6. Poor machine precision, which makes the relative roundoff errors larger.

Some of these factors are unavoidable in many applications. Keep in mind that they do not *necessarily* make the Kalman filtering problem hopeless. However, they are cause for concern—and for considering alternative implementation methods.

6.3 EFFECTS OF ROUNDOFF ERRORS ON KALMAN FILTERS

Quantifying the Effects of Roundoff Errors on Kalman Filtering. Although there was early experimental evidence of divergence due to roundoff errors, it has been difficult to obtain general principles describing how it is related to characteristics of the implementation. There are some general (but somewhat weak) principles relating roundoff errors to characteristics of the computer on which the filter is implemented and to properties of the filter parameters. These include the results of Verhaegen and Van Dooren [232] on the numerical analysis of various implementation methods in Kalman filtering. These results provide upper bounds on the propagation of roundoff errors as functions of the norms and singular values of key matrix variables. They show that some implementations have better bounds than others. In particular, they show that certain "symmetrization" procedures are provably beneficial and that the so-called square-root filter implementations have generally better error propagation bounds than the conventional Kalman filter equations.

Let us examine the ways that roundoff errors propagate in the computation of the Kalman filter variables and how they influence the accuracy of results in the Kalman filter. Finally, we provide some examples that demonstrate common failure modes.

6.3.1 Roundoff Error Propagation in Kalman Filters

Heuristic Analysis. We begin with a heuristic look at roundoff error propagation, from the viewpoint of the data flow in the Kalman filter, to show how roundoff errors in the Riccati equation solution are not controlled by feedback like roundoff errors in the estimate. Consider the matrix-level data flow diagram of the Kalman filter that is shown in Figure 6.2. This figure shows the data flow at the level of vectors and

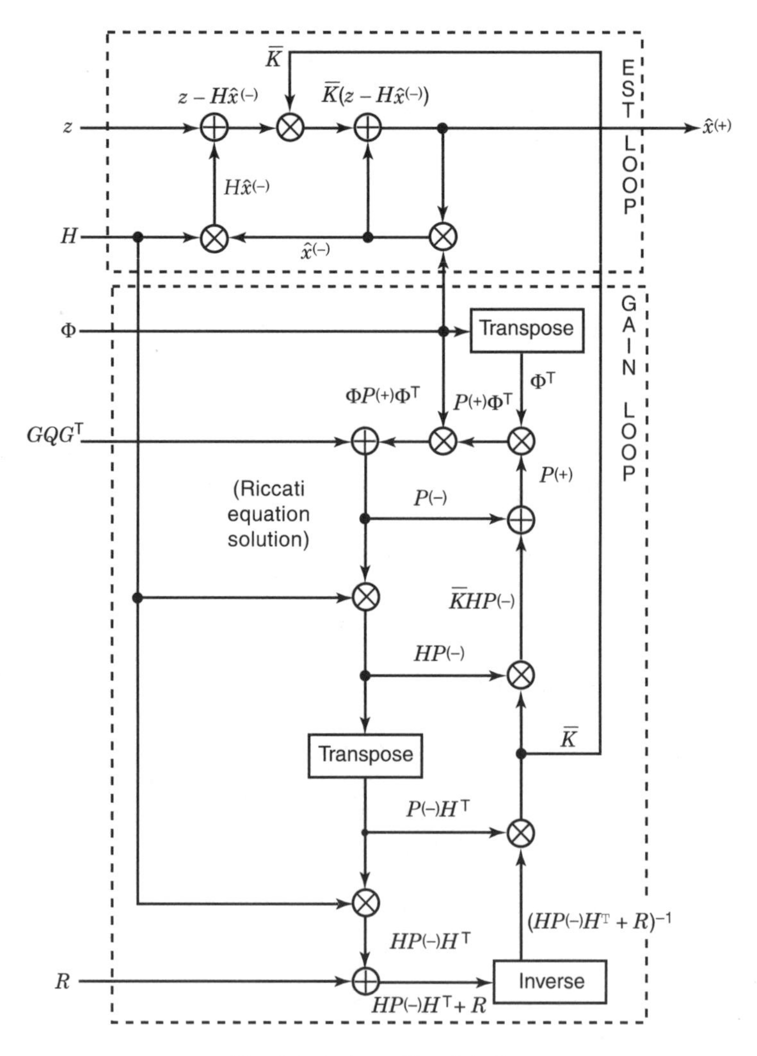

Fig. 6.2 *Kalman filter data flow.*

matrices, with operations of addition (⊕), multiplication (⊗), and inversion ($I\div$). Matrix transposition need not be considered a data operation in this context, because it can be implemented by index changes in subsequent operations. This data flow diagram is fairly representative of the straightforward Kalman filter algorithm, the way it was originally presented by Kalman, and as it might be implemented in MATLAB by a moderately conscientious programmer. That is, the diagram shows how partial results (including the Kalman gain, $\overline{K}$) might be saved and reused. Note that the internal data flow can be separated into two, semi-independent loops within the dashed boxes. The variable propagated around one loop is the state estimate. The variable propagated around the other loop is the covariance matrix of estimation uncertainty. (The diagram also shows some of the loop "shortcuts" resulting from reuse of partial results, but the basic data flows are still loops.)

Feedback in the Estimation Loop. The uppermost of these loops, labeled EST. LOOP, is essentially a feedback error correction loop with gain ($\overline{K}$) computed in the other loop (labeled GAIN LOOP). The difference between the expected value $H\hat{x}$ of the observation z (based on the current estimate $\hat{x}$ of the state vector) and the observed value is used in correcting the estimate $\hat{x}$. Errors in $\hat{x}$ will be corrected by this loop, so long as the gain is correct. This applies to errors in $\hat{x}$ introduced by roundoff as well as those due to noise and a priori estimation errors. Therefore, roundoff errors in the estimation loop are compensated by the feedback mechanism, so long as the loop gain is correct. That gain is computed in the other loop.

No Feedback in the Gain Loop. This is the loop in which the Riccati equation is solved for the covariance matrix of estimation uncertainty (P), and the Kalman gain is computed as an intermediate result. It is not stabilized by feedback, the way that the estimation loop is stabilized. There is no external reference for correcting the "estimate" of P. Consequently, there is no way of detecting and correcting the effects of roundoff errors. They propagate and accumulate unchecked. This loop also includes many more roundoff operations than the estimation loop, as evidenced by the greater number of matrix multiplies (⊗) in the loop. The computations involved in evaluating the filter gains are, therefore, more suspect as sources of roundoff error propagation in this "conventional" implementation of the Kalman filter. It has been shown by Potter [209] that the gain loop, by itself, is not unstable. However, even bounded errors in the computed value of P may momentarily destabilize the estimation loop.

EXAMPLE 6.4 An illustration of the effects that negative characteristic values of the computed covariance matrix P can have on the estimation errors is shown below:

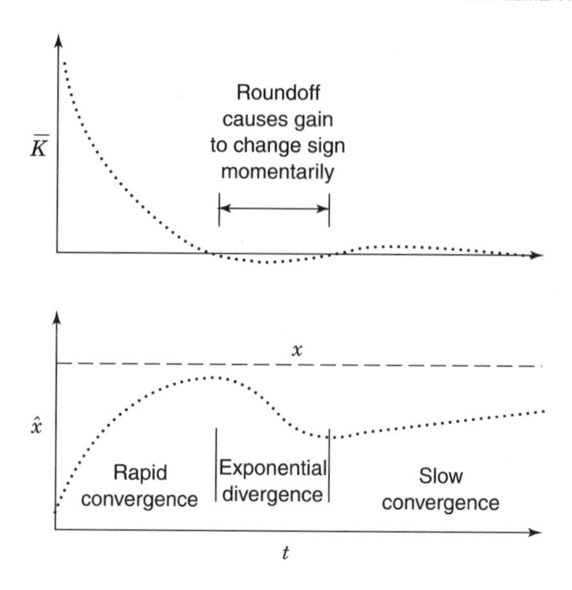

Roundoff errors can cause the computed value of P to have a negative characteristic value. The Riccati equation is stable, and the problem will eventually rectify itself. However, the effect on the actual estimation error can be a more serious problem.

Because P is a factor in the Kalman gain $\overline{K}$, a negative characteristic value of P can cause the gain in the prediction error feedback loop to have the *wrong sign*. However, in this transient condition, the estimation loop is momentarily destabilized. In this illustration, the estimate $\hat{x}$ converges toward the true value x until the gain changes sign. Then the error diverges momentarily. The gain computations may eventually recover with the correct sign, but the accumulated error due to divergence is not accounted for in the gain computations. The gain is not as big as it should be, and convergence is slower than it should be.

6.3.1.1 Numerical Analysis. Because the a priori value of P is the one used in computing the Kalman gain, it suffices to consider just the error propagation of that value. It is convenient, as well, to consider the roundoff error propagation for $x(-)$.

A first-order roundoff error propagation model is of the form

$$\delta x_{k+1}(-) = f_1(\delta x_k(-), \delta P_k(-)) + \Delta x_{k+1}, \tag{6.4}$$

$$\delta P_{k+1}(-) = f_2(\delta P_k(-)) + \Delta P_{k+1}(-), \tag{6.5}$$

where the δ term refers to the accumulated error and the Δ term refers to the added roundoff errors on each recursion step. This model ignores higher order terms in the error variables. The forms of the appropriate error propagation functions are given in Table 6.1. Error equations for the Kalman gain are also given, although the errors in $\overline{K}_k$ depend only on the errors in x and P—they are not propagated independently. These error propagation function values are from the paper by Verhaegen and Van

TABLE 6.1 First-Order Error Propagation Models

Roundoff Error in Filter Variable	Error Model (by Filter Type)	
	Conventional Implementation	Square-Root Covariance
$\delta x_{k+1}(-)$	$A_1[\delta x_k(-) + \delta P_k(-)A_2(z - Hx_k(-))] + \Delta x_{k+1}$	
$\delta \overline{K}_k$	$A_1 \delta P_k(-)$	
$\delta P_{k+1}(-)$	$A_1 \delta P_k(-) A_1^T + \Delta P_{k+1}$ $+ \Phi(\delta P_k(-) - \delta P_k^T(-))\Phi^T$ $- \Phi(\delta P_k(-) - \delta P_k^T(-))A_1^T$	$A_1 \delta P_k(-) A_1^T$ $+ \Delta P_{k+1}$

Notes: $A_1 = \Phi - \overline{K}_k H$; $A_2 = H^T[HP_k H^T + R]^{-1}$.

Dooren [232]. (Many of these results have also appeared in earlier publications.) These expressions represent the first-order error in the updated a prior variables on the $(k+1)$th temporal epoch in terms of the first-order errors in the kth temporal epoch and the errors added in the update process.

Roundoff Error Propagation. Table 6.1 compares two filter implementation types, in terms of their first-order error propagation characteristics. One implementation type is called "conventional." That corresponds to the straightforward implementation of the equations as they were originally derived in previous chapters, excluding the "Joseph-stabilized" implementation mentioned in Chapter 4. The other type is called "square root," the type of implementation presented in this chapter. A further breakdown of these implementation types will be defined in later sections.

Propagation of Antisymmetry Errors. Note the two terms in Table 6.1 involving the antisymmetry error $\delta P_k(-) - \delta P_k^T(-)$ in the covariance matrix P, which tends to confirm in theory what had been discovered in practice. Early computers had very little memory capacity, and programmers had learned to save time and memory by computing only the unique parts of symmetric matrix expressions such as $\Phi P \Phi^T$, HPH^T, $HPH^T + R$, or $(HPH^T + R)^{-1}$. To their surprise and delight, this was also found to improve error propagation. It has also been found to be beneficial in MATLAB implementations to maintain symmetry of P by evaluating the MATLAB expression P=.5*(P + P') on every cycle of the Riccati equation.

Added Roundoff Error. The roundoff error (Δ) that is added on each cycle of the Kalman filter is considered in Table 6.2. The tabulated formulas are upper bounds on these random errors.

The important points which these tables demonstrate are the following:

1. These expressions show the same first-order error propagation in the state update errors for both filter types (covariance and square-root forms). These

include terms coupling the errors in the covariance matrix into the state estimate and gain.

2. The error propagation expression for the conventional Kalman filter includes aforementioned terms proportional to the antisymmetric part of P. One must consider the effects of roundoff errors added in the computation of x, $\overline{K}$ and P as well as those propagated from the previous temporal epoch. In this case, Verhaegen and Van Dooren have obtained upper bounds on the norms of the added errors Δx, $\Delta\overline{K}$, and ΔP, as shown in Table 6.2. These upper bounds give a crude approximation of the dependence of roundoff error propagation on the characteristics of the unit roundoff error (ε) and the parameters of the Kalman filter model. Here, the bounds on the added state estimation error are similar for the two filter types, but the bounds on the added covariance error ΔP are better for the square-root filter. (The factor is something like the condition number of the matrix E.) In this case, one cannot relate the difference in performance to such factors as asymmetry of P.

The efficacy of various implementation methods for reducing the effects of roundoff errors have also been studied experimentally for some applications. The paper by Verhaegen and Van Dooren [232] includes results of this type as well as numerical analyses of other implementations (information filters and Chandrasekhar filters). Similar comparisons of square-root filters with conventional Kalman filters (and Joseph-stabilized filters) have been made by Thornton and Bierman [125].

TABLE 6.2 Upper Bounds on Added Roundoff Errors

Norm of Roundoff Errors	Upper Bounds (by Filter Type)	
	Conventional Implementation	Square-Root Covariance
$\lvert\Delta x_{k+1}(-)\rvert$	$\varepsilon_1(\lvert A_1\rvert\lvert x_k(-)\rvert + \lvert\overline{K}_k\rvert\lvert z_k\rvert)$ $+\lvert\Delta\overline{K}_k\rvert(\lvert H\rvert\lvert x_k(-)\rvert + \lvert z_k\rvert)$	$\varepsilon_4(\lvert A_1\rvert\lvert x_k(-)\rvert + \lvert\overline{K}_k\rvert\lvert z_k\rvert)$ $+\lvert\Delta\overline{K}_k\rvert(\lvert H\rvert\lvert x_k(-)\rvert + \lvert z_k\rvert)$
$\lvert\Delta\overline{K}_k\rvert$	$\varepsilon_2\kappa^2(R^\star)\lvert\overline{K}_k\rvert$	$\varepsilon_5\kappa(R^\star)[\lambda_m^{-1}(R^\star)\lvert C_{P(\overline{K}+1)}\rvert$ $+\lvert\overline{K}_k C_{R^\star}\rvert + \lvert A_3\rvert/\lambda_1(R^\star)]$
$\lvert\Delta P_{k+1}(-)\rvert$	$\varepsilon_3\kappa^2(R^\star)\lvert P_{k+1}(-)\rvert$	$\dfrac{\varepsilon_6[1+\kappa(R^\star)]\lvert P_{k+1}\rvert\lvert A_3\rvert}{\lvert C_{P(k+1)}\rvert}$

Notes: $\varepsilon_1, \ldots, \varepsilon_6$ are constant multiples of ε, the unit roundoff error; $A_1 = \Phi - \overline{K}_k H$; $A_3 = [(\overline{K}_k C_{R^\star})\lvert C_{P(k+1)}]$; $R^* = HP_k(-)H^{\mathrm{T}} + R$; $R^\star = C_{R^\star} C_{R^\star}^{\mathrm{T}}$ (triangular Cholesky decomposition); $P_{k+1}(-) = C_{P(k+1)} C_{P(k+1)}^{\mathrm{T}}$ (triangular Cholesky decomposition); $\lambda_1(R^*) \geq \lambda_2(R^*) \geq \cdots \geq \lambda_m(R^*) \geq 0$ are the characteristic values of R^*; $\kappa(R^*) = \lambda_1(R^\star)/\lambda_m(R^\star)$ is the condition number of R^*.

6.3.2 Examples of Filter Divergence

The following simple examples show how roundoff errors can cause the Kalman filter results to diverge from their expected values.

EXAMPLE 6.5: Roundoff Errors Due to Large a Priori Uncertainty If users have very little confidence in the a priori estimate for a Kalman filter, they tend to make the initial covariance of estimation uncertainty very large. This has its limitations, however.

Consider the scalar parameter estimation problem ($\Phi = I$, $Q = 0$, $\ell = n = 1$) in which the initial variance of estimation uncertainty $P_0 \gg R$, the variance of measurement uncertainty. Suppose that the measurement sensitivity $H = 1$ and that P_0 is so much greater than R that, in the floating-point machine precision, the result of adding R to P_0—with roundoff—is P_0. That is, $R < \varepsilon P_0$. In that case, the values computed in the Kalman filter calculations will be as shown in the table and plot below:

Observation Number	Expression	Value	
		Exact	Rounded
1	$P_0 H^T$	P_0	P_0
1	$HP_0 H^T$	P_0	P_0
1	$HP_0 H^T + R$	$P_0 + R$	P_0
1	$\overline{K}_1 = P_0 H^T (HP_0 H^T + R)^{-1}$	$\dfrac{P_0}{P_0 + R}$	1
1	$P_1 = P_0 - \overline{K}_1 H P_0$	$\dfrac{P_0 R}{P_0 + R}$	0
⋮	⋮	⋮	⋮
k	$\overline{K}_k = P_{k-1} H^T (HP_{k-1} H^T + R)^{-1}$	$\dfrac{P_0}{kP_0 + R}$	0
	$P_k = P_{k-1} - \overline{K}_k H P_{k-1}$	$\dfrac{P_0 R}{kP_0 + R}$	0

The rounded value of the calculated variance of estimation uncertainty is *zero* after the first measurement update, and remains zero thereafter. As a result, the calculated value of the Kalman gain is also zero after the first update. The exact (roundoff-free) value of the Kalman gain is $\approx 1/k$, where k is the observation number. After 10 observations,

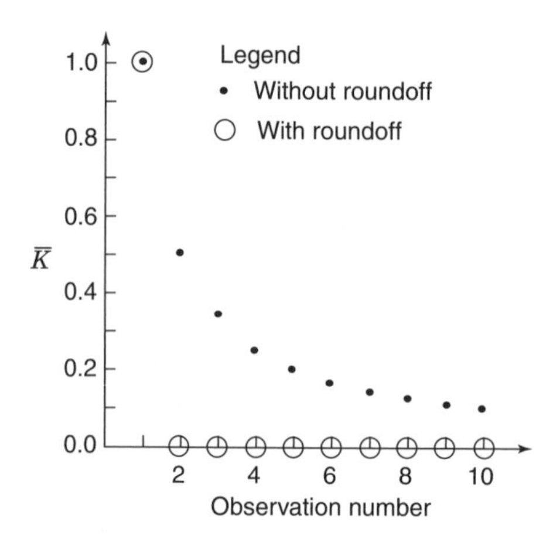

1. the *calculated* variance of estimation uncertainty is zero;
2. the *actual* variance of estimation uncertainty is $P_0R/(P_0 + R) \approx R$ (the value after the first observation and after which the computed Kalman gains were zeroed), and
3. the *theoretical* variance in the exact case (no roundoff) would have been $P_0R/(10P_0 + R) \approx \frac{1}{10}R$.

The ill-conditioning in this example is due to the misscaling between the a priori state estimation uncertainty and the measurement uncertainty.

6.4 FACTORIZATION METHODS FOR KALMAN FILTERING

Basic methods for factoring matrices are described in Sections B.6 and 6.4.2. This section describes how these methods are applied to Kalman filtering.

6.4.1 Overview of Matrix Factorization Tricks

Matrix Factoring and Decomposition. The terms *decomposition* and *factoring* (or *factorization*) are used interchangeably to describe the process of transforming a matrix or matrix expression into an equivalent product of factors.[3]

[3]The term decomposition is somewhat more general. It is also used to describe nonproduct representations, such as the *additive decomposition* of a square matrix into its symmetric and antisymmetric parts:

$$A = \tfrac{1}{2}(A + A^{\mathrm{T}}) + \tfrac{1}{2}(A - A^{\mathrm{T}}).$$

Another distinction between *decomposition* and *factorization* is made by Dongarra et al. [84], who use the term factorization to refer to an arithmetic process for performing a product decomposition of a matrix in which not all factors are preserved. The term *triangularization* is used in this book to indicate a *QR* factorization (in the sense of Dongarra et al.) involving a triangular factor that is preserved and an orthogonal factor that is not preserved.

Applications to Kalman Filtering. The more numerically stable implementations of the Kalman filter use one or more of the following techniques to solve the associated Riccati equation:

1. Factoring the *covariance matrix of state estimation uncertainty* P (the dependent variable of the Riccati equation) into Cholesky factors (see Section B.6) or into modified Cholesky factors (unit triangular and diagonal factors).
2. Factoring the *covariance matrix of measurement noise* R to reduce the computational complexity of the observational update implementation. (These methods effectively "decorrelate" the components of the measurement noise vector.)
3. Taking the symmetric *matrix square roots of elementary matrices*. A symmetric elementary matrix has the form $I - \sigma vv^{\mathrm{T}}$, where I is the $n \times n$ identity matrix, σ is a scalar, and v is an n-vector. The symmetric square root of an elementary matrix is also an elementary matrix with the same v but a different value for σ.
4. Factoring *general matrices* as products of triangular and orthogonal matrices. Two general methods are used in Kalman filtering:

   (a) *Triangularization (QR decomposition)* methods were originally developed for more numerically stable solutions of systems of linear equations. They factor a matrix into the product of an orthogonal matrix Q and a triangular matrix R. In the application to Kalman filtering, only the triangular factor is needed. We will call the QR decomposition triangularization, because Q and R already have special meanings in Kalman filtering. The two triangularization methods used in Kalman filtering are:

      i. *Givens rotations* [164] triangularize a matrix by operating on one element at a time. (A *modified Givens method* due to Gentleman [163] generates *diagonal* and *unit triangular* factors.)

      ii. *Householder transformations* triangularize a matrix by operating on one row or column at a time.

   (b) *Gram–Schmidt orthonormalization* is another general method for factoring a general matrix into a product of an orthogonal matrix and a triangular matrix. Usually, the triangular factor is not saved. In the application to Kalman filtering, only the triangular factor is saved.

5. *Rank 1 modification algorithms*. A "rank 1 modification" of a symmetric positive-definite $n \times n$ matrix M has the form $M \pm vv^{\mathrm{T}}$, where v is an n-vector (and therefore has matrix rank equal to 1). The algorithms compute a Cholesky factor of the modification $M \pm vv^{\mathrm{T}}$, given v and a Cholesky factor of M.
6. *Block matrix factorizations* of matrix expressions in the Riccati equation. The general approach uses two different factorizations to represent the two sides of

an equation, such as

$$\begin{aligned} CC^{\mathrm{T}} &= AA^{\mathrm{T}} + BB^{\mathrm{T}} \\ &= \begin{bmatrix} A & B \end{bmatrix} \begin{bmatrix} A^{\mathrm{T}} \\ B^{\mathrm{T}} \end{bmatrix}. \end{aligned}$$

The alternative Cholesky factors C and $\begin{bmatrix} A & B \end{bmatrix}$ must then be related by orthogonal transformations (triangularizations). A QR decomposition of $\begin{bmatrix} A & B \end{bmatrix}$ will yield a corresponding solution of the Riccati equation in terms of a Cholesky factor of the covariance matrix.

In the example used above, $\begin{bmatrix} A & B \end{bmatrix}$ would be called a "1×2" block partitioned matrix, because there are one row and two columns of blocks (matrices) in the partitioning. Different block dimensions are used to solve different problems:

(a) The *discrete-time temporal update* equation is solved in "square-root" form by using alternative 1×2 block-partitioned Cholesky factors.
(b) The *observational update* equation is solved in square-root form by using alternative 2×2 block-partitioned Cholesky factors and modified Cholesky factors representing the observational update equation.
(c) The *combined temporal/observational update* equations are solved in square-root form by using alternative 2×3 block-partitioned Cholesky factors of the combined temporal and observational update equations.

The different implementations of the Kalman filter based on these approaches are presented in Sections 6.5.2–6.6.2 and 6.6. They make use of the general numerical procedures presented in Sections 6.4.2–6.4.5.

6.4.2 Cholesky Decomposition Methods and Applications

Symmetric Products and Cholesky Factors. The product of a matrix C with its own transpose in the form $CC^{\mathrm{T}} = M$ is called the *symmetric product* of C, and C is called a Cholesky factor of M (Section B.6). Strictly speaking, a Cholesky factor is not a matrix square root, although the terms are often used interchangeably in the literature. (A matrix square root S of M is a solution of $M = SS = S^2$, without the transpose.)

All symmetric nonnegative definite matrices (such as covariance matrices) have Cholesky factors, but the Cholesky factor of a given symmetric nonnegative definite matrix is *not unique.* For any *orthogonal matrix* $\mathscr{T}$ (i.e., such that $\mathscr{T}\mathscr{T}^{\mathrm{T}} = I$), the product $\Gamma = C\mathscr{T}$ satisfies the equation

$$\Gamma\Gamma^{\mathrm{T}} = C\mathscr{T}\mathscr{T}^{\mathrm{T}}C^{\mathrm{T}} = CC^{\mathrm{T}} = M.$$

That is, $\Gamma = C\mathscr{T}$ is also a Cholesky factor of M. Transformations of one Cholesky factor into another are important for alternative Kalman filter implementations.

Applications to Kalman Filtering. Cholesky decomposition methods produce triangular matrix factors (Cholesky factors), and the sparseness of these factors can be exploited in the implementation of the Kalman filter equations. These methods are used for the following purposes:

1. in the decomposition of covariance matrices (P, R, and Q) for implementation of square-root filters;
2. in "decorrelating" measurement errors between components of vector-valued measurements, so that the components may be processed sequentially as independent scalar-valued measurements (Section 6.4.2.2);
3. as part of a numerically stable method for computing matrix expressions containing the factor $(HPH^{\mathrm{T}} + R)^{-1}$ in the conventional form of the Kalman filter (this matrix inversion can be obviated by the decorrelation methods, however); and
4. in Monte Carlo analysis of Kalman filters by simulation, in which Cholesky factors are used for generating independent random sequences of vectors with pre-specified means and covariance matrices (see Section 3.4.8).

6.4.2.1 Cholesky Decomposition Algorithms

Triangular Matrices. Recall that the *main diagonal* of an $n \times m$ matrix C is the set of elements $\{C_{ii} \mid 1 \leq i \leq \min(m, n)\}$ and that C is called *triangular* if the elements on one side of its main diagonal are zero. The matrix is called *upper triangular* if its nonzero elements are on and above its main diagonal and *lower triangular* if they are on or below the main diagonal.

A Cholesky decomposition algorithm is a procedure for calculating the elements of a triangular Cholesky factor of a symmetric, nonnegative definite matrix. It solves the Cholesky decomposition equation $P = CC^{\mathrm{T}}$ for a triangular matrix C, given the matrix P, as illustrated in the following example.

EXAMPLE 6.6 Consider the 3×3 example for finding a *lower triangular* Cholesky factor $P = CC^{\mathrm{T}}$ for symmetric P:

$$\begin{bmatrix} p_{11} & p_{21} & p_{31} \\ p_{21} & p_{22} & p_{32} \\ p_{31} & p_{32} & p_{33} \end{bmatrix} = \begin{bmatrix} c_{11} & 0 & 0 \\ c_{21} & c_{22} & 0 \\ c_{31} & c_{32} & c_{33} \end{bmatrix} \begin{bmatrix} c_{11} & 0 & 0 \\ c_{21} & c_{22} & 0 \\ c_{31} & c_{32} & c_{33} \end{bmatrix}^{\mathrm{T}}$$

$$= \begin{bmatrix} c_{11}^2 & c_{11}c_{21} & c_{11}c_{31} \\ c_{11}c_{21} & c_{21}^2 + c_{22}^2 & c_{21}c_{31} + c_{22}c_{32} \\ c_{11}c_{31} & c_{21}c_{31} + c_{22}c_{32} & c_{31}^2 + c_{32}^2 + c_{33}^2 \end{bmatrix}.$$

The corresponding matrix elements of the left- and right-hand sides of the last matrix equation can be equated as nine scalar equations. However, due to symmetry, only six of these are independent. The six scalar equations can be solved in sequence, making use of previous results. The following solution order steps down the rows and across the columns:

Six Independent Scalar Equations	Solutions Using Prior Results
$p_{11} = c_{11}^2$	$c_{11} = \sqrt{p_{11}}$
$p_{21} = c_{11}c_{21}$	$c_{21} = p_{21}/c_{11}$
$p_{22} = c_{21}^2 + c_{22}^2$	$c_{22} = \sqrt{p_{22} - c_{21}^2}$
$p_{31} = c_{11}c_{31}$	$c_{31} = p_{31}/c_{11}$
$p_{32} = c_{21}c_{31} + c_{22}c_{32}$	$c_{32} = (p_{32} - c_{21}c_{31})/c_{22}$
$p_{33} = c_{31}^2 + c_{32}^2 + c_{33}^2$	$c_{33} = \sqrt{p_{33} - c_{31}^2 - c_{32}^2}$

A solution can also be obtained by stepping across the rows and then down the rows, in the order c_{11}, c_{21}, c_{31}, c_{22}, c_{32}, c_{33}.

The general solutions can be put in the form of algorithms looping through the rows and columns of C and using prior results. The example above suggests two algorithmic solutions, one looping in row–column order and one looping in column–row order. There is also the choice of whether the solution C should be lower triangular or upper triangular.

Algorithmic solutions are given in Table 6.3. The one on the left can be implemented as C = chol(M)', using the built-in MATLAB function chol. The one in the right column is implemented in the m-file chol2.m.

Programming note: MATLAB automatically assigns the value zero to all the unassigned matrix locations. This would not be necessary if subsequent processes treat the resulting Cholesky factor matrix C as triangular and do not bother to add or multiply the zero elements.

6.4.2.2 Modified Cholesky (UD) Decomposition Algorithms

Unit Triangular Matrices. An upper triangular matrix U is called *unit upper triangular* if its diagonal elements are all 1 (unity). Similarly, a lower triangular matrix L is called *unit lower triangular* if all of its diagonal elements are unity.

UD Decomposition algorithm. The *modified* Cholesky decomposition of a symmetric positive-definite matrix M is a decomposition into products $M = UDU^{\mathrm{T}}$ such that U is unit upper triangular and D is diagonal. It is also called *UD decomposition.*

TABLE 6.3 Cholesky Decomposition Algorithms

Given an $m \times m$ symmetric positive definite matrix M, a triangular matrix C such that $M = CC^{\mathrm{T}}$ is computed.

Lower Triangular Result	Upper Triangular Result

`for j=1:m,`
`  for i=1:j,`
`    sigma=M(i,j);`
`    for k=1:j-1,`
`      sigma=sigma-C(i,k)*C(j,k);`
`    end;`
`    if i==j`
`      C(i,j)=sqrt(sigma);`
`    else`
`      C(i,j)=sigma/C(j,j)`
`    end;`
`  end;`
`end;`
	`for j=m:-1:1,`
	`  for i=j:-1:1,`
	`    sigma=M(i,j);`
	`    for k=j+1:m,`
	`      sigma=sigma-C(i,k)*C(j,k);`
	`    end;`
	`    if i==j`
	`      C(i,j)=sqrt(sigma);`
	`    else`
	`      C(i,j)=sigma/C(j,j)`
	`    end;`
	`  end;`
	`end;`

Computational complexity: $\frac{1}{6}m(m-1)(m+4)$ flops $+ m\sqrt{\ }$.

A procedure for implementing *UD* decomposition is presented in Table 6.4. This algorithm is implemented in the m-file modchol.m. It takes M as input and returns U and D as output. The decomposition can also be implemented in place, overwriting the input array containing M with D (on the diagonal of the array containing M) and U (in the strictly upper triangular part of the array containing M). This algorithm is only slightly different from the upper triangular Cholesky decomposition algorithm presented in Table 6.3. The big difference is that the modified Cholesky decomposition does not require taking square roots.

6.4.2.3 Decorrelating Measurement Noise. The decomposition methods developed for factoring the covariance matrix of estimation uncertainty may also be applied to the covariance matrix of measurement uncertainty, R. This operation redefines the measurement vector (via a linear transform of its components) such that its measurement errors are uncorrelated from component to component. That is, the new covariance matrix of measurement uncertainty is a *diagonal* matrix. In that case, the components of the redefined measurement vector can be processed serially as uncorrelated scalar measurements. The reduction in the computational complexity[4] of the Kalman filter from this approach will be covered in Section 6.6.1.

Suppose, for example, that

$$z = Hx + v \tag{6.6}$$

[4]The methodology used for determining the computational complexities of algorithms in this chapter is presented in Section 6.4.2.6.

TABLE 6.4 *UD* Decomposition Algorithm

Given M, a symmetric, positive-definite $m \times m$ matrix, U and D, modified Cholesky factors of M, are computed, such that U is a unit upper triangular matrix, D is a diagonal matrix, and $M = UDU^T$.

```
for j=m:-1:1,
   for i=j:-1:1,
      sigma=M(i,j);
      for k=j+1:m,
         sigma=sigma-U(i,k)*D(k,k)*U(j,k);
      end;
      if i==j
         D(j,j)=sigma;
         U(j,j)=1;
      else
         U(i,j)=sigma/D(j,j);
      end;
   end;
end;
```

Computational complexity: $\frac{1}{6}m(m-1)(m+4)$ flops.

is an observation with measurement sensitivity matrix H and noise v that is correlated from component to component of v. That is, the covariance matrix

$$\mathrm{E}\langle vv^{\mathrm{T}}\rangle = R \tag{6.7}$$

is not a diagonal matrix. Then the scalar components of z cannot be processed serially as scalar observations with statistically independent measurement errors.

However, R can always be factored in the form

$$R = UDU^{\mathrm{T}}, \tag{6.8}$$

where D is a diagonal matrix and U is an upper triangular matrix. Unit triangular matrices have some useful properties:

- The determinant of a unit triangular matrix is 1. Unit triangular matrices are, therefore, always *nonsingular*. In particular, they always have a matrix inverse.
- The inverse of a unit triangular matrix is a unit triangular matrix. The inverse of a unit upper triangular matrix is unit upper triangular, and the inverse of a unit lower triangular matrix is a unit lower triangular matrix.

It is not necessary to compute U^{-1} to perform measurement decorrelation, but it is useful for pedagogical purposes to use U^{-1} to redefine the measurement as

$$\acute{z} = U^{-1}z \tag{6.9}$$

$$= U^{-1}(Hx + v) \tag{6.10}$$

$$= (U^{-1}H)x + (U^{-1}v) \tag{6.11}$$

$$= \acute{H}\mathrm{x} + \acute{v}. \tag{6.12}$$

That is, this "new" measurement $\acute{z}$ has measurement sensitivity matrix $\acute{H} = U^{-1}H$ and observation error $\acute{v} = U^{-1}v$. The covariance matrix R' of the observation error $\acute{v}$ will be the expected value

$$R' = \mathrm{E}\langle \acute{v}\acute{v}^{\mathrm{T}} \rangle \tag{6.13}$$

$$= \mathrm{E}\langle (U^{-1}v)(U^{-1}v)^{\mathrm{T}} \rangle \tag{6.14}$$

$$= \mathrm{E}\langle U^{-1}vv^{\mathrm{T}}U^{T-1} \rangle \tag{6.15}$$

$$= U^{-1}\mathrm{E}\langle vv^{\mathrm{T}} \rangle U^{\mathrm{T}-1} \tag{6.16}$$

$$= U^{-1}RU^{\mathrm{T}-1} \tag{6.17}$$

$$= U^{-1}(UDU^{\mathrm{T}})U^{\mathrm{T}-1} \tag{6.18}$$

$$= D. \tag{6.19}$$

That is, this redefined measurement has uncorrelated components of its measurement errors, which is what was needed for serializing the processing of the components of the new vector-valued measurement.

In order to decorrelate the measurement errors, one must solve the unit upper triangular system of equations

$$U\acute{z} = z \tag{6.20}$$

$$U\acute{H} = H \tag{6.21}$$

for $\acute{z}$ and $\acute{H}$, given z, H, and U. As noted previously, *it is not necessary to invert U to* solve for $\acute{z}$ and $\acute{H}$.

Solving Unit Triangular Systems. It was mentioned above that it is not necessary to invert U to decorrelate measurement errors. In fact, it is only necessary to solve equations of the form $UX = Y$, where U is a unit triangular matrix and X and Y have conformable dimensions. The objective is to solve for X, given Y. It can be done by what is called "back substitution." The algorithms listed in Table 6.5 perform the solutions by back substitution. The one on the right overwrites Y with $U^{-1}Y$. This feature is useful when several procedures are composed into one special-purpose procedure, such as the decorrelation of vector-valued measurements.

Specialization for Measurement Decorrelation. A complete procedure for measurement decorrelation is listed in Table 6.6. It performs the UD decomposition and upper triangular system solution in place (overwriting H with $U^{-1}H$ and z with $U^{-1}z$), after decomposing R as $R = UDU^{\mathrm{T}}$ in place (overwriting the diagonal of R with $\acute{R} = D$ and overwriting the strictly upper triangular part of R with the strictly upper triangular part of U^{-1}).

6.4.2.4 Symmetric Positive-Definite System Solution. Cholesky decomposition provides an efficient and numerically stable method for solving equations of

TABLE 6.5 Unit Upper Triangular System Solution

Input: U, $m \times m$ unit upper triangular matrix; Y, $m \times p$ matrix Output: $X := U^{-1}Y$	Input: U, $m \times m$ unit upper triangular matrix; Y, $m \times p$ matrix Output: $Y := U^{-1}Y$

`for j=1:p,`
`  for i=m:-1:1,`
`    X(i,j)=Y(i,j);`
`    for k=i+1:m,`
`      X(i,j)=X(i,j)-U(i,k)*X(k,j);`
`    end;`
`  end;`
`end;`
	`for j=1:p,`
	`  for i=m:-1:1,`
	`    for k=i+1:m,`
	`      Y(i,j)=Y(i,j)-U(i,k)*Y(k,j);`
	`    end;`
	`  end;`
	`end;`

Computational complexity: $pm(m-1)/2$ flops.

the form $AX = Y$ when A is a symmetric, positive-definite matrix. The modified Cholesky decomposition is even better, because it avoids taking scalar square roots. It is the recommended method for forming the term $[HPH^{T} + R]^{-1}H$ in the conventional Kalman filter without explicitly inverting a matrix. That is, if one decomposes $HPH^{T} + R$ as UDU^{T}, then

$$[UDU^{T}][HPH^{T} + R]^{-1}H = H. \tag{6.22}$$

It then suffices to solve

$$UDU^{T}X = H \tag{6.23}$$

TABLE 6.6 Measurement Decorrelation Procedure

The vector-valued measurement $z = Hx + v$, with correlated components of the measurement error $E(vv^{T}) = R$, is transformed to the measurement $\acute{z} = \acute{H}x + \acute{v}$ with uncorrelated components of the measurement error $\acute{v}$ [$E(\acute{v}\acute{v}$T$) =$ D, a diagonal matrix], by overwriting H with $\acute{H} = U^{-1}H$ and z with $\acute{z} = U^{-1}z$, after decomposing R to UDU^{T}, overwriting the diagonal of R with D.

Symbol	Definition
R	Input: $\ell \times \ell$ covariance matrix of measurement uncertainty Output: D (on diagonal), U (above diagonal)
H	Input: $\ell \times n$ measurement sensitivity matrix Output: overwritten with $\acute{H} = U^{-1}H$
z	Input: measurement ℓ-vector Output: overwritten with $\acute{z} = U^{-1}z$

Procedure:

1. Perform UD decomposition of R in place.
2. Solve $U\acute{z} = z$ and $U\acute{H} = H$ in place.

Computational complexity: $\frac{1}{6}\ell(\ell - 1)(\ell + 4) + \frac{1}{2}\ell(\ell - 1)(n + 1)$ flops.

for X. This can be done by solving the three problems

$$UX_{[1]} = H \text{ for } X_{[1]}, \tag{6.24}$$

$$DX_{[2]} = X_{[1]} \text{ for } X_{[2]}, \tag{6.25}$$

$$U^{T}X = X_{[2]} \text{ for } X. \tag{6.26}$$

The first of these is a unit upper triangular system, which was solved in the previous subsection. The second is a system of independent scalar equations, which has a simple solution. The last is a unit lower triangular system, which can be solved by "forward substitution"—a simple modification of back substitution. The computational complexity of this method is m^2p, where m is the row and column dimension of A and p is the column dimension of X and Y.

6.4.2.5 Transforming Covariance Matrices to Information Matrices.

The information matrix is the inverse of the covariance matrix—and vice versa. Although matrix inversion is generally to be avoided if possible, it is just not possible to avoid it forever. This is one of those problems that require it.

The inversion is not possible unless one of the matrices (either P or Y) is positive definite, in which case both will be positive-definite and they will have the same condition number. If they are sufficiently well conditioned, they can be inverted in place by UD decomposition, followed by inversion and recomposition in place. The in-place UD decomposition procedure is listed in Table 6.4. A procedure for inverting the result in place is shown in Table 6.7. A matrix inversion procedure using these two is outlined in Table 6.8. It should be used with caution, however.

6.4.2.6 Computational Complexities.

Using the general methods outlined in [85] and [89], one can derive the complexity formulas shown in Table 6.9 for methods using Cholesky factors.

TABLE 6.7 Unit Upper Triangular Matrix Inversion

Input/output: U, an $m \times m$ unit upper triangular matrix (U is overwritten with U^{-1})

```
for i=m:-1:1,
   for j=m:-1:i+1,
      U(i,j)=-U(i,j);
      for k=i+1:j-1,
         U(i,j)=U(i,j)-U(i,k)*U(k,j);
      end;
   end;
end;
```

Computational complexity: $m(m-1)(m-2)/6$ flops.

TABLE 6.8 Symmetric Positive-Definite Matrix Inversion Procedure[a]

Symbol	Description
M	Input: $m \times m$ symmetric positive definite matrix Output: M is overwritten with M^{-1}

Procedure:
1. Perform UD decomposition of M in place.
2. Invert U in place (in the M-array).
3. Invert D in place: for $i = 1 : m$, $M(i, i) = 1/M(i, i)$; end;
4. Recompose $M^{-1} = (U^{-T}D^{-1})U^{-1}$ in place:

```
for i=m:-1:1,
   for j=1:i-1,
      M(i,j)=M(j,i)*M(j,j);
   end;
   for j=m:-1:i,
      if i<j
         M(i,j)=M(i,j)*M(i,i);
      end;
      for k=1:i-1,
         M(i,j)=M(i,j)+M(k,j)*M(i,k);
      end;
      M(j,i)=M(i,j);
   end;
end;
```

Computational complexity: $m^3 + \frac{1}{2}m^2 + \frac{1}{2}m$ flops.

[a]Inverts a symmetric positive-definite matrix in place.

6.4.3 Kalman Implementation with Decorrelation

It was pointed out by Kaminski [115] that the computational efficiency of the conventional Kalman observational update implementation can be improved by processing the components of vector-valued observations sequentially using the error decorrelation algorithm in Table 6.6, if necessary. The computational savings with the measurement decorrelation approach can be evaluated by comparing the rough operations counts of the two approaches using the operations counts for the sequential approach given in Table 6.10. One must multiply by ℓ, the number of operations required for the implementation of the scalar observational update equations, and add the number of operations required for performing the decorrelation.

The computational advantage of the decorrelation approach is

$$\tfrac{1}{3}\ell^3 - \tfrac{1}{2}\ell^2 + \tfrac{7}{6}\ell - \ell n + 2\ell^2 n + \ell n^2 \text{ flops.}$$

That is, it requires that many fewer flops to decorrelate vector-valued measurements and process the components serially.

6.4.4 Symmetric Square Roots of Elementary Matrices

Historical Background. Square-root filtering was introduced by James Potter [5] to overcome an ill-conditioned Kalman filtering problem for the Apollo moon

TABLE 6.9 Computational Complexity Formulas

Cholesky decomposition of an $m \times m$ matrix:

$$\begin{aligned}\mathscr{C}_{\text{Cholesky}} &= \sum_{j=1}^{m}\left[m - j + \sum_{i=j}^{m}(m-j)\right] \\ &= \tfrac{1}{3}m^3 + \tfrac{1}{2}m^2 - \tfrac{5}{6}m\end{aligned}$$

UD decomposition of an $m \times m$ matrix:

$$\begin{aligned}\mathscr{C}_{UD} &= \sum_{j=1}^{m}\left[m - j + \sum_{i=j}^{m}2(m-j)\right] \\ &= \tfrac{2}{3}m^3 + \tfrac{1}{2}m^2 - \tfrac{7}{6}m\end{aligned}$$

Inversion of an $m \times m$ unit triangular matrix:

$$\begin{aligned}\mathscr{C}_{\text{UTINV}} &= \sum_{i=1}^{m-1}\sum_{j=i+1}^{m}(j-i-1) \\ &= \tfrac{1}{6}m^3 - \tfrac{1}{2}m^2 + \tfrac{1}{3}m\end{aligned}$$

Measurement decorrelation ($\ell \times n$ H-matrix):

$$\begin{aligned}\mathscr{C}_{\text{DeCorr}} &= \mathscr{C}_{UD} + \sum_{i=1}^{\ell-1}\sum_{k=i+1}^{\ell}(n+1) \\ &= \tfrac{2}{3}\ell^3 + \ell^2 - \tfrac{5}{3}\ell + \tfrac{1}{2}\ell^2 n - \tfrac{1}{2}\ell n\end{aligned}$$

Inversion of an $m \times m$ covariance matrix:

$$\begin{aligned}\mathscr{C}_{\text{COVINV}} &= \mathscr{C}_{UD} + \mathscr{C}_{\text{UTINV}} + m + \sum_{i=1}^{m}[i(i-1)(m-i+1)] \\ &= m^3 + \tfrac{1}{2}m^2 + \tfrac{1}{2}m\end{aligned}$$

project. The mission used an onboard sextant to measure the angles between stars and the limb of the earth or moon. These are scalar measurements, and Potter was able to factor the resulting measurement update equations of the Riccati equation into Cholesky factors of the covariance matrix and an *elementary matrix* of the type used by Householder [172]. Potter was able to factor the elementary matrix into a product of its square roots using the approach presented here. Potter's application of this result to Kalman filtering is presented in Section 6.6.1.3.

Elementary Matrices. An elementary matrix is a matrix of the form $I - s\mathbf{v}\mathbf{w}^{\mathrm{T}}$, where I is an identity matrix, s is a scalar, and $\mathbf{v}$, $\mathbf{w}$ are column vectors of the same row dimension as I. Elementary matrices have the property that their products are also elementary matrices. Their squares are also elementary matrices, with the same vector values ($\mathbf{v}$, $\mathbf{w}$) but with different scalar values (s).

TABLE 6.10 Operations for Sequential Processing of Measurements

Operation	Flops
$H \times P(-)$	n^2
$H \times [HP(-)]^{\mathrm{T}} + R$	n
$\{H[HP(-)]^{\mathrm{T}} + R\}^{-1}$	1
$\{H[HP(-)]^{\mathrm{T}} + R\}^{-1} \times [HP(-)]$	n
$P(-) - [HP(-)] \times \{H[HP(-)]^{\mathrm{T}} + R\}^{-1}[HP(-)]$	$\frac{1}{2}n^2 + \frac{1}{2}n$
Total (per component) $\times \ell$ components	$(\frac{1}{2}n^2 + \frac{5}{2}n + 1) \times \ell$
+ decorrelation complexity	$\frac{2}{3}\ell^3 + \ell^2 - \frac{5}{3}\ell + \frac{1}{2}\ell^2 n - \frac{1}{2}\ell n$
Total	$\frac{2}{3}\ell^3 + \ell^2 - \frac{2}{3}\ell + \frac{1}{2}\ell^2 n + 2\ell + n\frac{1}{2}\ell n^2$

Symmetric Elementary Matrices. An elementary matrix is symmetric if $v = w$. The squares of such matrices have the same format:

$$(I - \sigma \mathbf{v}\mathbf{v}^{\mathrm{T}})^2 = (I - \sigma \mathbf{v}\mathbf{v}^{\mathrm{T}})(I - \sigma \mathbf{v}\mathbf{v}^{\mathrm{T}}) \tag{6.27}$$

$$= I - 2\sigma \mathbf{v}\mathbf{v}^{\mathrm{T}} + \sigma^2 |\mathbf{v}|^2 \mathbf{v}\mathbf{v}^{\mathrm{T}} \tag{6.28}$$

$$= I - (2\sigma - \sigma^2 |\mathbf{v}|^2)\mathbf{v}\mathbf{v}^{\mathrm{T}} \tag{6.29}$$

$$= I - s\mathbf{v}\mathbf{v}^{\mathrm{T}} \tag{6.30}$$

$$s = (2\sigma - \sigma^2 |\mathbf{v}|^2). \tag{6.31}$$

Symmetric Square Root of a Symmetric Elementary Matrix. One can also invert the last equation above and take the square root of the symmetric elementary matrix $(I - s\mathbf{v}\mathbf{v}^{\mathrm{T}})$. This is done by solving the scalar quadratic equation

$$s = 2\sigma - \sigma^2 |\mathbf{v}|^2, \tag{6.32}$$

$$\sigma^2 |\mathbf{v}|^2 - 2\sigma + s = 0 \tag{6.33}$$

to obtain the solution

$$(I - s\mathbf{v}\mathbf{v}^{\mathrm{T}})^{1/2} = I - \sigma \mathbf{v}\mathbf{v}^{\mathrm{T}}, \tag{6.34}$$

$$\sigma = \frac{1 + \sqrt{1 - s|\mathbf{v}|^2}}{|\mathbf{v}|^2}. \tag{6.35}$$

In order that this square root be a real matrix, it is necessary that the radicand

$$1 - s|\mathbf{v}|^2 \geq 0. \tag{6.36}$$

6.4.5 Triangularization Methods

Triangularization Methods for Least-Squares Problems. These techniques were originally developed for solving least-squares problems. The overdetermined system

$$Ax = b$$

can be solved efficiently and relatively accurately by finding an orthogonal matrix T such that the product $B = TA$ is a triangular matrix. In that case, the solution to the triangular system of equations

$$Bx = Tb$$

can be solved by backward substitution.

Triangularization (QR Decomposition) of A. It is a theorem of linear algebra that any general matrix A can be represented as a product[5]

$$A = C_{k+1}(-)T \tag{6.37}$$

of a triangular matrix $C_{k+1}(-)$ and an orthogonal matrix T. This type of decomposition is called QR decomposition or triangularization. By means of this triangularization, the symmetric matrix product factorization

$$P_{k+1}(-) = AA^{\mathrm{T}} \tag{6.38}$$

$$= [C_{k+1}(-)T][C_{k+1}(-)T]^{\mathrm{T}} \tag{6.39}$$

$$= C_{k+1}(-)TT^{\mathrm{T}}C_{k+1}^{\mathrm{T}}(-) \tag{6.40}$$

$$= C_{k+1}(-)(TT^{\mathrm{T}})C_{k+1}^{\mathrm{T}}(-) \tag{6.41}$$

$$= C_{k+1}(-)C_{k+1}^{\mathrm{T}}(-) \tag{6.42}$$

[5]This is the so-called "QR" decomposition in disguise. It is customarily represented as $A = QR$ (whence the name), where Q is orthogonal and R is triangular. However, as mentioned earlier, we have already committed the symbols Q and R to play other roles in this book. In this instance of the QR decomposition, it has the transposed form $A^{\mathrm{T}} = T^{\mathrm{T}}C_{k+1}^{\mathrm{T}}(-)$, where T^{T} is the stand-in for the original Q (the orthogonal factor) and $C_{k+1}^{\mathrm{T}}(-)$ is the stand-in for the original R (the triangular factor).

also defines a triangular Cholesky decomposition of $C_{k+1}(-)$ of $P_{k+1}(-)$. This is the basis for performing temporal updates of Cholesky factors of P.

Uses of Triangularization in Kalman Filtering. Matrix triangularization methods were originally developed for solving least-squares problems. They are used in Kalman filtering for

- temporal updates of Cholesky factors of the covariance matrix of estimation uncertainty, as described above;
- observational updates of Cholesky factors of the estimation information matrix, as described in Section 6.6.3.5; and
- combined updates (observational and temporal) of Cholesky factors of the covariance matrix of estimation uncertainty, as described in Section 6.6.2.

A modified Givens rotation due to W. Morven Gentleman [163] is used for the temporal updating of modified Cholesky factors of the covariance matrix.

In these applications, as in most least-squares applications, the orthogonal matrix factor is unimportant. The resulting triangular factor is the intended result, and numerically stable methods have been developed for computing it.

Triangularization Algorithms. Two of the more stable methods for matrix triangularization are presented in the following subsections. These methods are based on orthogonal transformations (matrices) that, when applied to (multiplied by) general matrices, reduce them to triangular form. Both were published in the same year (1958). Both define the requisite transformation as a product of "elementary" orthogonal transformations:

$$T = T_1 T_2 T_3 \cdots T_m. \tag{6.43}$$

These elementary transformations are either *Givens rotations* or *Householder reflections.* In each case, triangularization is achieved by zeroing of the nonzero elements on one side of the main diagonal. Givens rotations zero these elements one by one. Householder reflections zero entire subrows of elements (i.e., the part of a row left of the triangularization diagonal) on each application. The order in which such transformations may be applied must be constrained so that they do not "unzero" previously zeroed elements.

6.4.5.1 Triangularization by Givens Rotations. This method for triangularization, due to Wallace Givens [164], uses a *plane rotation matrix* $T_{ij}(\theta)$ of the following form:

$$T_{ij}(\theta) = \begin{array}{c} \\ \\ \\ j \\ \\ \\ \\ i \\ \\ \\ \\ \end{array} \begin{pmatrix} 1 & \cdots & 0 & 0 & 0 & \cdots & 0 & 0 & 0 & \cdots & 0 \\ \vdots & \ddots & \vdots & \vdots & \vdots & & \vdots & \vdots & \vdots & & \vdots \\ 0 & \cdots & 1 & 0 & 0 & \cdots & 0 & 0 & 0 & \cdots & 0 \\ 0 & \cdots & 0 & \cos(\theta) & 0 & \cdots & 0 & \sin(\theta) & 0 & \cdots & 0 \\ 0 & \cdots & 0 & 0 & 1 & \cdots & 0 & 0 & 0 & \cdots & 0 \\ \vdots & & \vdots & \vdots & \vdots & \ddots & \vdots & \vdots & \vdots & & \vdots \\ 0 & \cdots & 0 & 0 & 0 & \cdots & 1 & 0 & 0 & \cdots & 0 \\ 0 & \cdots & 0 & -\sin(\theta) & 0 & \cdots & 0 & \cos(\theta) & 0 & \cdots & 0 \\ 0 & \cdots & 0 & 0 & 0 & \cdots & 0 & 0 & 1 & \cdots & 0 \\ \vdots & & \vdots & \vdots & \vdots & & \vdots & \vdots & \vdots & \ddots & \vdots \\ 0 & \cdots & 0 & 0 & 0 & \cdots & 0 & 0 & 0 & \cdots & 1 \end{pmatrix}. \tag{6.44}$$

(Column j is the column containing $\cos(\theta)$ in row j; column i is the column containing $\cos(\theta)$ in row i.)

It is also called a *Givens rotation matrix* or *Givens transformation matrix*. Except for the ith and jth rows and columns, the plane rotation matrix has the appearance of an identity matrix. When it is multiplied on the right-hand side of another matrix, it affects only the ith and jth columns of the matrix product. It rotates the ith and jth elements of a row or column vector, as shown in Figure 6.3. It can be used to rotate one of the components all the way to zero, which is how it is used in triangularization.

Triangularization of a matrix A by Givens rotations is achieved by successive multiplications of A on one side by Givens rotation matrices, as illustrated by the following example.

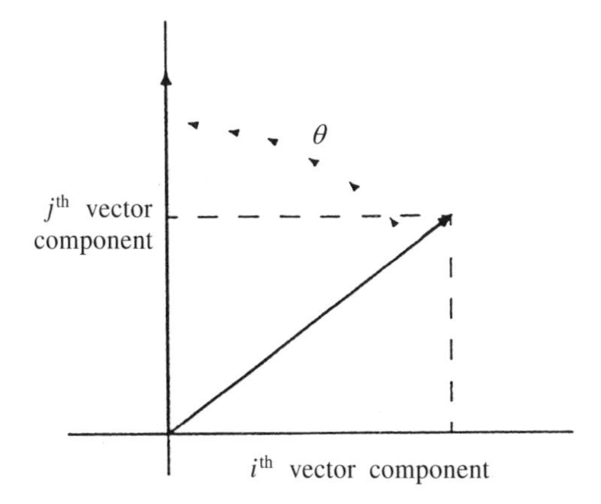

Fig. 6.3 *Component transformations by plane rotation.*

EXAMPLE 6.7 Consider the problem of upper triangularizing the 2×3 symbolic matrix

$$A = \begin{bmatrix} a_{11} & a_{12} & a_{13} \\ a_{21} & a_{22} & a_{23} \end{bmatrix} \tag{6.45}$$

by multiplying with Givens rotation matrices on the right. The first product

$$\begin{aligned} \acute{A}(\theta) &= A T_{23}(\theta) \\ &= \begin{bmatrix} a_{11} & a_{12} & a_{13} \\ a_{21} & a_{22} & a_{23} \end{bmatrix} \begin{bmatrix} 1 & 0 & 0 \\ 0 & \cos(\theta) & \sin(\theta) \\ 0 & -\sin(\theta) & \cos(\theta) \end{bmatrix} \\ &= \begin{bmatrix} a_{11} & a_{12}\cos(\theta) - a_{13}\sin(\theta) & a_{12}\sin(\theta) + a_{13}\cos(\theta) \\ a_{21} & \boxed{a_{22}\cos(\theta) - a_{23}\sin(\theta)} & a_{22}\sin(\theta) + a_{23}\cos(\theta) \end{bmatrix}. \end{aligned}$$

The framed element in the product will be zero if $a_{22}^2 + a_{23}^2 = 0$, and if $a_{22}^2 + a_{23}^2 > 0$, the values

$$\cos(\theta) = \frac{a_{23}}{\sqrt{a_{22}^2 + a_{23}^2}}, \qquad \sin(\theta) = \frac{a_{22}}{\sqrt{a_{22}^2 + a_{23}^2}}.$$

will force it to zero. The resulting matrix $\acute{A}$ can be multiplied again on the right by the Givens rotation matrix $T_{13}(\acute{\theta})$ to yield yet a second intermediate matrix form

$$\begin{aligned} \grave{A}(\acute{\theta}) &= A T_{23}(\theta) T_{13}(\acute{\theta}) \\ &= \begin{bmatrix} a_{11} & \acute{a}_{12} & \acute{a}_{13} \\ a_{21} & 0 & \acute{a}_{23} \end{bmatrix} \begin{bmatrix} \cos(\acute{\theta}) & & \sin(\acute{\theta}) \\ 0 & 1 & 0 \\ -\sin(\acute{\theta}) & 0 & \cos(\acute{\theta}) \end{bmatrix} \\ &= \begin{bmatrix} \grave{a}_{11} & \acute{a}_{12} & \grave{a}_{13} \\ 0 & 0 & \grave{a}_{23} \end{bmatrix} \end{aligned}$$

for $\acute{\theta}$ such that

$$\cos(\acute{\theta}) = \frac{\acute{a}_{23}}{\sqrt{\acute{a}_{21}^2 + \acute{a}_{23}^2}}, \qquad \sin(\acute{\theta}) = \frac{\acute{a}_{21}}{\sqrt{\acute{a}_{21}^2 + \acute{a}_{23}^2}}.$$

A third Givens rotation yields the final matrix form

$$
\begin{aligned}
\breve{A}(\grave{\theta}) &= A T_{23}(\theta) T_{13}(\acute{\theta}) T_{12}(\grave{\theta}) \\
&= \begin{bmatrix} \grave{a}_{11} & \acute{a}_{12} & \grave{a}_{13} \\ 0 & 0 & \grave{a}_{23} \end{bmatrix} \begin{bmatrix} \cos(\grave{\theta}) & \sin(\grave{\theta}) & 0 \\ -\sin(\grave{\theta}) & \cos(\grave{\theta}) & 0 \\ 0 & 0 & 1 \end{bmatrix} \\
&= \begin{bmatrix} 0 & \breve{a}_{12} & \grave{a}_{13} \\ 0 & 0 & \grave{a}_{23} \end{bmatrix}
\end{aligned}
$$

for $\grave{\theta}$ such that

$$
\cos(\grave{\theta}) = \frac{\acute{a}_{12}}{\sqrt{\grave{a}_{11}^2 + \acute{a}_{12}^2}}, \qquad \sin(\grave{\theta}) = \frac{\grave{a}_{11}}{\sqrt{\grave{a}_{11}^2 + \acute{a}_{12}^2}}.
$$

The remaining nonzero part of this final result is an upper triangular submatrix right adjusted within the original array dimensions.

The order in which successive Givens rotation matrices are applied is constrained to avoid "unzeroing" elements of the matrix that have already been zeroed by previous Givens rotations. Figure 6.4 shows the constraints that guarantee such noninterference. If we suppose that the element to be annihilated (designated by x in the figure) is in the ith column and kth row and the corresponding diagonal element of the soon-to-be triangular matrix is in the jth column, then it is sufficient if the elements *below* the kth rows in those two columns have already been annihilated by Givens rotations. The reason for this is simple: the Givens rotations can only form linear combinations of row elements in those two columns. If those row elements are already zero, then any linear combination of them will also be zero. The result: no effect.

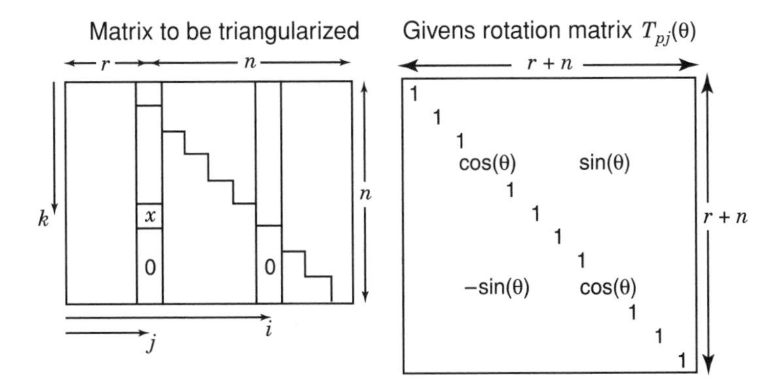

Fig. 6.4 *Constraints on Givens triangularization order.*

A Givens Triangularization Algorithm. The method used in the previous example can be generalized to an algorithm for upper triangularization of an $n \times (n+r)$ matrix, as listed below:

```
Input: A, a n-by(n+r) matrix

Output: A is overwritten by an upper triangular matrix C,
        right-adjusted in the array, such that output value of
        CC' equals input value of AA'.

        for i=n:-1:1,
            for j=1:r+i,
               rho=sqrt(A(i,r+i)^2+A(i,j)^2);
               s=A(i,j) / rho;
               c=A(i,r+i)/rho;
               for k=1:i,
                           x =c*A(k,j)-s*A(k,r+i);
                 A(k,r+i)=s*A(k,j)+c*A(k,r+i);
                    A(k,j)=x;
               end;
            end;
        end;
```

In this particular form of the algorithm, the outermost loop (the loop with index i in the listing) zeros elements of A one row at a time. An analogous algorithm can be designed in which the outermost loop is by columns.

6.4.5.2 Triangularization by Householder Reflections. This method of triangularization was discovered by Alston S. Householder [172]. It uses an elementary matrix of the form

$$T(v) = I - \frac{2}{v^{\mathrm{T}} v} v v^{\mathrm{T}}, \tag{6.46}$$

where v is a column vector and I is the identity matrix of the same dimension. This particular form of the elementary matrix is called a *Householder reflection, Householder transformation*, or *Householder matrix.*

Note that Householder transformation matrices are always *symmetric.* They are also *orthogonal*, for

$$T(v)T^{\mathrm{T}}(v) = \left(I - \frac{2}{v^{\mathrm{T}} v} v v^{\mathrm{T}}\right)\left(I - \frac{2}{v^{\mathrm{T}} v} v v^{\mathrm{T}}\right) \tag{6.47}$$

$$= I - \frac{4}{v^{\mathrm{T}} v} v v^{\mathrm{T}} + \frac{4}{(v^{\mathrm{T}} v)^2} v (v^{\mathrm{T}} v) v^{\mathrm{T}} \tag{6.48}$$

$$= I. \tag{6.49}$$

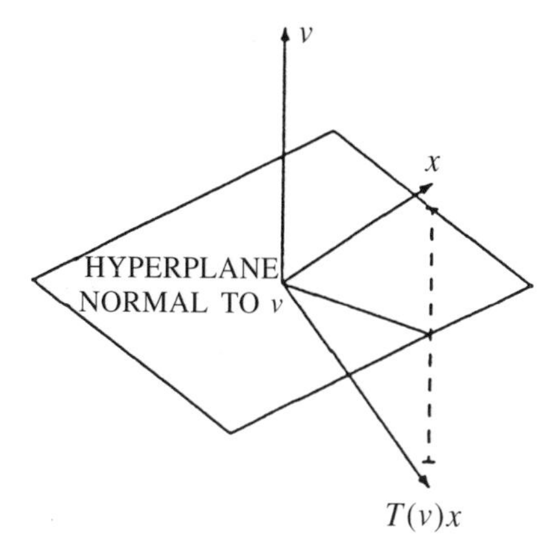

Fig. 6.5 Householder reflection of a vector x

They are called "reflections" because they transform any matrix x into its "mirror reflection" in the plane (or hyperplane[6]) normal to the vector v, as illustrated in Figure 6.5 (for three-dimensional v and x). By choosing the proper mirror plane, one can place the reflected vector $T(v)x$ along any direction whatsoever, including parallel to any coordinate axis.

EXAMPLE 6.8: Householder Reflection Along One Coordinate Axis Let x be any n-dimensional *row vector*, and let

$$e_k \stackrel{\text{def}}{=} \begin{pmatrix} 0 & 0 & 0 & \cdots & 0 & \overset{k}{1} & 0 & \cdots & 0 \end{pmatrix}$$

be the kth row of the $n \times n$ identity matrix. If the vector v of the Householder reflection $T(v)$ is defined as

$$v = x^{\mathrm{T}} + \alpha e_k^{\mathrm{T}},$$

where α is a scalar, then the inner products

$$\begin{aligned} v^{\mathrm{T}} v &= |x|^2 + 2\alpha x_k + \alpha^2, \\ xv &= |x|^2 + \alpha x_k, \end{aligned}$$

[6]The dimension of the hyperplane normal to the vector v will be one less than the dimension of the space containing v. When, as in the illustration, v is a three-dimensional vector (i.e., the space containing v is three-dimensional), the hyperplane normal to v is a two-dimensional plane.

where x_k is the kth component of x. The Householder reflection $xT(v)$ of x will be

$$
\begin{aligned}
xT(v) &= x\left(I - \frac{2}{v^{\mathrm{T}}v}vv^{\mathrm{T}}\right) \\
&= x\left(I - \frac{2}{x^{\mathrm{T}}x + 2\alpha x_k + \alpha^2}vv^{\mathrm{T}}\right) \\
&= x - \frac{2xv}{|x|^2 + 2\alpha x_k + \alpha^2}v^{\mathrm{T}} \\
&= x - \frac{2(|x|^2 + \alpha x_k)}{|x|^2 + 2\alpha x_k + \alpha^2}(x + \alpha e_k) \\
&= \left[1 - \frac{2(|x|^2 + \alpha x_k)}{|x|^2 + 2\alpha x_k + \alpha^2}\right]x - \left[\frac{2\alpha(|x|^2 + \alpha x_k)}{|x|^2 + 2\alpha x_k + \alpha^2}\right]e_k \\
&= \left[\frac{\alpha^2 - |x|^2}{|x|^2 + 2\alpha x_k + \alpha^2}\right]x - \left[\frac{2\alpha(|x|^2 + \alpha x_k)}{|x|^2 + 2\alpha x_k + \alpha^2}\right]e_k.
\end{aligned}
$$

Consequently, if one lets

$$\alpha = \mp|x|, \tag{6.50}$$

then

$$xT(v) = \pm|x|e_k. \tag{6.51}$$

That is, $xT(v)$ is parallel to the kth coordinate axis.

If, in the above example, x were the last row vector of an $n \times (n + r)$ matrix

$$M = \begin{bmatrix} Z \\ x \end{bmatrix},$$

and letting $k = 1$, then

$$MT(v) = \begin{bmatrix} ZT(v) \\ xT(v) \end{bmatrix} \tag{6.52}$$

$$= \begin{bmatrix} & & & ZT(v) & & \\ 0 & 0 & 0 & \cdots & 0 & |x| \end{bmatrix}, \tag{6.53}$$

the first step in upper triangularizing a matrix by Householder reflections.

Upper Triangularization by Successive Householder Reflections. A single Householder reflection can be used to zero all the elements to the left of the diagonal in an entire row of a matrix, as shown in Figure 6.6. In this case, the vector x to be operated upon by the Householder reflection is a row vector composed of the first k components of a row of a matrix. Consequently, the dimension of the Householder reflection matrix $T(v)$ need only be k, which may be strictly less than the number of columns in the matrix to be triangularized. The "undersized" Householder matrix is placed in the upper left corner of the transformation matrix, and the remaining diagonal block is filled with an (appropriately dimensioned) identity matrix I, such that the row dimension of the resulting transformation matrix equals the column dimension of the matrix to be triangularized. The resulting composite matrix will always be orthogonal, so long as $T(v)$ is orthogonal.

There are two important features of this construction. The first is that the presence of the identity matrix has the effect of leaving the columns to the right of the kth column undisturbed by the transformation. The second is that the transformation does not "unzero" previously zeroed rows below. Together, these features allow the matrix to be triangularized by the sequence of transformations shown in Figure 6.7. Note that the *size* of the Householder transformation shrinks with each step.

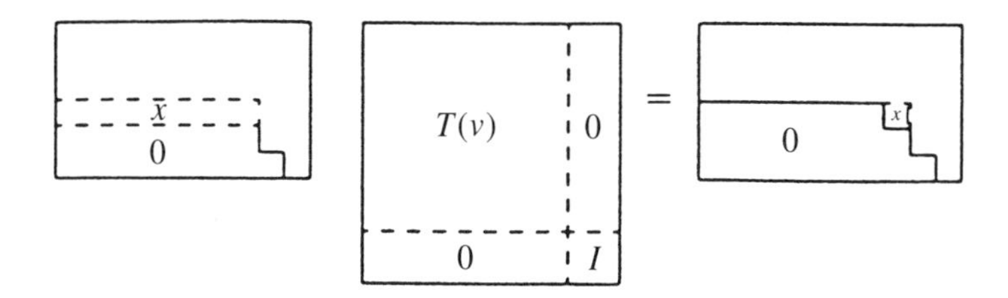

Fig. 6.6 *Zeroing one subrow with a Householder reflection.*

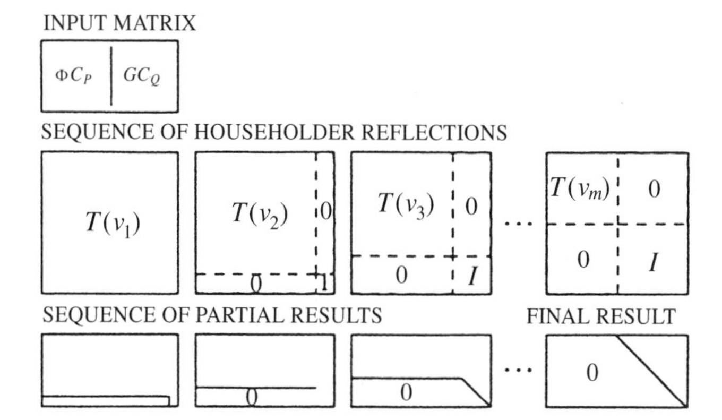

Fig. 6.7 *Upper triangularization by Householder reflections.*

Householder Triangularization Algorithm. The algorithm listed below performs upper triangularization of a rectangular matrix $n \times (n+r)$ A in place using a scratchpad $(n+r)$-vector v. The result is an upper triangular $n \times n$ matrix, right adjusted in the array A.

This algorithm includes a rescaling computation (involving the absolute value function abs) for better numerical stability. It is modified after the one given by Golub and Van Loan [89] for Householder triangularization. The modification is required for applying the Householder transformations from the right, rather than from the left, and for the particular form of the input matrix used in the Kalman filter implementation. Further specialization of this algorithm for Kalman filtering is presented later in Table 6.14.

```
    for k=n:-1:1,
       sigma=0;
       for j=1:r+k,
          sigma=sigma+A(k,j)^2;
       end;
       a=sqrt(sigma);
       sigma=0;
       for j=1:r+k,
          if j==r+k
             v(j)=A(k,j)-a;
          else
             v(j)=A(k,j);
          end;
          sigma=sigma+v(j)^2;
       end;
       a=2/sigma;
       for i=1:k,
       sigma=0;
       for j=1:r+k,
          sigma=sigma+A(i,j)*v(j);
       end;
       b=a*sigma;
       for j=1:r+k,
          A(i,j)=A(i,j)-b*v(j);
       end;
    end;
 end;
```

6.5 SQUARE-ROOT AND *UD* FILTERS

Square-root filtering uses a reformulation of the Riccati equations such that the dependent variable is a Cholesky factor (or modified Cholesky factor) of the state estimation uncertainty matrix P. We present here just two of the many forms of square-root Kalman filtering, with other forms presented in the following section. The two selected forms of square-root filtering are

1. Carlson–Schmidt square-root filtering, which uses Cholesky factors of P, and
2. Bierman–Thornton *UD* filtering, which uses modified Cholesky factors of P.

These are perhaps the more favored implementation forms (after the conventional Kalman filter), because they have been used extensively and successfully on many problems that would be too poorly conditioned for conventional Kalman filter implementation. The *UD* filter, in particular, has been used successfully on problems with thousands of state variables.

This does not necessarily imply that these two methods are to be preferred above all others, however. It may be possible that the Morf–Kailath combined square-root filter (Section 6.6.2), for example, performs as well or better, but we are currently not aware of any comparable experience using that method.

6.5.1 Carlson–Schmidt Square-Root Filtering

This is a matched pair of algorithms for the observational and temporal update of the covariance matrix P in terms of its Cholesky factors. If the covariance matrices R (measurement noise) and Q (dynamic process noise) are not diagonal matrices, the implementation also requires UD or Cholesky factorization of these matrices.

6.5.1.1 Carlson "Fast Triangular" Update. This algorithm is implemented in the MATLAB m-file carlson.m on the accompanying diskette. The algorithm is due to Neal A. Carlson [149]. It generates an upper triangular Cholesky factor W for the Potter factorization and has generally lower computational complexity than the Potter algorithm. It is a specialized and simplified form of an algorithm used by Agee and Turner [106] for Kalman filtering. It is a rank 1 modification algorithm, like the Potter algorithm, but it produces a triangular Cholesky factor. It can be derived from Problem 6.14.

In the case that $m = j$, the summation on the left-hand side of Equation 6.157 has but one term:

$$W_{ij}W_{jj} = \Delta_{ij} - \frac{\mathbf{v}_i\mathbf{v}_j}{R + \sum_{k=1}^{j}\mathbf{v}_k^2}, \tag{6.54}$$

which can be solved in terms of the elements of the upper triangular Cholesky factor W:

$$W_{ij} = \begin{cases} 0, & i > j, \\ \sqrt{\dfrac{R + \sum_{k=1}^{j-1}\mathbf{v}_k^2}{R + \sum_{k=1}^{j}\mathbf{v}_k^2}}, & i = j, \\ \dfrac{-\mathbf{v}_i\mathbf{v}_j}{(R + \sum_{k=1}^{j-1}\mathbf{v}_k^2)(R + \sum_{k=1}^{j}\mathbf{v}_k^2)}, & i < j. \end{cases} \tag{6.55}$$

Given the above formula for the elements of W, one can derive the formula for the elements of $C(+) = C(-)W$, the upper triangular Cholesky factor of the a posteriori covariance matrix of estimation uncertainty $P(+)$, given $C(-)$, the upper triangular Cholesky factor of the a priori covariance matrix $P(-)$.

Because both C and W are upper triangular, the elements $C_{ik} = 0$ for $k < i$ and the elements $W_{kj} = 0$ for $k > j$. Consequently, for $1 \leq i \leq j \leq n$, the element in the ith row and jth column of the matrix product $C(-)W = C(+)$ will be

$$C_{ij}(+) = \sum_{k=i}^{j} C_{ik}(-)W_{kj} + \text{ terms with zero factors} \tag{6.56}$$

$$= C_{ij}(-)W_{jj} + \sum_{k=i}^{j-1} C_{ik}(-)W_{kj} \tag{6.57}$$

$$= C_{ij}(-)\sqrt{\frac{R + \sum_{k=1}^{j-1}\mathbf{v}_k^2}{R + \sum_{k=1}^{j}\mathbf{v}_k^2}} - \sum_{k=i}^{j-1} \frac{C_{ik}(-)\mathbf{v}_k\mathbf{v}_j}{(R + \sum_{k=1}^{j-1}\mathbf{v}_k^2)(R + \sum_{k=1}^{j}\mathbf{v}_k^2)} \tag{6.58}$$

$$= \left(R + \sum_{k=1}^{j}\mathbf{v}_k^2\right)^{-1/2} \times \left[C_{ij}(-)\sqrt{R + \sum_{k=1}^{j-1}\mathbf{v}_k^2} - \frac{\mathbf{v}_j\sum_{k=i}^{j-1} C_{ik}(-)\mathbf{v}_k}{(R + \sum_{k=1}^{j-1}\mathbf{v}_k^2)^{1/2}}\right]. \tag{6.59}$$

This is a general formula for the upper triangular a posteriori Cholesky factor of the covariance matrix of estimation uncertainty, in terms of the upper triangular a priori Cholesky factor $C(-)$ and the vector $\mathbf{v} = C^{\mathrm{T}}H^{\mathrm{T}}$, where H is the measurement sensitivity matrix (a row vector). An algorithm implementing the formula is given in Table 6.11. This algorithm performs the complete observational update, including the update of the state estimate, in place. [Note that this algorithm forms the product $\mathbf{v} = C^{\mathrm{T}}(-)H^{\mathrm{T}}$ internally, computing and using the components $\sigma = \mathbf{v}_j$ as needed, without storing the vector $\mathbf{v}$. It does store and use the vector $\mathbf{w} = C(-)\mathbf{v}$, the unscaled Kalman gain, however.]

It is possible—and often desirable—to save array space by storing triangular matrices in singly subscripted arrays. An algorithm (in FORTRAN) for such an implementation of this algorithm is given in Carlson's original paper [149].

6.5.1.2 Schmidt Temporal Update

Nonsquare, Nontriangular Cholesky Factor of $P_k(-)$. If C_P is a Cholesky factor of $P_{k-1}(+)$ and C_Q is a Cholesky factor of Q_k, then the partitioned $n \times (n + q)$ matrix

$$A = [\,G_k C_Q \quad | \quad \Phi_k C_P] \tag{6.60}$$

TABLE 6.11 Carlson's Fast Triangular Observational Update

Symbol	Definition
z	Value of scalar measurement
R	Variance of scalar measurement uncertainty
H	Scalar measurement sensitivity vector ($1 \times n$ matrix)
C	Cholesky factors of $P(-)$ (input) and $P(+)$ (output)
x	State estimates $x(-)$ (input) and $x(+)$ (output)
$\mathbf{w}$	Unscaled Kalman gain (internal)

```
alpha=R;
delta=z;
for j=1:n,
   delta=delta-H(j)*x(j);
   sigma=0;
   for i=1:j,
      sigma=sigma+C(i,j)*H(i);
   end;
   beta=alpha;
   alpha=alpha+sigma^2;
   gamma=sqrt(alpha*beta);
   eta=beta / gamma;
   zeta=sigma/ gamma;
   w(j)=0;
   for i=1:j,
      tau=C(i,j);
      C(i,j)=eta*C(i,j)-zeta*w(i);
      w(i)=w(i)+tau*sigma;
   end;
end;
epsilon=delta / alpha;
for i=1:n,
   x(i)=x(i)+w(i)*epsilon;
end;
```

Computational complexity: $(2n^2 + 7n + 1)$ flops $+ n\sqrt{\ }$.

has the $n \times n$ symmetric matrix product value

$$
\begin{aligned}
AA^{\mathrm{T}} &= [G_{k-1}C_Q|\Phi_{k-1}C_P][G_{k-1}C_Q|\Phi_{k-1}C_P]^{\mathrm{T}} && (6.61)\\
&= \Phi_{k-1}C_PC_P^{\mathrm{T}}\Phi_{k-1}^{\mathrm{T}} + G_{k-1}C_QC_Q^{\mathrm{T}}G_{k-1}^{\mathrm{T}} && (6.62)\\
&= \Phi_{k-1}P_{k-1}(+)\Phi_{k-1}^{\mathrm{T}} + G_{k-1}Q_{k-1}G_{k-1}^{\mathrm{T}} && (6.63)\\
&= P_k(-) && (6.64)
\end{aligned}
$$

That is, A is a nonsquare, nontriangular Cholesky factor of $P_k(-)$. If *only* it were square and triangular, it would be the kind of Cholesky factor we are looking for. It is not, but fortunately there are algorithmic procedures for modifying A to that format.

Programming note: Computation of GC_Q and ΦC_P *in place. This should be attempted only if memory limitations demand it.* The product ΦC_P can be computed in place by overwriting Φ with the product ΦC_P. This is not desirable if Φ is constant, however. (It is possible to overwrite C_P with the product ΦC_P, but this requires the use of an additional n-vector of storage. This option is left as an exercise for the truly needy.) Similarly, the product GC_Q can be computed in place by overwriting G with the product GC_Q if $r \leq n$. Algorithms for doing the easier in-place operations when the Cholesky factors C_Q and C_P are *upper* triangular are shown in Table 6.12. Note that these have roughly half the computational complexities of the general matrix products.

Triangularization using Givens Rotations. The Givens triangularization algorithm is presented in Section 6.4.5. The specialization of this algorithm to use GC_Q and ΦC_P in place, without having to stuff them into a common array, is shown in Table 6.13.

The computational complexity of Givens triangularization is greater than that of Householder triangularization, which is covered next. There are two attributes of the Givens method that might recommend it for certain applications, however:

1. The Givens triangularization procedure can exploit sparseness of the matrices to be triangularized. Because it zeros elements one at a time, it can skip over elements that are already zero. (The procedure may "unzero" some elements that are already zero in the process, however.) This will tend to decrease the computational complexity of the application.

TABLE 6.12 Algorithms Performing Matrix Products in Place

<table>
<tr><th>Overwriting Φ by ΦC_P</th><th>Overwriting G by GC_Q</th></tr>
<tr><td><pre>
for j=n:-1:1,
  for i=1:n,
    sigma=0;
    for k=1:j,
      sigma=sigma+Phi(i,k)*CP(k,j);
    end;
    Phi(i,j)=sigma;
  end;
end;
</pre></td><td><pre>
for j=r:-1:1,
  for i=1:n,
    sigma=0;
    for k=1:j,
      sigma=sigma+G(i,k)*CQ(kj);
    end;
    G(i,j)=sigma;
  end;
end;
</pre></td></tr>
<tr><td colspan="2">Computational complexities</td></tr>
<tr><td>$n^2(n+1)/2$</td><td>$nr(r+1)/2$</td></tr>
</table>

TABLE 6.13 Temporal Update by Givens Rotations

Symbol	Description
A	Input: $G_k C_{Q_k}$, an $n \times r$ matrix. Output: A is overwritten by intermediate results.
B	Input: $\Phi_k C_{P_k}(+)$, an $n \times n$ matrix. Output: B is overwritten by the upper triangular matrix $C_{P_{k+1}}(-)$.

```
for i=n:-1:1,
   for j=1:r,
      rho=sqrt(B(i,i)^2+A(i,j)^2);
      s=A(i,j)/rho;
      c=B(i,i)/rho;
      for k=1:i,
         tau=c*A(k,j)-s*B(k,i);
         B(k,i)=s*A(k,j)+c*B(k,i);
         A(k,j)=tau;
      end;
   end;
   for j=1:i-1,
      rho=sqrt(B(i,i)^2+B(i,j)^2);
      s=B(i,j)/rho;
      c=B(i,i)/rho;
      for k=1:i,
         tau=c*B(k,j)-s*B(k,i);
         B(k,i)=s*B(k,j)+c*B(k,i);
         B(k,j)=tau;
      end;
   end;
end;
```

Computational complexity: $\frac{2}{3}n^2(2n+3r+6)+6nr+\frac{8}{3}n$ flops $+\frac{1}{2}n(n+2r+1)\surd$.

2. The Givens method can be "parallelized" to exploit concurrent processing capabilities, if they are available. The parallelized Givens method has no data contention among concurrent processes—they are working with data in different parts of the matrix as it is being triangularized.

Schmidt Temporal Updates Using Householder Reflections. The basic Householder triangularization algorithm (see Section 6.4.5.2) operates on a single $n \times (n+r)$ matrix. For the method of Dyer and McReynolds, this matrix is composed of two blocks containing the matrices $GC_Q(n \times r)$ and $\Phi P(n \times n)$.

The specialization of the Householder algorithm to use the matrices GC_Q and ΦP directly, without having to place them into a common array first, is described and listed in Table 6.14. Algorithms for computing GC_Q and ΦP in place were given in the previous subsection.

TABLE 6.14 Schmidt–Householder Temporal Update Algorithm

This modification of the Householder algorithm performs upper triangularization of the partitioned matrix $[\Phi C_P(+), GC_Q]$ by modifying $\Phi C_P(+)$ and GC_Q in place using Householder transformations of the (effectively) partitioned matrix.

Symbol	Description
	Input Variables
A	$n \times n$ matrix $\Phi C_P(+)$
B	$n \times r$ matrix GC_Q
	Output Variables
A	Array is overwritten by upper triangular result $C_P(-)$ such that $C_P(-)C_P^T(-) = \Phi C_P(+)C_P^T(+)\Phi^T + GC_Q C_Q^T G^T$.
B	Array is zeroed during processing.
	Intermediate Variables
α, β, σ	Scalars
v	Scratchpad n-vector
w	Scratchpad $n + r$-vector

```
for k=n:-1:1,
   sigma=0;
   for j=1:r,
      sigma=sigma+B(k,j)^2;
   end;
   for j=1:k,
      sigma=sigma+A(k,j)^2;
   end;
   alpha=sqrt(sigma);
   sigma=0;
   for j=1:r,
      w(j)=B(k,j);
      sigma=sigma+w(j)^2;
   end;
   for j=1:k,
      if j==k
         v(j)=A(k,j)-alpha;
      else
         v(j)=A(k,j);
      end;
      sigma=sigma+v(j)^2;
   end;
   alpha=2/sigma;
   for i=1:k,
      sigma=0;
      for j=1:r,
         sigma=sigma+B(i,j)*w(j);
      end;
      for j=1:k,
         sigma=sigma+A(i,j)*v(j);
      end;
      beta=alpha*sigma;
      for j=1:r,
         B(i,j)=B(i,j)-beta*w(j);
      end;
      for j=1:k,
         A(i,j)=A(i,j)-beta*v(j);
      end;
   end;
end;
```

Computational complexity: $n^3 r + \frac{1}{2}(n+1)^2 r + 5 + \frac{1}{3}(2n+1)$ flops.

6.5.2 Bierman–Thornton *UD* Filtering

This is a pair of algorithms, including the Bierman algorithm for the observational update of the modified Cholesky factors U and D of the covariance matrix $P = UDU^{\mathrm{T}}$, and the corresponding Thornton algorithm for the temporal update of U and D.

6.5.2.1 Bierman UD Observational Update. Bierman's algorithm is one of the more stable implementations of the Kalman filter observational update. It is similar in form and computational complexity to the Carlson algorithm but avoids taking scalar square roots. (It has been called "square-root filtering without square roots.") The algorithm was developed by the late Gerald J. Bierman (1941–1987), who made many useful contributions to optimal estimation theory, especially in implementation methods.

Partial UD Factorization of Covariance Equations. In a manner similar to the case with Cholesky factors for scalar-valued measurements, the conventional form of the observational update of the covariance matrix

$$P(+) = P(-) - \frac{P(-)H^{\mathrm{T}}HP(-)}{R + HP(-)H^{\mathrm{T}}}$$

can be partially factored in terms of UD factors:

$$P(-) \stackrel{\text{def}}{=} U(-)D(-)U^{\mathrm{T}}(-), \tag{6.65}$$

$$P(+) \stackrel{\text{def}}{=} U(+)D(+)U^{\mathrm{T}}(+), \tag{6.66}$$

$$\begin{aligned} U(+)D(+)U^{\mathrm{T}}(+) &= U(-)D(-)U^{\mathrm{T}}(-) \qquad (6.67)\\ &\quad - \frac{U(-)D(-)U^{\mathrm{T}}(-)H^{\mathrm{T}}HU(-)D(-)}{R + HU(-)D(-)U^{\mathrm{T}}(-)H^{\mathrm{T}}}U^{\mathrm{T}}(-) \\ &= U(-)D(-)U^{\mathrm{T}}(-) - \frac{U(-)D(-)\mathbf{v}\mathbf{v}^{\mathrm{T}}D(-)U^{\mathrm{T}}(-)}{R + \mathbf{v}^{\mathrm{T}}D(-)\mathbf{v}} \qquad (6.68)\\ &= U(-)\left[D(-) - \frac{D(-)\mathbf{v}\mathbf{v}^{\mathrm{T}}D(-)}{R + \mathbf{v}^{\mathrm{T}}D(-)\mathbf{v}}\right]U^{\mathrm{T}}(-), \qquad (6.69) \end{aligned}$$

where

$$\mathbf{v} = U^{\mathrm{T}}(-)H^{\mathrm{T}} \tag{6.70}$$

is an n-vector and n is the dimension of the state vector.

Equation 6.69 contains the unfactored expression

$$D(-) - D(-)\mathbf{v}[\mathbf{v}^{\mathrm{T}}D(-)\mathbf{v} + R]^{-1}\mathbf{v}^{\mathrm{T}}D(-).$$

If one were able to factor it with a unit triangular factor B in the form

$$D(-) - D(-)\mathbf{v}[\mathbf{v}^{\mathrm{T}}D(-)\mathbf{v} + R]^{-1}\mathbf{v}^{\mathrm{T}}D(-) = BD(+)B^{\mathrm{T}}, \tag{6.71}$$

then $D(+)$ would be the a posteriori D factor of P because the resulting equation

$$U(+)D(+)U^{\mathrm{T}}(+) = U(-)\{BD(+)B^{\mathrm{T}}\}U^{\mathrm{T}}(-) \tag{6.72}$$

$$= \{U(-)B\}D(+)\{U(-)B\}^{\mathrm{T}} \tag{6.73}$$

can be solved for the a posteriori U factor as

$$U(+) = U(-)B. \tag{6.74}$$

Therefore, for the observational update of the UD factors of the covariance matrix P, it suffices to find a numerically stable and efficient method for the UD factorization of a matrix expression of the form $D - D\mathbf{v}[\mathbf{v}^{\mathrm{T}}D\mathbf{v} + R]^{-1}\mathbf{v}^{\mathrm{T}}D$, where $\mathbf{v} = U^{\mathrm{T}}H^{\mathrm{T}}$ is a column vector. Bierman [7] found such a solution, in terms of a rank 1 modification algorithm for modified Cholesky factors.

Bierman UD Factorization. Derivations of the Bierman UD observational update can be found in [7]. It is implemented in the accompanying MATLAB m-file bierman.m.

An alternative algorithm implementing the Bierman UD observational update in place is given in Table 6.15. One can also store $\mathbf{w}$ over $\mathbf{v}$, to save memory requirements. It is possible to reduce the memory requirements even further by storing D on the diagonal of U, or, better yet, storing just the strictly upper triangular part of U in a singly-subscripted array. These programming tricks do little to enhance readability of the algorithms, however. They are best avoided unless one is truly desperate for memory.

6.5.2.2 Thornton *UD* Temporal Update. This UD factorization of the temporal update in the discrete-time Riccati equation is due to Catherine Thornton [124]. It is also called *modified weighted Gram–Schmidt* (*MWGS*) orthogonalization.[7] It uses a factorization algorithm due to Björck [141] that is actually quite different from the conventional Gram–Schmidt orthogonalization algorithm and

[7]Gram–Schmidt *orthonormalization* is a procedure for generating a set of "unit normal" vectors as linear combinations of a set of linearly independent vectors. That is, the resulting vectors are mutually orthogonal and have unit length. The procedure without the unit-length property is called Gram–Schmidt *orthogonalization.* These algorithmic methods were derived by Jorgen Pedersen Gram (1850–1916) and Erhard Schmidt (1876–1959).

TABLE 6.15 Bierman Observational Update

Symbol	Definition
z	Value of scalar measurement
R	Variance of scalar measurement uncertainty
H	Scalar measurement sensitivity vector ($1 \times n$ matrix)
U, D	UD factors of $P(-)$ (input) and $P(+)$ (output)
x	State estimates $x(-)$ (input) and $x(+)$ (output)
$\mathbf{v}$	scratchpad n-vector
$\mathbf{w}$	scratchpad n-vector

```
delta=z;
for j=1:n,
   delta=delta-H(j)*x(j);
   v(j)=H(j);
   for i=1:j-1,
      v(j)=v(j)+U(i,j)*H(i);
   end;
end;
sigma=R;
for j=1:n,
   nu=v(j);
   v(j)=v(j)*D(j,j);
   w(j)=nu;
   for i=1:j-1,
      tau=U(i,j)*nu;
      U(i,j)=U(i,j)-nu*w(i)/sigma;
      w(i)=w(i)+tau;
   end;
   D(j,j)=D(j,j)*sigma;
   sigma=sigma+nu*v(j);
   D(j,j)=D(j,j)*sigma;
end;
epsilon=delta / sigma;
for i=1:n,
   x(i)=x(i)+v(i)*epsilon;
end;
```

Computational complexity: $(2n^2 + 7n + 1)$ flops.

more robust against roundoff errors. However, the algebraic properties of Gram–Schmidt orthogonalization are useful for deriving the factorization.

Gram–Schmidt orthogonalization is an algorithm for finding a set of n mutually orthogonal m-vectors $b_1, b_2, b_3, \ldots, b_m$ that are linear combinations of a set of n *linearly independent* m-vectors $a_1, a_2, a_3, \ldots, a_m$. That is, the inner products

$$b_i^{\mathrm{T}} b_j = \begin{cases} |b_i|^2 & \text{if } i = j, \\ 0 & \text{if } i \neq j. \end{cases} \tag{6.75}$$

The Gram–Schmidt algorithm defines a unit lower[8] triangular $n \times n$ matrix L such that $A = BL$, where

$$A = [a_1 \quad a_2 \quad a_3 \quad \cdots \quad a_n] \tag{6.76}$$

$$= BL \tag{6.77}$$

$$= [b_1 \quad b_2 \quad b_3 \quad \cdots \quad b_n] \begin{bmatrix} 1 & 0 & 0 & \cdots & 0 \\ \ell_{21} & 1 & 0 & \cdots & 0 \\ \ell_{31} & \ell_{32} & 1 & \cdots & 0 \\ \vdots & \vdots & \vdots & \ddots & \vdots \\ \ell_{n1} & \ell_{n2} & \ell_{n3} & \cdots & 1 \end{bmatrix} \tag{6.78}$$

where the vectors b_i are column vectors of B and the matrix product

$$B^{\mathrm{T}}B = \begin{bmatrix} |b_1|^2 & 0 & 0 & \cdots & 0 \\ 0 & |b_2|^2 & 0 & \cdots & 0 \\ 0 & 0 & |b_3|^2 & \cdots & 0 \\ \vdots & \vdots & \vdots & \ddots & \vdots \\ 0 & 0 & 0 & \cdots & |b_n|^2 \end{bmatrix} \tag{6.79}$$

$$= \mathrm{diag}_{1 \le i \le n}\{|b_i|^2\} \tag{6.80}$$

$$= D_\beta, \tag{6.81}$$

a diagonal matrix with positive diagonal values $\beta_i = |b_i|^2$.

Weighted Gram–Schmidt Orthogonalization. The m-vectors x and y are said to be orthogonal *with respect to the weights* $w_1, w_2, w_3, \ldots, w_m$ if the *weighted* inner product

$$\sum_{i=1}^{m} x_i w_i y_i = x^{\mathrm{T}} D_w y \tag{6.82}$$

$$= 0, \tag{6.83}$$

where the diagonal *weighting matrix*

$$D_w = \mathrm{diag}_{1 \le i \le n}\{w_i\}. \tag{6.84}$$

[8]In its original form, the algorithm produced a unit upper triangular matrix U by processing the a_i in the order $i = 1, 2, 3, \ldots, n$. However, if the order is reversed, the resulting coefficient matrix will be lower triangular and the resulting vectors b_i will still be mutually orthogonal.

The Gram–Schmidt orthogonalization procedure can be extended to include orthogonality of the column vectors b_i and b_j with respect to the weighting matrix D_w:

$$b_i^{\mathrm{T}} D_w b_j = \begin{cases} \beta_i > 0 & \text{if } i = j, \\ 0 & \text{if } i \neq j. \end{cases} \tag{6.85}$$

The resulting weighted Gram–Schmidt orthogonalization of a set of n linearly independent m-vectors $a_1, a_2, a_3, \ldots, a_m$ with respect to a weighting matrix D_w defines a unit lower triangular $n \times n$ matrix L_w such that the product $AL_w = B_w$ and

$$B_w^{\mathrm{T}} D_w B_w = \mathrm{diag}_{1 \leq i \leq n}\{\beta_i\}, \tag{6.86}$$

where $D_w = I$ for conventional orthogonalization.

Modified Weighted Gram–Schmidt Orthogonalization. The standard Gram–Schmidt orthogonalization algorithms are not reliably stable numerical processes when the column vectors of A are close to being linearly dependent or the weighting matrix has a large condition number. An alternative algorithm due to Björck has better overall numerical stability. Although L is not an important result for the orthogonalization problem (B is), its inverse turns out to be more useful for the UD filtering problem.

The *Thornton temporal update for UD factors* uses triangularization of the Q matrix (if it is not already diagonal) in the form $Q = GD_Q G^{\mathrm{T}}$, where D_Q is a diagonal matrix. If we let the matrices

$$A = \begin{bmatrix} U_{k-1}^{\mathrm{T}}(+)\Phi_k^{\mathrm{T}} \\ G_k^{\mathrm{T}} \end{bmatrix}, \tag{6.87}$$

$$D_w = \begin{bmatrix} D_{k-1}(+) & 0 \\ 0 & D_{Q_k} \end{bmatrix}, \tag{6.88}$$

then the MWGS orthogonalization procedure will produce a unit lower triangular $n \times n$ matrix L^{-1} and a diagonal matrix D_β such that

$$A = BL, \tag{6.89}$$

$$L^{\mathrm{T}} D_\beta L = L^{\mathrm{T}} B^{\mathrm{T}} D_w BL \tag{6.90}$$

$$= (BL)^{\mathrm{T}} D_w BL \tag{6.91}$$

$$= A^{\mathrm{T}} D_w A \tag{6.92}$$

$$= \begin{bmatrix} \Phi_{k-1}U_{k-1}(+) & G_{k-1} \end{bmatrix} \begin{bmatrix} D_{k-1}(+) & 0 \\ 0 & D_{Q_{k-1}} \end{bmatrix} \begin{bmatrix} U_{k-1}^{\mathrm{T}}(+)\Phi_{k-1}^{\mathrm{T}} \\ G_{k-1}^{\mathrm{T}} \end{bmatrix} \quad (6.93)$$

$$= \Phi_{k-1}U_{k-1}(+)D_{k-1}(+)U_{k-1}^{\mathrm{T}}(+)\Phi_{k-1}^{\mathrm{T}} + G_{k-1}D_{Q_{k-1}}G_{k-1}^{\mathrm{T}} \quad (6.94)$$

$$= \Phi_{k-1}P_{k-1}(+)\Phi_{k-1}^{\mathrm{T}} + Q_{k-1} \quad (6.95)$$

$$= P_k(-). \quad (6.96)$$

Consequently, the factors

$$U_k(-) = L^{\mathrm{T}}, \quad (6.97)$$

$$D_k(-) = D_\beta \quad (6.98)$$

are the solutions of the temporal update problem for the UD filter.

Diagonalizing Q. It is generally worth the effort to "diagonalize" Q (if it is not already a diagonal matrix), because the net computational complexity is reduced by this procedure. The formula given for total computational complexity of the Thornton algorithm in Table 6.16 includes the computational complexity for the UD decomposition of Q as $U_Q\acute{Q}U_Q^{\mathrm{T}}$ for $\acute{Q}$ diagonal [$\frac{1}{6}p(p-1)(p+4)$ flops] plus the computational complexity for multiplying G by the resulting $p \times p$ unit upper triangular factor U_Q [$\frac{1}{2}np(p-1)$ flops] to obtain $\acute{G}$.

The algorithm listed in Table 6.16 operates with the matrix blocks ΦU, $\acute{G}$, D, and $\acute{Q}$ by name and not as submatrices of some larger matrix. It is not necessary to physically relocate these matrices into larger arrays in order to apply the Björck orthogonalization procedure. The algorithm is written to find them in their original arrays. (It makes the listing a little longer, but the computational complexity is the same.)

Multiplying ΦU in Place. The complexity formulas in Table 6.16 also include the computations required for forming the product ΦU of the $n \times n$ state transition matrix Φ and the $n \times n$ unit upper triangular matrix U. This matrix product can be performed in place—overwriting Φ with the product ΦU—by the following algorithm:

```
for i=1:n,
   for j=n:-1:1,
      sigma=Phi(i,j);
      for k=1:j-1,
         sigma=sigma+Phi(i,k)*U(k,j);
      end;      Phi(i,j)=sigma;
  end;
end;
```

TABLE 6.16 Thornton *UD* Temporal Update Algorithm[a]

Symbol	Description
	Inputs
D	The $n \times n$ diagonal matrix. Can be stored as an n-vector.
ΦU	Matrix product of $n \times n$ state transition matrix Φ and $n \times n$ unit upper triangular matrix U such that $UDU^{\mathrm{T}} = P(+)$, the covariance matrix of a posteriori state estimation uncertainty.
$\acute{G}$	$= GU_Q$. The modified $n \times p$ process noise coupling matrix, where $Q = U_Q D_Q U_Q^{\mathrm{T}}$
D_Q	Diagonalized $p \times p$ covariance matrix of process noise. Can be stored as a p-vector.
	Outputs
ΦU	Is overwritten by intermediate results.
$\acute{G}$	Is overwritten by intermediate results.
$\acute{D}$	The $n \times n$ diagonal matrix. Can be stored as an n-vector.
$\acute{U}$	The $n \times n$ unit upper triangular matrix such that $\acute{U}\acute{D}\acute{U}^{\mathrm{T}} = \Phi UDU^{\mathrm{T}}\Phi^{\mathrm{T}} + GQG^{\mathrm{T}}$.

```
for i=n:-1:1,
   sigma=0;
   for j=1:n,
      sigma=sigma+PhiU(i,j)^2*D(j,j);
   end;
   for j=1:p,
      sigma=sigma+G(i,j)^2 *DQ(j,j);
   end;
   D(i,i)=sigma;
   U(i,i)=1;
   for j=1:i-1,
      sigma=0;
      for k=1:n,
         sigma=sigma+PhiU(i,k)*D(k,k)*PhiU(j,k);
      end;
      for k=1:p,
         sigma=sigma+G(i,k)*DQ(k,k)*G(j,k);
      end;
      U(j,i)=sigma / D(i,i);
      for k=1:n,
         PhiU(j,k)=PhiU(j,k)-U(j,i)*PhiU(i,k);
      end;
      for k=1:p,
         G(j,k)=G(j,k)-U(j,i)*G(i,k);
      end;
   end;
end;
```

	Computational Complexity Breakdown (in flops)
Matrix product ΦU	$n^2(n-1)/2$
Solve $U_Q D_Q U_Q^{\mathrm{T}} = Q$, $\acute{G} = GU_Q$	$p(p-1)(3n+p+4)/6$
Thornton algorithm	$3n(n-1)(n+p)/2$
Total	$n(n-1)(4n+3p-1)/2 + p(p-1)(3n+p+4)/6$

[a]Performs the temporal update of the modified Cholesky factors (*UD* factors) of the covariance matrix of state estimation uncertainty for the Kalman filter.

The computational complexity of this specialized matrix multiplication algorithm is $n^2(n-1)/2$, which is less than half of the computational complexity of the general $n \times n$ matrix product (n^3). Performing this multiplication in place also frees up the array containing $U_k(+)$ to accept the updated value $U_{k+1}(-)$. In some applications, Φ will have a sparse structure that can be exploited to reduce the computational requirements even more.

6.6 OTHER ALTERNATIVE IMPLEMENTATION METHODS

The main thrust of this chapter is the square-root filtering methods presented in the previous section. Although factorization methods are probably the most numerically stable implementation methods currently available, there are other alternative methods that may perform adequately on some applications, and there are some earlier alternatives that bear mentioning for their historical significance and for comparisons with the factorization methods.

6.6.1 Earlier Implementation Methods

6.6.1.1 Swerling Inverse Formulation. This is *not* recommended as an implementation method for the Kalman filter, because its computational complexity and numerical stability place it at a disadvantage relative to the other methods. Its computational complexity is derived here for the purpose of demonstrating this.

Recursive Least Mean Squares. This form of the covariance update for recursive least mean squares estimation was published by Peter Swerling [227]. Swerling's estimator was essentially the Kalman filter but with the observational update equation for the covariance matrix in the form

$$P(+) = \left[P^{-1}(-) + H^{\mathrm{T}} R^{-1} H\right]^{-1}.$$

This implementation requires three matrix inversions and two matrix products. If the observation is scalar valued ($m = 1$), the matrix inversion R^{-1} requires only one divide operation. One can also take advantage of diagonal and symmetric matrix forms to make the implementation more efficient.

Computational Complexity of Swerling Inverse Formulation.[9] For the case that the state dimension $n = 1$ and the measurement dimension $\ell = 1$, it can be done with 4 flops. In the case that $n > 1$, the dependence of the computational complexity

[9]See Section 6.4.2.6 for an explanation of how computational complexity is determined.

TABLE 6.17 Operation Summary for Swerling Inverse Formulation

Operation	Flops
R^{-1}	$\ell^3 + \frac{1}{2}\ell^2 + \frac{1}{2}\ell$ if $\ell > 1$,
	1 if $\ell = 1$
$(R^{-1})H$	$n\ell^2$
$H^T(R^{-1}H)$	$\frac{1}{2}n^2\ell + \frac{1}{2}n\ell$
$P^{-1}(-) + (H^T R^{-1} H)$	$n^3 + \frac{1}{2}n^2 + \frac{1}{2}n$
$[P^{-1}(-) + H^T R^{-1} H]^{-1}$	$n^3 + \frac{1}{2}n^2 + \frac{1}{2}n$
Total	$2n^3 + n^2 + n + \frac{1}{2}n^2\ell + n\ell^2 + \frac{1}{2}n\ell + \ell^3 + \frac{1}{2}\ell^2 + \ell$

on ℓ and n is shown in Table 6.17.[10] This is the most computationally intensive method of all. The number of arithmetic operations increases as n^3. The numbers of operations for the other methods increase as $n^2\ell + \ell^3$, but usually $n > \ell$ in Kalman filtering applications.

6.6.1.2 Kalman Formulation

Data Flows. A data flow diagram of the implementation equations defined by Kalman [179] is shown in Figure 6.8. This shows the points at which the measurements (z) and model parameters (H, R, Φ, and Q) are introduced in the matrix operations. There is some freedom to choose exactly how these operations shown will be implemented, however, and the computational requirements can be reduced substantially by reuse of repeated subexpressions.

Reuse of Partial Results. In this "conventional" form of the observational update equations, the factor $\{HP\}$ occurs several times in the computation of $\overline{K}$ and P:

$$\overline{K} = P(-)H^T[HP(-)H^T + R]^{-1} \tag{6.99}$$

$$= \{HP(-)\}^T[\{HP(-)\}H^T + R]^{-1}, \tag{6.100}$$

$$P(+) = P(-) - \overline{K}\{HP(-)\}. \tag{6.101}$$

The factored form shows the reusable partial results $[HP(-)]$ and $\overline{K}$ (the Kalman gain). Using these partial results where indicated, the implementation of the factored

[10]There is an alternative matrix inversion method (due to Strassen [228]) that reduces the number of multiplies in an $n \times n$ matrix inversion from n^3 to $n^{\log_2 7}$ but increases the number of additions significantly.

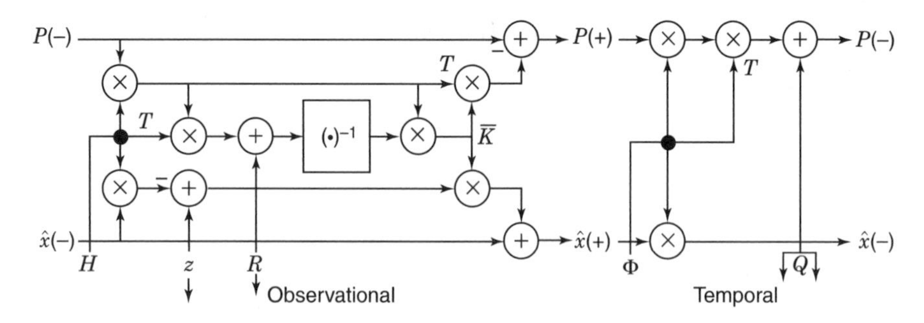

Fig. 6.8 Data flows of Kalman update implementations.

form requires four matrix multiplies and one matrix inversion. As in the case of the Swerling formulation, the matrix inversion can be accomplished by a divide if the dimension of the observation (ℓ) is 1, and the efficiency of all operations can be improved by computing only the unique elements of symmetric matrices. The number of floating-point arithmetic operations required for the observational update of the Kalman filter, using these methods, is summarized in Table 6.18. Note that the total number of operations required does not increase as n^3, as was the case with the Swerling formulation.

6.6.1.3 Potter Square-Root Filter.

The inventor of square-root filtering is James E. Potter. Soon after the introduction of the Kalman filter, Potter introduced the idea of factoring the covariance matrix and provided the first of the square-root methods for the observational update (see Battin [5, pp. 338–340] and Potter [208]).

Potter defined the Cholesky factor of the covariance matrix P as

$$P(-) \stackrel{\text{def}}{=} C(-)C^{\mathrm{T}}(-), \tag{6.102}$$

$$P(+) \stackrel{\text{def}}{=} C(+)C(+)^{\mathrm{T}}, \tag{6.103}$$

TABLE 6.18 Operation Summary for Conventional Kalman Filter

Operation	Flops
$H \times P(-)$	$n^2\ell$
$H \times [HP(-)]^{\mathrm{T}} + R$	$\frac{1}{2}n\ell^2 + \frac{1}{2}n\ell$
$\{H[HP(-)]^{\mathrm{T}} + R\}^{-1}$	$\ell^3 + \frac{1}{2}\ell^2 + \frac{1}{2}\ell$
$\{H[HP(-)]^{\mathrm{T}} + R\}^{-1} \times [HP(-)]$	$n\ell^2$
$P(-) - [HP(-)] \times \{H[HP(-)]^{\mathrm{T}} + R\}^{-1}[HP(-)]$	$\frac{1}{2}n^2\ell + \frac{1}{2}n\ell$
Total	$\frac{3}{2}n^2\ell + \frac{3}{2}n\ell^2 + n\ell + \ell^3 + \frac{1}{2}\ell^2 + \frac{1}{2}\ell$

so that the observational update equation

$$P(+) = P(-) - P(-)H^{\mathrm{T}}[HP(-)H^{\mathrm{T}} + R]^{-1}HP(-) \tag{6.104}$$

could be partially factored as

$$\begin{aligned} C(+)C^{\mathrm{T}}(+) &= C(-)C^{\mathrm{T}}(-) \\ &\quad - C(-)C^{\mathrm{T}}(-)H^{\mathrm{T}}[HC(-)C^{\mathrm{T}}(-)H^{\mathrm{T}} + R]^{-1}HC(-)C^{\mathrm{T}}(-) && (6.105) \\ &= C(-)C^{\mathrm{T}}(-) - C(-)V[V^{\mathrm{T}}V + R]^{-1}V^{\mathrm{T}}C^{\mathrm{T}}(-) && (6.106) \\ &= C(-)\{I_n - V[V^{\mathrm{T}}V + R]^{-1}V^{\mathrm{T}}\}C^{\mathrm{T}}(-), && (6.107) \end{aligned}$$

where

$I_n =$ $n \times n$ identity matrix
$V = C^{\mathrm{T}}(-)H^{\mathrm{T}}$ is an $n \times \ell$ general matrix
$n =$ dimension of state vector
$\ell =$ dimension of measurement vector

Equation 6.107 contains the unfactored expression $\{I_n - V[V^{\mathrm{T}}V + R]^{-1}V^{\mathrm{T}}\}$. For the case that the measurement is a scalar ($\ell = 1$), Potter was able to factor it in the form

$$I_n - V[V^{\mathrm{T}}V + R]^{-1}V^{\mathrm{T}} = WW^{\mathrm{T}}, \tag{6.108}$$

so that the resulting equation

$$\begin{aligned} C(+)C(+)^{\mathrm{T}} &= C(-)\{WW^{\mathrm{T}}\}C^{\mathrm{T}}(-) && (6.109) \\ &= \{C(-)W\}\{C(-)W\}^{\mathrm{T}} && (6.110) \end{aligned}$$

could be solved for the a posteriori Cholesky factor of $P(+)$ as

$$C(+) = C(-)W. \tag{6.111}$$

When the measurement is a scalar, the expression to be factored is a symmetric *elementary matrix* of the form[11]

$$I_n - \frac{\mathbf{v}\mathbf{v}^{\mathrm{T}}}{R + |\mathbf{v}|^2}, \tag{6.112}$$

where R is a positive scalar and $\mathbf{v} = C^{\mathrm{T}}(-)H^{\mathrm{T}}$ is a column n-vector.

[11]This expression—or something very close to it—is used in many of the square-root filtering methods for observational updates. The Potter square-root filtering algorithm finds a symmetric factor W, which does not preserve triangularity of the product $C(-)W$. The Carlson observational update algorithm (in Section 6.5.1.1) finds a triangular factor W, which preserves triangularity of $C(+) = C(-)W$ if both factors $C(+)$ and W are of the same triangularity (i.e., if both $C(-)$ and W are upper triangular or both lower triangular). The Bierman observational update algorithm uses a related UD factorization. Because the rank of the matrix $\mathbf{v}\mathbf{v}^{\mathrm{T}}$ is 1, these methods are referred to as rank 1 modification methods.

The formula for the symmetric square root of a symmetric elementary matrix is given in Equation 6.35. For the elementary matrix format in 6.112, the scalar s of Equation 6.35 has the value

$$s = \frac{1}{R + |\mathbf{v}|^2}, \tag{6.113}$$

so that the radicand

$$\begin{aligned} 1 - s|\mathbf{v}|^2 &= 1 - \frac{|\mathbf{v}|^2}{R + |\mathbf{v}|^2} & (6.114) \\ &= \frac{R}{R + |\mathbf{v}|^2} & (6.115) \\ &\geq 0 & (6.116) \end{aligned}$$

because the variance $R \geq 0$. Consequently, the matrix expression 6.112 will always have a real matrix square root.

Potter Formula for Observational Updates. Because the matrix square roots of symmetric elementary matrices are also symmetric matrices, they are also Cholesky factors. That is,

$$\begin{aligned} (I - s\mathbf{v}\mathbf{v}^{\mathrm{T}}) &= (I - \sigma\mathbf{v}\mathbf{v}^{\mathrm{T}})(I - \sigma\mathbf{v}\mathbf{v}^{\mathrm{T}}) & (6.117) \\ &= (I - \sigma\mathbf{v}\mathbf{v}^{\mathrm{T}})(I - \sigma\mathbf{v}\mathbf{v}^{\mathrm{T}})^{\mathrm{T}}. & (6.118) \end{aligned}$$

Following the approach leading to Equation 6.111, the solution for the a posteriori Cholesky factor $C(+)$ of the covariance matrix P can be expressed as the product

$$\begin{aligned} C(+)C^{\mathrm{T}}(+) &= P(+) & (6.119) \\ &= C(-)(I - s\mathbf{v}\mathbf{v}^{\mathrm{T}})C^{\mathrm{T}}(-) & (6.120) \\ &= C(-)(I - \sigma\mathbf{v}\mathbf{v}^{\mathrm{T}})(I - \sigma\mathbf{v}\mathbf{v}^{\mathrm{T}})^{\mathrm{T}}C^{\mathrm{T}}(-), & (6.121) \end{aligned}$$

which can be factored as[12]

$$C(+) = C(-)(I - \sigma\mathbf{v}\mathbf{v}^{\mathrm{T}}) \tag{6.122}$$

[12]Note that, as $R \to \infty$ (no measurement), $\sigma \to 2/|v|^2$ and $I - \sigma vv^{\mathrm{T}}$ becomes a Householder matrix.

with

$$\sigma = \frac{1 + \sqrt{1 - s|\mathbf{v}|^2}}{|\mathbf{v}|^2} \tag{6.123}$$

$$= \frac{1 + \sqrt{R/(R + |\mathbf{v}|^2)}}{|\mathbf{v}|^2}. \tag{6.124}$$

Equations 6.122 and 6.124 define the Potter square-root observational update formula, which is implemented in the accompanying MATLAB m-file potter.m. The Potter formula can be implemented *in place* (i.e., by overwriting C).

This algorithm updates the state estimate x and a Cholesky factor C of P in place. This Cholesky factor is a general $n \times n$ matrix. That is, it is not maintained in any particular form by the Potter update algorithm. The other square-root algorithms maintain C in triangular form.

6.6.1.4 Joseph-Stabilized Implementation. This variant of the Kalman filter is due to Joseph [15], who demonstrated improved numerical stability by rearranging the standard formulas for the observational update (given here for scalar measurements) into the formats

$$\acute{z} = R^{-1/2} z, \tag{6.125}$$

$$\acute{H} = \acute{z} H, \tag{6.126}$$

$$\overline{K} = (\acute{H} P(-) \acute{H}^{\mathrm{T}} + 1)^{-1} P(-) \acute{H}^{\mathrm{T}}, \tag{6.127}$$

$$P(+) = (I - \overline{K}\acute{H}) P(-) (I - \overline{K}\acute{H})^{\mathrm{T}} + \overline{K}\,\overline{K}^{\mathrm{T}}, \tag{6.128}$$

taking advantage of partial results and the redundancy due to symmetry. The mathematical equivalence of Equation 6.128 to the conventional update formula for the covariance matrix was shown as Equation 4.23. This formula, by itself, does not uniquely define the Joseph implementation, however. As shown, it has $\sim n^3$ computational complexity.

Bierman Implementation. This is a slight alteration due to Bierman [7] that reduces computational complexity by measurement decorrelation (if necessary) and the parsimonious use of partial results. The data flow diagram shown in Figure 6.9 is for a scalar measurement update, with data flow from top (inputs) to bottom (outputs) and showing all intermediate results. Calculations at the same level in this diagram may be implemented in parallel. Intermediate (temporary) results are labeled as $\mathscr{T}_1, \mathscr{T}_2, \ldots, \mathscr{T}_8$, where $\mathscr{T}_6 = \overline{K}$, the Kalman gain. If the result (left-hand side) of an $m \times m$ array is symmetric, then only the $\frac{1}{2}m(m+1)$ unique elements need be computed. Bierman [7] has made the implementation more memory efficient by the reuse of memory locations for these intermediate results. Bierman's implementation does not eliminate the redundant memory from symmetric arrays, however.

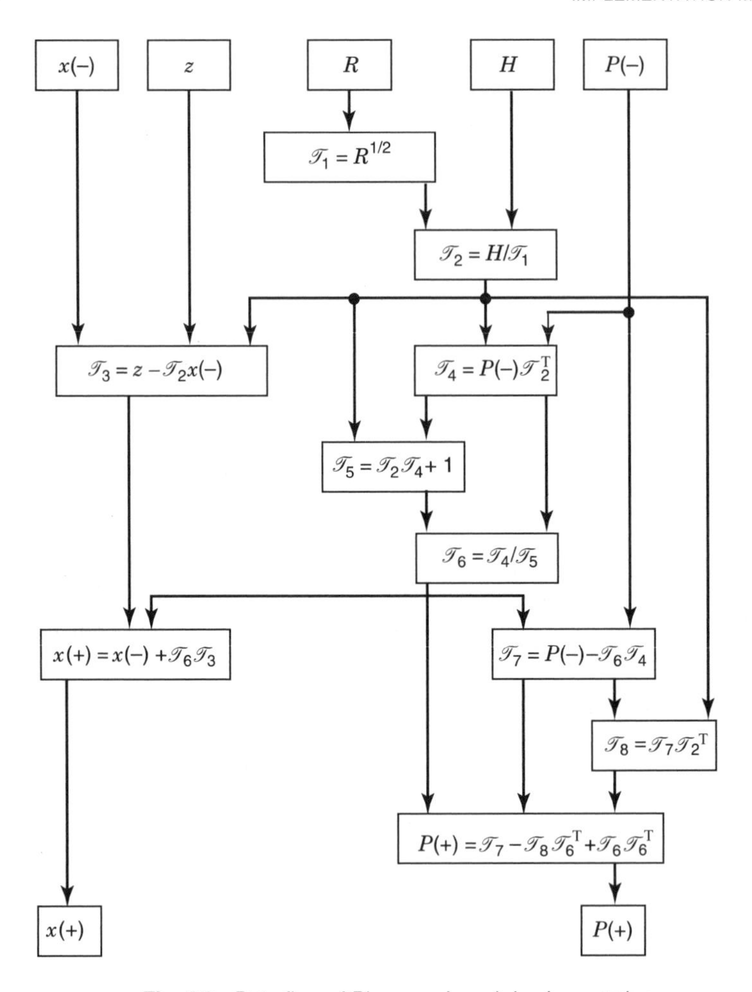

Fig. 6.9 *Data flow of Bierman–Joseph implementation.*

The computational complexity of this implementation grows as $3\ell n(3n + 5)/2$ flops, where n is the number of components in the state vector and ℓ is the number of components in the measurement vector [7]. However, this formulation does require that R be a diagonal matrix. Otherwise, an additional computational complexity of $(4\ell^3 + \ell^2 - 10\ell + 3\ell^2 n - 3\ell n)/6$ flops for measurement decorrelation is incurred.

De Vries Implementation. This implementation, which was shown to the authors by Thomas W. De Vries at Rockwell International, is designed to reduce the computational complexity of the Joseph formulation by judicious rearrangement of the matrix expressions and reuse of intermediate results. The fundamental operations—and their computational complexities—are summarized in Table 6.19.

TABLE 6.19 De Vries–Joseph Implementation of Covariance Update

Operation	Complexity
Without Using Decorrelation	
$\mathcal{T}_1 = P(-)H^{\mathrm{T}}$	ℓn^2
$\mathcal{T}_2 = H\mathcal{T}_1 + R$	$n\ell(\ell+1)/2$
$\mathcal{U}\mathcal{D}\mathcal{U}^{\mathrm{T}} = \mathcal{T}_2$	$\frac{1}{6}\ell(\ell+1)(\ell+2)$ (*UD* factorization)
$\mathcal{U}\mathcal{D}\mathcal{U}^{\mathrm{T}}K^{\mathrm{T}} = \mathcal{T}_1^{\mathrm{T}}$	$\ell^2 n$ [to solve for K]
$\mathcal{T}_3 = \frac{1}{2}K\mathcal{T}_2 - \mathcal{T}_1$	$\ell^2(n+1)$
$\mathcal{T}_4 = \mathcal{T}_3 K^{\mathrm{T}}$	ℓn^2
$P(+) = P(-) + \mathcal{T}_4 + \mathcal{T}_4^{\mathrm{T}}$	(included above)
Total	$\frac{1}{6}\ell^3 + \frac{3}{2}\ell^2 + \frac{1}{3}\ell + \frac{1}{2}\ell n + \frac{5}{2}\ell^2 n + 2\ell n^2$
Using Decorrelation	
Decorrelation	$\frac{2}{3}\ell^3 + \ell^2 - \frac{5}{3}\ell - \frac{1}{2}\ell n + \frac{1}{2}\ell^2 n$
ℓ repeats:	$\ell \times \{$
$\mathcal{T}_1 = P(-)H^{\mathrm{T}}$	n^2
$\mathcal{T}_2 = H\mathcal{T}_1 + R$	n
$K = \mathcal{T}_1/\mathcal{T}_2$	n
$\mathcal{T}_3 = \frac{1}{2}K\mathcal{T}_2 - \mathcal{T}_1$	$n+1$
$\mathcal{T}_4 = \mathcal{T}_3 K^{\mathrm{T}}$	n^2
$P(+) = P(-) + \mathcal{T}_4 + \mathcal{T}_4^{\mathrm{T}}$	(included above)$\}$
Total	$\frac{2}{3}\ell^3 + \ell^2 - \frac{2}{3}\ell + \frac{5}{3}\ell n + \frac{1}{2}\ell^2 n + 2\ell n^2$

Negative Evaluations of Joseph-Stabilized Implementation. In comparative evaluations of several Kalman filter implementations on orbit estimation problems, Thornton and Bierman [125] found that the Joseph-stabilized implementation failed on some ill-conditioned problems for which square-root methods performed well.

6.6.2 Morf–Kailath Combined Observational/Temporal Update

The lion's share of the computational effort in Kalman filtering is spent in solving the Riccati equation. This effort is necessary for computing the Kalman gains. However, only the a priori value of the covariance matrix is needed for this purpose. Its a posteriori value is only used as an intermediate result on the way to computing the next a priori value.

Actually, it is not necessary to compute the a posteriori values of the covariance matrix explicitly. It is possible to compute the a priori values from one temporal epoch to the next temporal epoch, without going through the intermediate a posteriori values. This concept, and the methods for doing it, were introduced by Martin Morf and Thomas Kailath [204].

6.6.2.1 Combined Updates of Cholesky Factors. The direct computation of $C_{P(k+1)}(-)$, the triangular Cholesky factor of $P_{k+1}(-)$, from $C_{P(k)}(-)$, the triangular Cholesky factor of $P_k(-)$, can be implemented by triangularization of the $(n+m) \times (p+n+m)$ partitioned matrix

$$A_k = \begin{bmatrix} GC_{Q(k)} & \Phi_k C_{P(k)} & 0 \\ 0 & H_k C_{P(k)} & C_{R(k)} \end{bmatrix}, \tag{6.129}$$

where $C_{R(k)}$ is a Cholesky factor of R_k and $C_{Q(k)}$ is a Cholesky factor of Q_k. Note that the $(n+m) \times (n+m)$ symmetric product

$$A_k A_k^{\mathrm{T}} = \begin{bmatrix} \Phi_k P_k(-)\Phi_k^{\mathrm{T}} + G_k Q_k G_k^{\mathrm{T}} & \Phi_k P_k(-) H_k^{\mathrm{T}} \\ H_k P_k(-)\Phi_k^{\mathrm{T}} & H_k P_k(-) H_k^{\mathrm{T}} \end{bmatrix}. \tag{6.130}$$

Consequently, if A_k is upper triangularized in the form

$$A_k T = C_k \tag{6.131}$$

$$= \begin{bmatrix} 0 & C_{P(k+1)} & \Psi_k \\ 0 & 0 & C_{E(k)} \end{bmatrix} \tag{6.132}$$

by an orthogonal transformation T, then the matrix equation

$$C_k C_k^{\mathrm{T}} = A_k A_k^{\mathrm{T}}$$

implies that the newly created block submatrices $C_{E(k)}$, Ψ_k, and $C_{P(k+1)}$ satisfy the equations

$$C_{E(k)} C_{E(k)}^{\mathrm{T}} = H_k P_k(-) H_k^{\mathrm{T}} + R_k \tag{6.133}$$

$$= E_k, \tag{6.134}$$

$$\Psi_k \Psi_k^{\mathrm{T}} = \Phi_k P_k(-) H_k^{\mathrm{T}} E_k^{-1} H_k P_k(-) \Phi_k, \tag{6.135}$$

$$\Psi_k = \Phi_k P_k(-) H_k^{\mathrm{T}} C_{E(k)}^{-1}, \tag{6.136}$$

$$C_{P(k+1)} C_{P(k+1)}^{\mathrm{T}} = \Phi_k P_k(-)\Phi_k^{\mathrm{T}} + G_k Q_k G_k^{\mathrm{T}} - \Psi_k \Psi_k^{\mathrm{T}} \tag{6.137}$$

$$= P_{k+1}(-), \tag{6.138}$$

and the Kalman gain can be computed as

$$\overline{K}_k = \Psi_k C_{E(k)}^{-1}. \tag{6.139}$$

The computation of C_k from A_k can be done by Householder or Givens triangularization.

6.6.2.2 Combined Updates of UD Factors. This implementation uses the UD factors of the covariance matrices P, R, and Q,

$$P_k = U_{P(k)} D_{P(k)} U_{P(k)}^{\mathrm{T}}, \tag{6.140}$$

$$R_k = U_{R(k)} D_{R(k)} U_{R(k)}^{\mathrm{T}}, \tag{6.141}$$

$$Q_k = U_{Q(k)} D_{Q(k)} U_{Q(k)}^{\mathrm{T}}, \tag{6.142}$$

in the partitioned matrices

$$B_k = \begin{bmatrix} GU_{Q(k)} & \Phi_k U_{P(k)} & 0 \\ 0 & H_k U_{P(k)} & U_{R(k)} \end{bmatrix}, \tag{6.143}$$

$$D_k = \begin{bmatrix} D_{Q(k)} & 0 & 0 \\ 0 & D_{P(k)} & 0 \\ 0 & 0 & D_{R(k)} \end{bmatrix}, \tag{6.144),}$$

which satisfy the equation

$$B_k D_k B_k^{\mathrm{T}} = \begin{bmatrix} \Phi_k P_k(-)\Phi_k^{\mathrm{T}} + G_k Q_k G_k^{\mathrm{T}} & \Phi_k P_k(-)H_k^{\mathrm{T}} \\ H_k P_k(-)\Phi_k^{\mathrm{T}} & H_k P_k(-)H_k^{\mathrm{T}} \end{bmatrix}. \tag{6.145}$$

The MWGS orthogonalization of the rows of B_k with respect to the weighting matrix D_k will yield the matrices

$$\acute{B}_k = \begin{bmatrix} U_{P(k+1)} & U_{\Psi(k)} \\ 0 & U_{E(k)} \end{bmatrix}, \tag{6.146}$$

$$\acute{D}_k = \begin{bmatrix} D_{P(k+1)} & 0 \\ 0 & D_{E(k)} \end{bmatrix}, \tag{6.147}$$

where $U_{P(k+1)}$ and $D_{P(k+1)}$ are the UD factors of $P_{k+1}(-)$, and

$$U_{\Psi(k)} = \Phi_k P_k(-) H_k^{\mathrm{T}} U_{E(k)}^{-\mathrm{T}} D_{E(k)}^{-1} \tag{6.148}$$

$$= \overline{K}_k U_{E(k)}. \tag{6.149}$$

Consequently, the Kalman gain

$$\overline{K}_k = U_{\Psi(k)} U_{E(k)}^{-1} \tag{6.150}$$

can be computed as a by-product of the MWGS procedure, as well.

6.6.3 Information Filtering

6.6.3.1 Information Matrix of an Estimate. The inverse of the covariance matrix of estimation uncertainty is called the *information matrix*[13]:

$$Y \stackrel{\text{def}}{=} P^{-1}. \tag{6.151}$$

Implementations using Y (or its Cholesky factors) rather than P (or its Cholesky factors) are called *information filters*. (Implementations using P are also called *covariance filters*.)

6.6.3.2 Uses of Information Filtering

Problems without Prior Information. Using the information matrix, one can express the idea that an estimation process may start with no a priori information whatsoever, expressed by

$$Y_0 = 0, \tag{6.152}$$

a matrix of zeros. An information filter starting from this condition will have absolutely no bias toward the a priori estimate. Covariance filters cannot do this.

One can also represent a priori estimates with no information in specified subspaces of state space by using information matrices with characteristic values equal to zero. In that case, the information matrix will have an eigenvalue–eigenvector decomposition of the form

$$Y_0 = \sum_i \lambda_i e_i e_i^{\mathrm{T}}, \tag{6.153}$$

where some of the eigenvalues $\lambda_i = 0$ and the corresponding eigenvectors e_i represent directions in state space with zero a priori information. Subsequent estimates will have no bias toward these components of the a priori estimate.

Information filtering cannot be used if P is singular, just as covariance filtering cannot be used if Y is singular. However, one may switch representations if both conditions do not occur simultaneously. For example, an estimation problem with zero initial information can be started with an information filter and then switched to a covariance implementation when Y becomes nonsingular. Conversely, a filtering problem with zero initial uncertainty may be started with a covariance filter, then switched to an information filter when P becomes nonsingular.

[13]This is also called the *Fisher information matrix*, named after the English statistician Ronald Aylmer Fisher (1890–1962). More generally, for distributions with differentiable probability density functions, the information matrix is defined as the matrix of second-order derivatives of the logarithm of the probability density with respect to the variates. For Gaussian distributions, this equals the inverse of the covariance matrix.

Robust Observational Updates. The observational update of the *uncertainty matrix* is less robust against roundoff errors than the temporal update. It is more likely to cause the uncertainty matrix to become indefinite, which tends to destabilize the estimation feedback loop.

The observational update of the information matrix is more robust against roundoff errors. This condition is the result of a certain duality between information filtering and covariance filtering, by which the algorithmic structures of the temporal and observational updates are switched between the two approaches. The downside of this duality is that the temporal update of the information matrix is less robust than the observational update against roundoff errors and is a more likely cause of degradation. Therefore, information filtering may not be a panacea for all conditioning problems, but in those cases for which the observational update of the uncertainty matrix is the culprit, information filtering offers a possible solution to the roundoff problem.

Disadvantages of Information Filtering. The greatest objection to information filtering is the loss of "transparency" of the representation. Although information is a more practical concept than uncertainty for some problems, it can be more difficult to interpret its physical significance and to use it in our thinking. With a little practice, it is relatively easy to visualize how σ (the square root of variance) is related to probabilities and to express uncertainties as "3σ" values. One must invert the information matrix before one can interpret its values in this way.

Perhaps the greatest impediment to widespread acceptance of information filtering is the loss of physical significance of the associated state vector components. These are linear combinations of the original state vector components, but the coefficients of these linear combinations change with the state of information/uncertainty in the estimates.

6.6.3.3 Information States. Information filters do not use the same state vector representations as covariance filters. Those that use the information matrix in the filter implementation use the *information state*

$$d \stackrel{\text{def}}{=} Yx, \tag{6.154}$$

and those that use its Cholesky factors C_Y such that

$$C_Y C_Y^{\mathrm{T}} = Y \tag{6.155}$$

use the square-root information state

$$s \stackrel{\text{def}}{=} C_Y x. \tag{6.156}$$

6.6.3.4 Information Filter Implementation. The implementation equations for the "straight" information filter (i.e., using Y, rather than its Cholesky factors) are shown in Table 6.20. These can be derived from the Kalman filter equations and the definitions of the information matrix and information state. Note the similarities in form between these equations and the Kalman filter equations, with respective observational and temporal equations switched.

6.6.3.5 Square-Root Information Filtering. The *square-root information filter* is usually abbreviated as *SRIF.* (The conventional square root filter is often abbreviated as *SRCF,* which stands for *square-root covariance filter.*) Like the SRCF, the SRIF is more robust against roundoff errors than the "straight" form of the filter.

Historical note: A complete formulation (i.e., including both updates) of the SRIF was developed by Dyer and McReynolds [156], using the square-root least-squares methods (triangularization) developed by Golub [165] and applied to sequential least-squares estimation by Lawson and Hanson [91]. The form developed by Dyer and McReynolds is shown in Table 6.21.

TABLE 6.20 Information Filter Equations

Observational update:

$$\hat{d}_k(+) = \hat{d}_k(-) + H_k^{\mathrm{T}} R_k^{-1} z_k$$
$$Y_k(+) = Y_k(-) + H_k^{\mathrm{T}} R_k^{-1} H_k$$

Temporal update:

$$A_k \stackrel{\text{def}}{=} \Phi_k^{-\mathrm{T}} Y_k(+) \Phi_k^{-1}$$
$$Y_{k+1}(-) = \left\{ I - A_k G_k [G_k^{\mathrm{T}} A_k G_k + Q_k^{-1}]^{-1} G_k^{\mathrm{T}} \right\} A_k$$
$$\hat{d}_{k+1}(-) = \left\{ I - A_k G_k [G_k^{\mathrm{T}} A_k G_k + Q_k^{-1}]^{-1} G_k^{\mathrm{T}} \right\} \Phi_k^{-\mathrm{T}} \hat{d}_k(+)$$

TABLE 6.21 Square-Root Information Filter Using Triangularization

Observational update:

$$\begin{bmatrix} C_{Y_k}(-) & H_k^{\mathrm{T}} C_{R_k^{-1}} \\ \hat{s}_k^{\mathrm{T}}(-) & z_k^{\mathrm{T}} C_{R_k^{-1}} \end{bmatrix} T_{\text{obs}} = \begin{bmatrix} C_{Y_k}(+) & 0 \\ \hat{s}_k^{\mathrm{T}}(+) & \epsilon \end{bmatrix}$$

Temporal update:

$$\begin{bmatrix} C_{Q_k^{-1}} & -G_k \Phi_k^{-\mathrm{T}} C_{Y_k}(+) \\ 0 & \Phi_k^{-\mathrm{T}} C_{Y_k}(+) \\ 0 & \hat{s}_k^{\mathrm{T}}(+) \end{bmatrix} T_{\text{temp}} = \begin{bmatrix} \Theta & 0 \\ \Gamma & C_{Y_{k+1}}(-) \\ \tau^{\mathrm{T}} & \hat{s}_{k+1}^{\mathrm{T}}(-) \end{bmatrix}$$

Note: T_{obs} and T_{temp} are orthogonal matrices (composed of Householder or Givens transformations), which lower triangularize the left-hand-side matrices. The submatrices other than s and C_Y on the right-hand sides are extraneous.

6.7 SUMMARY

Although Kalman filtering has been called "ideally suited to digital computer implementation" [21], the digital computer is not ideally suited to the task. The conventional implementation of the Kalman filter—in terms of covariance matrices—is particularly sensitive to roundoff errors.

Many methods have been developed for decreasing the sensitivity of the Kalman filter to roundoff errors. The most successful approaches use alternative representations for the covariance matrix of estimation uncertainty, in terms of symmetric products of triangular factors. These fall into three general classes:

1. *Square-root covariance filters*, which use a decomposition of the covariance matrix of estimation uncertainty as a symmetric product of triangular Cholesky factors:
$$P = CC^{\mathrm{T}}.$$
2. *UD covariance filters*, which use a modified (square-root-free) Cholesky decomposition of the covariance matrix:
$$P = UDU^{\mathrm{T}}.$$
3. *Square root information filters*, which use a symmetric product factorization of the information matrix, P^{-1}.

The alternative Kalman filter implementations use these factors of the covariance matrix (or it inverse) in three types of filter operations:

1. *temporal updates*,
2. *observational updates*, and
3. *combined updates* (temporal and observational).

The basic algorithmic methods used in these alternative Kalman filter implementations fall into four general categories. The first three of these categories of methods are concerned with decomposing matrices into triangular factors and maintaining the triangular form of the factors through all the Kalman filtering operations:

1. *Cholesky decomposition methods*, by which a symmetric positive-definite matrix M can be represented as symmetric products of a triangular matrix C:
$$M = CC^{\mathrm{T}} \quad \text{or} \quad M = UDU^{\mathrm{T}}.$$
The Cholesky decomposition algorithms compute C (or U and D), given M.
2. *Triangularization methods*, by which a symmetric product of a general matrix A can be represented as a symmetric product of a triangular matrix C:
$$AA^{\mathrm{T}} = CC^{\mathrm{T}} \quad \text{or} \quad A\acute{D}A^{\mathrm{T}} = UDU^{\mathrm{T}}.$$

These methods compute C (or U and D), given A (or A and $\acute{D}$).

3. *Rank 1 modification methods*, by which the sum of a symmetric product of a triangular matrix $\acute{C}$ and scaled symmetric product of a vector (rank 1 matrix) $\mathbf{v}$ can be represented by a symmetric product of a new triangular matrix C:

$$\acute{C}\acute{C}^{\mathrm{T}} + s\mathbf{v}\mathbf{v}^{\mathrm{T}} = CC^{\mathrm{T}} \quad \text{or} \quad \acute{U}\acute{D}\acute{U}^{\mathrm{T}} + s\mathbf{v}\mathbf{v}^{\mathrm{T}} = UDU^{\mathrm{T}}.$$

These methods compute C (or U and D), given $\acute{C}$ (or $\acute{U}$ and $\acute{D}$), s, and $\mathbf{v}$.

The fourth category of methods includes standard matrix operations (multiplications, inversions, etc.) that have been specialized for triangular matrices.

These implementation methods have succeeded where the conventional Kalman filter implementation has failed.

It would be difficult to overemphasize the importance of good numerical methods in Kalman filtering. Limited to finite precision, computers will always make approximation errors. They are not infallible. One must always take this into account in problem analysis. The effects of roundoff may be thought to be minor, but overlooking them could be a major blunder.

PROBLEMS

6.1 An $n \times n$ *Moler matrix* M has elements

$$M_{ij} = \begin{cases} i & \text{if } i = j, \\ \min(i, j) & \text{if } i \neq j. \end{cases}$$

Calculate the 3×3 Moler matrix and its lower triangular Cholesky factor.

6.2 Write a MATLAB script to compute and print out the $n \times n$ Moler matrices and their lower triangular Cholesky factors for $2 \leq n \leq 20$.

6.3 Show that the condition number of a Cholesky factor C of $P = CC^{\mathrm{T}}$ is the square root of the condition number of P.

6.4 Show that, if A and B are $n \times n$ upper triangular matrices, then their product AB is also upper triangular.

6.5 Show that a square, triangular matrix is singular if and only if one of its diagonal terms is zero. (*Hint*: What is the determinant of a triangular matrix?)

6.6 Show that the inverse of an upper (lower) triangular matrix is also an upper (lower) triangular matrix.

6.7 Show that, if the upper triangular Cholesky decomposition algorithm is applied to the matrix product

$$\begin{bmatrix} H & z \end{bmatrix}^{\mathrm{T}} \begin{bmatrix} H & z \end{bmatrix} = \begin{bmatrix} H^{\mathrm{T}}H & H^{\mathrm{T}}z \\ z^{\mathrm{T}}H & z^{\mathrm{T}}z \end{bmatrix}$$

and the upper triangular result is similarly partitioned as $\begin{bmatrix} U & y \\ 0 & \varepsilon \end{bmatrix}$, then the solution $\hat{x}$ to the equation $U\hat{x} = y$ (which can be computed by back substitution) solves the least-squares problem $Hx \approx z$ with root summed square residual $\|H\hat{x} - z\| = \varepsilon$ (Cholesky's method of least squares).

6.8 The *singular-value decomposition* of a symmetric, nonnegative-definite matrix P is a factorization $P = EDE^{\mathrm{T}}$ such that E is an orthogonal matrix and $D = \mathrm{diag}(d_1, d_2, d_3, \ldots, d_n)$ is a diagonal matrix with nonnegative elements $d_i \geq 0, 1 \leq i \leq n$. For $D^{1/2} = \mathrm{diag}(d_1^{1/2}, d_2^{1/2}, d_3^{1/2}, \ldots, d_n^{1/2})$, show that the symmetric matrix $C = ED^{1/2}E^{\mathrm{T}}$ is both a Cholesky factor of P and a square root of P.

6.9 Show that the column vectors of the orthogonal matrix E in the singular value decomposition of P (in the above exercise) are the characteristic vectors (eigenvectors) of P, and the corresponding diagonal elements of D are the respective characteristic values (eigenvalues). That is, for $1 \leq i \leq n$, if e_i is the ith column of E, show that $Pe_i = d_i e_i$.

6.10 Show that, if $P = EDE^{\mathrm{T}}$ is a singular-value decomposition of P (defined above), then $P = \sum_{i=1}^{n} d_i e_i e_i^{\mathrm{T}}$, where e_i is the ith column vector of E.

6.11 Show that, if C is an $n \times n$ Cholesky factor of P, then, for any orthogonal matrix T, CT is also a Cholesky factor of P.

6.12 Show that $(I - vv^{\mathrm{T}})^2 = (I - vv^{\mathrm{T}})$ if $|v|^2 = 1$ and that $(I - vv^{\mathrm{T}})^2 = I$ if $|v|^2 = 2$.

6.13 Show that the following formula generalizes the Potter observational update to include vector-valued measurements:

$$C(+) = C(-)[I - VM^{-\mathrm{T}}(M + F)^{-1}V^{\mathrm{T}}],$$

where $V = C^{\mathrm{T}}(-)H^{\mathrm{T}}$ and F and M are Cholesky factors of R and $R + V^{\mathrm{T}}V$, respectively.

6.14 Prove the following lemma: If W is an upper triangular $n \times n$ matrix such that

$$WW^{\mathrm{T}} = I - \frac{\mathbf{v}\mathbf{v}^{\mathrm{T}}}{R + |\mathbf{v}|^2},$$

then[14]

$$\sum_{k=m}^{j} W_{ik} W_{mk} = \Delta_{im} - \frac{\mathbf{v}_i \mathbf{v}_m}{R + \sum_{k=1}^{j} \mathbf{v}_k^2} \tag{6.157}$$

for all i, m, j such that $1 \leq i \leq m \leq j \leq n$.

[14]Kronecker's delta (Δ_{ij}) is defined to equal 1 only if its subscripts are equal ($i = j$) and to equal zero otherwise.

6.15 Prove that the Björck "modified" Gram–Schmidt algorithm results in a set of mutually orthogonal vectors.

6.16 Suppose that

$$V = \begin{bmatrix} 1 & 1 & 1 \\ \epsilon & 0 & 0 \\ 0 & \epsilon & 0 \\ 0 & 0 & \epsilon \end{bmatrix},$$

where ϵ is so small that $1 + \epsilon^2$ (but not $1 + \epsilon$) rounds to 1 in machine precision. Compute the rounded result of Gram–Schmidt orthogonalization by the conventional and modified methods. Which result is closer to the theoretical value?

6.17 Show that, if A and B are orthogonal matrices, then

$$\begin{bmatrix} A & 0 \\ 0 & B \end{bmatrix}$$

is an orthogonal matrix.

6.18 What is the inverse of the Householder reflection matrix $I - 2vv^{\mathrm{T}}/v^{\mathrm{T}}v$?

6.19 How many Householder transformations are necessary for triangularization of an $n \times q$ matrix when $n < q$? Does this change when $n = q$?

6.20 (Continuous temporal update of Cholesky factors.) Show that all differentiable Cholesky factors $C(t)$ of the solution $P(t)$ to the linear dynamic equation $\dot{P} = F(t)P(t) + P(t)F^{\mathrm{T}}(t) + G(t)Q(t)G^{\mathrm{T}}(t)$, where Q is symmetric, are solutions of a nonlinear dynamic equation $\dot{C}(t) = F(t)C(t) + \frac{1}{2}[G(T)Q(t)G^{\mathrm{T}}(t) + A(t)]C^{-\mathrm{T}}(t)$, where $A(t)$ is a skew-symmetric matrix [130].

6.21 Prove that the condition number of the information matrix is equal to the condition number of the corresponding covariance matrix in the case that neither of them is singular. (The condition number is the ratio of the largest characteristic value to the smallest characteristic value.)

6.22 Prove the correctness of the triangularization equation for the observational update of the SRIF. (*Hint*: Multiply the partitioned matrices on the right by their respective transposes.)

6.23 Prove the correctness of the triangularization equation for the temporal update of the SRIF.

6.24 Prove to yourself that the conventional Kalman filter Riccati equation

$$P(+) = P(-) - P(-)H^{\mathrm{T}}[HP(-)H^{\mathrm{T}} + R]^{-1}HP(-)$$

for the observational update is equivalent to the information form

$$P^{-1}(+) = P^{-1}(-) + H^{\mathrm{T}}R^{-1}H$$

of Peter Swerling. (*Hint:* Try multiplying the form for $P(+)$ by the form for $P^{-1}(+)$ and see if it equals I, the identity matrix.)

6.25 Show that, if C is a Cholesky factor of P (i.e., $P = CC^{\mathrm{T}}$), then $C^{-\mathrm{T}} = (C^{-1})^{\mathrm{T}}$ is a Cholesky factor of $Y = P^{-1}$, provided that the inverse of C exists. Conversely, the *transposed* inverse of any Cholesky factor of the information matrix Y is a Cholesky factor of the covariance matrix P, provided that the inverse exists.

6.26 Write a MATLAB script to implement Example 4.4 using the Bierman–Thornton UD filter, plotting as a function of time the resulting RMS estimation uncertainty values of $P(+)$ and $P(-)$ and the components of $\overline{K}$. (You can use the scripts bierman.m and thornton., but you will have to compute UDU^{T} and take the square roots of its diagonal values to obtain RMS uncertainties.)

6.27 Write a MATLAB script to implement Example 4.4 using the Potter square-root filter and plotting the same values as in the problem above.

7

Practical Considerations

"The time has come," the Walrus said,
"To talk of many things:
Of shoes—and ships—and sealing wax—
Of cabbages—and kings—
And why the sea is boiling hot—
And whether pigs have wings."
From "The Walrus and the Carpenter," in Through the Looking Glass, 1872
Lewis Carroll [Charles Lutwidge Dodgson] (1832–1898)

7.1 CHAPTER FOCUS

The discussion turns now to what might be called Kalman filter *engineering*, which is that body of applicable knowledge that has evolved through practical experience in the use and misuse of the Kalman filter. The material of the previous two chapters (extended Kalman filtering and square-root filtering) has also evolved in this way and is part of the same general subject. Here, however, the discussion includes many more matters of practice than nonlinearities and finite-precision arithmetic.

7.1.1 Main Points to Be Covered

1. *Roundoff errors are not the only causes for the failure of the Kalman filter to achieve its theoretical performance.* There are diagnostic methods for identifying causes and remedies for other common patterns of misbehavior.
2. *Prefiltering to reduce computational requirements.* If the dynamics of the measured variables are "slow" relative to the sampling rate, then a simple prefilter can reduce the overall computational requirements without sacrificing performance.

3. *Detection and rejection of anomalous sensor data*. The inverse of the matrix $(HPH^{\mathrm{T}} + R)$ characterizes the probability distribution of the innovation $z - H\hat{x}$ and may be used to test for exogenous measurement errors, such as those resulting from sensor or transmission malfunctions.
4. *Statistical design of sensor and estimation systems*. The covariance equations of the Kalman filter provide an analytical basis for the predictive design of systems to estimate the state of dynamic systems. They may also be used to obtain suboptimal (but feasible) observation scheduling.
5. *Testing for asymptotic stability*. The relative robustness of the Kalman filter against minor modeling errors is due, in part, to the asymptotic stability of the Riccati equations defining performance.
6. *Model simplifications to reduce computational requirements*. A dual-state filter implementation can be used to analyze the expected performance of simplified Kalman filters, based on simplifying the dynamic system model and/or measurement model. These analyses characterize the trade-offs between performance and computational requirements.
7. *Memory and throughput requirements*. These computational requirements are represented as functions of "problem parameters" such as the dimensions of state and measurement vectors.
8. *Offline processing to reduce on-line computational requirements*. Except in extended (nonlinear) Kalman filtering, the gain computations do not depend upon the real-time data. Therefore, they can be precomputed to reduce the real-time computational load.
9. *Application to aided inertial navigation*, in which the power of Kalman filtering is demonstrated on a realistic problem (see also [22]).

7.2 DETECTING AND CORRECTING ANOMALOUS BEHAVIOR

7.2.1 Convergence, Divergence, and "Failure to Converge"

Definitions of Convergence and Divergence. A sequence $\{\eta_k | k = 1, 2, 3, \ldots\}$ of real vectors η_k is said to *converge* to a *limit* η_∞ if, for every $\varepsilon > 0$, for some n, for all $k > n$, the norm of the differences $\|\eta_k - \eta_\infty\| < \varepsilon$. Let us use the expressions

$$\lim_{k \to \infty} \eta_k = \eta_\infty \quad \text{or} \quad \eta_k \to \eta_\infty$$

to represent convergence. One vector sequence is said to converge to another vector sequence if their differences converge to the zero vector, and a sequence is said to converge[1] if, for every $\varepsilon > 0$, for some integer n, for all $k, \ell > n$, $\|\eta_k - \eta_\ell\| < \varepsilon$.

[1] Such sequences are called *Cauchy sequences*, after Augustin Louis Cauchy (1789–1857).

Divergence is defined as convergence to ∞: for every $\varepsilon > 0$, for some integer n, for all $k > n$, $|\eta_k| > \varepsilon$. In that case, $\|\eta_k\|$ is said to *grow without bound*.

Nonconvergence. This is a more common issue in the performance of Kalman filters than strict divergence. That is, the filter fails because it does not converge to the desired limit, although it does not necessarily diverge.

Dynamic and Stochastic Variables Subject to Convergence or Divergence. The operation of a Kalman filter involves the following sequences that may or may not converge or diverge:

x_k, the sequence of actual state values;
$\mathrm{E}\langle x_k x_k^{\mathrm{T}} \rangle$, the mean-squared state;
$\hat{x}_k$, the estimated state;
$\tilde{x}_k(-) = \hat{x}_k(-) - x_k$, the a priori estimation error;
$\tilde{x}_k(+) = \hat{x}_k(+) - x_k$, the a posteriori estimation error;
$P_k(-)$, the covariance of a priori estimation errors;
$P_k(+)$, the covariance of a posteriori estimation errors.

One may also be interested in whether or not the sequences $\{P_{k(-)}\}$ and $\{P_{k(+)}\}$ computed from the Riccati equations converge to the corresponding *true* covariances of estimation error.

7.2.2 Use of Riccati Equation to Predict Behavior

The covariance matrix of estimation uncertainty characterizes the theoretical performance of the Kalman filter. It is computed as an ancillary variable in the Kalman filter as the solution of a matrix Riccati equation with the given initial conditions. It is also useful for predicting performance. If its characteristic values are growing without bound, then the theoretical performance of the Kalman filter is said to be diverging. This can happen if the system state is unstable and unobservable, for example. This type of divergence is detectable by solving the Riccati equation to compute the covariance matrix.

The Riccati equation is not always well conditioned for numerical solution and one may need to use the more numerically stable methods of Chapter 6 to obtain reasonable results. One can, for example, use eigenvalue–eigenvector decomposition of solutions to test their characteristic roots (they should be positive) and condition numbers. Condition numbers within one or two orders of magnitude of ε^{-1} (the reciprocal of the unit roundoff error in computer precision) are considered probable cause for concern and reason to use square-root methods.

7.2.3 Testing for Unpredictable Behavior

Not all filter divergence is predictable from the Riccati equation solution. Sometimes the actual performance does not agree with theoretical performance.

One cannot measure estimation error directly, except in simulations, so one must find other means to check on estimation accuracy. Whenever the estimation error is deemed to differ significantly from its expected value (as computed by the Riccati equation), the filter is said to diverge from its predicted performance. We will now consider how one might go about detecting divergence.

Examples of typical behaviors of Kalman filters are shown in Figure 7.1, which is a multiplot of the estimation errors on 10 different simulations of a filter implementation with independent pseudorandom-error sequences. Note that each time the filter is run, different estimation errors $\tilde{x}(t)$ result, even with the same initial condition $\hat{x}(0)$. Also note that at any particular time the average estimation error (across the ensemble of simulations) is approximately zero,

$$\frac{1}{N}\sum_{i=1}^{N}[\hat{x}_i(t_k) - x(t_k)] \approx \mathop{\mathrm{E}}_{i}\langle\hat{x}_i(t_k) - x(t_k)\rangle = 0, \tag{7.1}$$

where N is the number of simulation runs and $\hat{x}_i(t_k) - x(t_k)$ is the estimation error at time t_k on the ith simulation run.

Monte Carlo analysis of Kalman filter performance uses many such runs to test that the ensemble mean estimation error is *unbiased* (i.e., has effectively zero mean) and that its ensemble covariance is in close agreement with the theoretical value computed as a solution of the Riccati equation.

Convergence of Suboptimal Filters. In the suboptimal filters discussed in Section 7.5, the estimates can be biased. Therefore, in the analysis of suboptimal filters, the behavior of $P(t)$ is not sufficient to define convergence. A suboptimal filter is said to converge if the covariance matrices converge,

$$\lim_{t\to\infty}[\mathrm{trace}(P_{\text{sub-opt}} - P_{\text{opt}})] = 0, \tag{7.2}$$

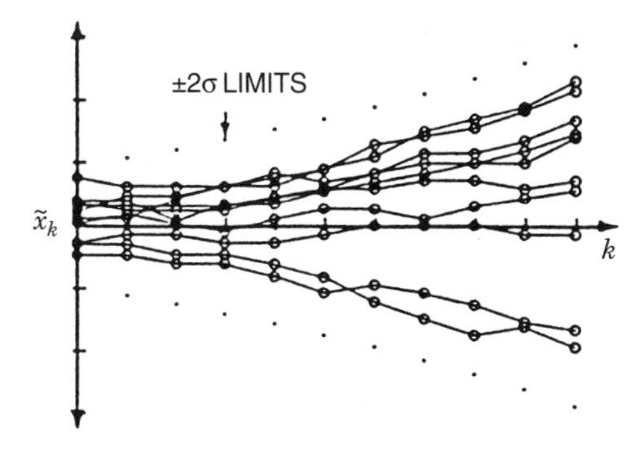

Fig. 7.1 Dispersion of multiple runs.

and the asymptotic estimation error is unbiased,

$$\lim_{t\to\infty} \mathrm{E}[\tilde{x}(t)] = 0. \tag{7.3}$$

Example 7.1: Some typical behavior patterns of suboptimal filter convergence are depicted by the plots of $P(t)$ in Figure 7.2*a*, and characteristics of systems with these symptoms are given here as examples.

Case A: Let a scalar continuous system equation be given by

$$\dot{x}(t) = Fx(t), \qquad F > 0, \tag{7.4}$$

in which the system is unstable, or

$$\dot{x}(t) = Fx(t) + w(t) \tag{7.5}$$

in which the system has driving noise and is unstable.

Case B: The system has constant steady-state uncertainty:

$$\lim_{t\to\infty} \dot{P}(t) = 0. \tag{7.6}$$

Case C: The system is stable and has no driving noise:

$$\dot{x}(t) = -Fx(t), \quad F > 0. \tag{7.7}$$

Example 7.2: Behaviors of Discrete-Time Systems Plots of P_k are shown in Figure 7.2*b* for the following system characteristics:

Case A: Effects of system driving noise and measurement noise are large relative to $P_0(t)$ (initial uncertainty).

Case B: $P_0 = P_\infty$ (Wiener filter).

Case C: Effects of system driving noise and measurement noise are small relative to $P_0(t)$.

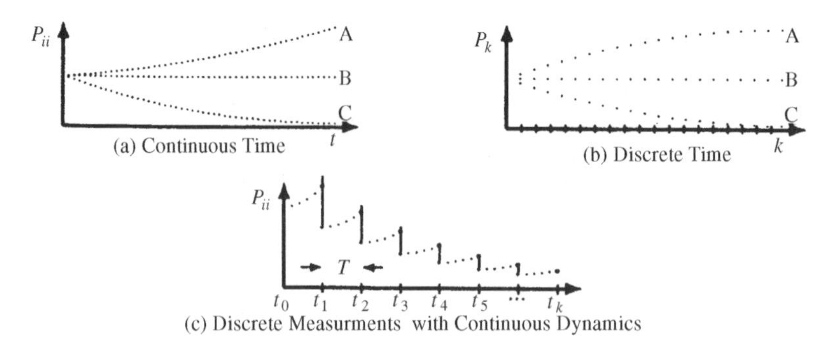

Fig. 7.2 *Asymptotic behaviors of estimation uncertainties.*

Example 7.3: Continuous System with Discrete Measurements A scalar example of a behavior pattern of the covariance propagation equation ($P_k(-)$, $\dot{P}(t)$) and covariance update equation $P_k(+)$,

$$\dot{x}(t) = Fx(t) + w(t), \quad F < 0,$$
$$z(t) = x(t) + v(t),$$

is shown in Figure 7.2*c*.

The following features may be observed in the behavior of $P(t)$:

1. Processing the measurement tends to reduce P.
2. Process noise covariance (Q) tends to increase P.
3. Damping in a stable system tends to reduce P.
4. Unstable system dynamics ($F > 0$) tend to increase P.
5. With white Gaussian measurement noise, the time between samples (T) can be reduced to decrease P.

The behavior of P represents a composite of all these effects (1–5) as shown in Figure 7.2*c*.

Causes of Predicted Nonconvergence. Nonconvergence of P predicted by the Riccati equation can be caused by

1. "natural behavior" of the dynamic equations or
2. nonobservability with the given measurements.

The following examples illustrate these behavioral patterns.

Example 7.4: The "natural behavior" for P in some cases is for

$$\lim_{t \to \infty} P(t) = P_\infty \text{ (a constant).} \tag{7.8}$$

For example,

$$\left.\begin{array}{ll} \dot{x} = w, & \text{cov}(w) = Q \\ z = x + v, & \text{cov}(v) = R \end{array}\right\} \Longrightarrow \begin{array}{l} F = 0 \\ G = H = 1 \end{array} \quad \text{in} \quad \begin{array}{l} \dot{x} = Fx + Gw, \\ z = Hx + v. \end{array} \tag{7.9}$$

Applying the continuous Kalman filter equations from Chapter 4, then

$$\dot{P} = FP + PF^{\mathrm{T}} + GQG^{\mathrm{T}} - \overline{K}R\overline{K}^{\mathrm{T}}$$

and

$$\overline{K} = PH^{\mathrm{T}}R^{-1}$$

become

$$\dot{P} = Q \; - \; \overline{K}^2 R$$

and

$$\overline{K} = \frac{P}{R}$$

or

$$\dot{P} = Q - \frac{P^2}{R}.$$

The solution is

$$P(t) = \alpha \left(\frac{P_0 \cosh(\beta t) + \alpha \sinh(\beta t)}{P_0 \sinh(\beta t) + \alpha \cosh(\beta t)} \right), \tag{7.10}$$

where

$$\alpha = \sqrt{RQ}, \qquad \beta = \sqrt{Q/R}. \tag{7.11}$$

Note that the solution of the Riccati equation converges to a finite limit:

1. $\lim_{t \to \infty} P(t) = \alpha > 0$, a finite, but nonzero, limit. (See Figure 7.3*a*.)
2. This is no cause for alarm, and there is no need to remedy the situation if the asymptotic mean-squared uncertainty is tolerable. If it is *not* tolerable, then the remedy must be found in the hardware (e.g., by attention to the physical sources of R or Q—or both) and not in software.

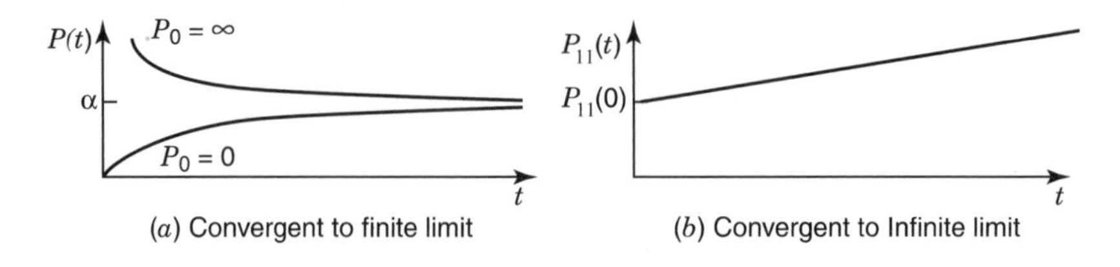

(*a*) Convergent to finite limit (*b*) Convergent to Infinite limit

Fig. 7.3 *Behavior patterns of P.*

Example 7.5: Divergence Due to "Structural" Unobservability The filter is said to diverge at infinity if its limit is unbounded:

$$\lim_{t \to \infty} P(t) = \infty. \tag{7.12}$$

As an example in which this occurs, consider the system

$$\begin{aligned} \dot{x}_1 &= w, \\ \dot{x}_2 &= 0, \\ z &= x_2 + v, \end{aligned} \qquad \begin{aligned} \operatorname{cov}(w) &= Q, \\ \operatorname{cov}(v) &= R, \end{aligned} \tag{7.13}$$

with initial conditions

$$P_0 = \begin{bmatrix} \sigma_1^2 & 0 \\ 0 & \sigma_2^2 \end{bmatrix} = \begin{bmatrix} P_{11}(0) & 0 \\ 0 & P_{22}(0) \end{bmatrix}. \tag{7.14}$$

The continuous Kalman filter equations

$$\begin{aligned} \dot{P} &= FP + PF^{\mathrm{T}} + GQG^{\mathrm{T}} - \overline{K}R\overline{K}^{\mathrm{T}}, \\ \overline{K} &= PH^{\mathrm{T}}R^{-1} \end{aligned}$$

can be combined to give

$$\dot{P} = FP + PF^{\mathrm{T}} + GQG^{\mathrm{T}} - PH^{\mathrm{T}}R^{-1}HP, \tag{7.15}$$

or

$$\dot{p}_{11} = Q - \frac{p_{12}^2}{R}, \qquad \dot{p}_{12} = -\frac{p_{12}p_{22}}{R}, \qquad \dot{p}_{22} = -\frac{p_{22}^2}{R}, \tag{7.16}$$

the solution to which is

$$p_{11}(t) = p_{11}(0) + Qt, \qquad p_{12}(t) = 0, \qquad p_{22}(t) = \frac{p_{22}(0)}{1 + [p_{22}(0)/R]t}, \tag{7.17}$$

as plotted in Figure 7.3*b*. The only remedy in this example is to alter or add measurements (sensors) to achieve observability.

Example 7.6: Nonconvergence Due to "Structural" Unobservability Parameter estimation problems have no state dynamics and no process noise. One might reasonably expect the estimation uncertainty to approach zero asymptotically as more and more measurements are made. However, it can still happen that the filter will not converge to absolute certainty. That is, the asymptotic limit of the estimation uncertainty

$$0 < \lim_{k \to \infty} P_k < \infty \tag{7.18}$$

is actually bounded away from zero uncertainty.

Parameter estimation model for continuous time. Consider the two-dimensional parameter estimation problem

$$\left.\begin{aligned} &\dot{x}_1 = 0, \qquad \dot{x}_2 = 0, \qquad P_0 = \begin{bmatrix} \sigma_1^2(0) & 0 \\ 0 & \sigma_2^2(0) \end{bmatrix}, \qquad H = [1 \quad 1], \\ &z = H\begin{bmatrix} x_1 \\ x_2 \end{bmatrix} + v, \qquad \text{cov}(v) = R, \end{aligned}\right\} \tag{7.19}$$

in which only the *sum* of the two state variables is measurable. The difference of the two state variables will then be unobservable.

Problem in discrete time. This example also illustrates a difficulty with a standard shorthand notation for discrete-time dynamic systems: the practice of using subscripts to indicate discrete time. Subscripts are more commonly used to indicate components of vectors. The solution here is to move the component indices "upstairs" and make them superscripts. (This approach only works here because the problem is linear. Therefore, one does not need superscripts to indicate powers of the components.) For these purposes, let x_k^i denote the ith component of the state vector at time t_k. The continuous form of the parameter estimation problem can then be "discretized" to a model for a discrete Kalman filter (for which the state transition matrix is the identity matrix; see Section 4.2):

$$x_k^1 = x_{k-1}^1 \quad (x^1 \text{ is constant}), \tag{7.20}$$

$$x_k^2 = x_{k-1}^2 \quad (x^2 \text{ is constant}), \tag{7.21}$$

$$z_k = [1 \quad 1]\begin{bmatrix} x_k^1 \\ x_k^2 \end{bmatrix} + v_k. \tag{7.22}$$

Let

$$\hat{x}_0 = 0.$$

The estimator then has two sources of information from which to form an optimal estimate of x_k:

1. the a priori information in $\hat{x}_0$ and P_0 and
2. the measurement sequence $z_k = x_k^1 + x_k^2 + v_k$ for $k = 1, 2, 3, \ldots$.

In this case, the best the optimal filter can do with the measurements is to "average out" the effects of the noise sequence $v_1, \ldots, v_k$. One might expect that an infinite number of measurements (z_k) would be equivalent to *one* noise-free measurement, that is,

$$z_1 = (x_1^1 + x_1^2), \quad \text{where } v_1 \to 0 \quad \text{and} \quad R = \text{cov}(v_1) \to 0. \tag{7.23}$$

Estimation uncertainty from a single noise-free measurement. By using the discrete filter equations with one stage of estimation on the measurement z_1, one can obtain the gain in the form

$$\overline{K}_1 = \begin{bmatrix} \dfrac{\sigma_1^2(0)}{(\sigma_1^2(0) + \sigma_2^2(0) + R)} \\ \dfrac{\sigma_2^2(0)}{\sigma_1^2(0) + \sigma_2^2(0) + R} \end{bmatrix}. \tag{7.24}$$

The estimation uncertainty covariance matrix can then be shown to be

$$P_1(+) = \begin{bmatrix} \dfrac{\sigma_1^2(0)\sigma_2^2(0) + R\sigma_1^2(0)}{\sigma_1^2(0) + \sigma_2^2(0) + R} & \dfrac{-\sigma_1^2(0)\sigma_2^2(0)}{\sigma_1^2(0) + \sigma_2^2(0) + R} \\ \dfrac{-\sigma_1^2(0)\sigma_2^2(0)}{\sigma_1^2(0) + \sigma_2^2(0) + R} & \dfrac{\sigma_1^2(0)\sigma_2^2(0) + R\sigma_2^2(0)}{\sigma_1^2(0) + \sigma_2^2(0) + R} \end{bmatrix} \equiv \begin{bmatrix} p_{11} & p_{12} \\ p_{12} & p_{22} \end{bmatrix}, \tag{7.25}$$

where the *correlation coefficient* (defined in Equation 3.138) is

$$\rho_{12} = \frac{p_{12}}{\sqrt{p_{11}p_{22}}} = \frac{-\sigma_1^2(0)\sigma_2^2(0)}{\sqrt{[\sigma_1^2(0)\sigma_2^2(0) + R\sigma_1^2(0)][\sigma_1^2(0)\sigma_2^2(0) + R\sigma_2^2(0)]}}, \tag{7.26}$$

and the state estimate is

$$\hat{x}_1 = \hat{x}_1(0) + \overline{K}_1[z_1 - H\hat{x}_1(0)] = [I - \overline{K}_1 H]\hat{x}_1(0) + \overline{K}_1 z_1. \tag{7.27}$$

However, for the *noise-free* case,

$$v_1 = 0, \quad R = \text{cov}(v_1) = 0,$$

the correlation coefficient is

$$\rho_{12} = -1, \tag{7.28}$$

and the estimates for $\hat{x}_1(0) = 0$,

$$\hat{x}_1^1 = \left(\frac{\sigma_1^2(0)}{\sigma_1^2(0) + \sigma_2^2(0)}\right)(x_1^1 + x_1^2),$$

$$\hat{x}_1^2 = \left(\frac{\sigma_2^2(0)}{\sigma_1^2(0) + \sigma_2^2(0)}\right)(x_1^1 + x_1^2),$$

are totally insensitive to the difference $x_1^1 - x_1^2$. As a consequence, the filter will *almost never get the right answer*! This is a fundamental characteristic of the problem, however, and not attributable to the design of the filter. There are *two* unknowns (x_1^1 and x_1^2) and *one* constraint:

$$z_1 = \left(x_1^1 + x_1^2\right). \tag{7.29}$$

Conditions for serendipitous design. The conditions under which the filter will still get the right answer can easily be derived. Because x_1^1 and x_1^2 are constants, their ratio constant

$$C \stackrel{\text{def}}{=} \frac{x_1^2}{x_1^1} \tag{7.30}$$

will also be a constant, such that the sum

$$\begin{aligned} x_1^1 + x_1^2 &= (1 + C)x_1^1 \\ &= \left(\frac{1 + C}{C}\right)x_1^2. \end{aligned}$$

Then

$$\hat{x}_1^1 = \left(\frac{\sigma_1^2(0)}{\sigma_1^2(0) + \sigma_2^2(0)}\right)\left[(1 + C)x_1^1\right] = x_1^1 \quad \text{only if} \quad \frac{\sigma_1^2(0)(1 + C)}{\sigma_1^2(0) + \sigma_2^2(0)} = 1,$$

$$\hat{x}_1^2 = \left(\frac{\sigma_2^2(0)}{\sigma_1^2(0) + \sigma_2^2(0)}\right)\left(\frac{1 + C}{C}\right)x_1^2 = x_1^2 \quad \text{only if} \quad \frac{\sigma_2^2(0)(1 + C)}{[\sigma_1^2(0) + \sigma_2^2(0)](C)} = 1.$$

Both these conditions are satisfied *only if*

$$\frac{\sigma_2^2(0)}{\sigma_1^2(0)} = C = \frac{x_1^2}{x_1^1} \geq 0, \tag{7.31}$$

because $\sigma_1^2(0)$ and $\sigma_2^2(0)$ are nonnegative numbers.

Likelihood of serendipitous design. For the filter to obtain the right answer, it would be necessary that

1. x_1^1 and x_1^2 have the *same sign* and
2. it is known that their ratio $C = x_1^2/x_1^1$.

Since both of these conditions are rarely satisfied, the filter estimates would rarely be correct.

What can be done about it. The following methods can be used to detect nonconvergence due to this type of structural unobservability:

- Test the system for observability using the "observability theorems" of Section 2.5.
- Look for perfect correlation coefficients ($\rho = \pm 1$) and be very suspicious of high correlation coefficients (e.g., $|\rho| > 0.9$).

- Perform eigenvalue–eigenvector decomposition of P to test for negative characteristic values or a large condition number. (This is a better test than correlation coefficients to detect unobservability.)
- Test the filter on a system simulator with *noise-free outputs* (measurements).

Remedies for this problem include

- attaining observability by adding another type of measurement or
- defining $\acute{x} \equiv x^1 + x^2$ as the only state variable to be estimated.

Example 7.7: Unobservability Caused by Poor Choice of Sampling Rate The problem in Example 7.6 might be solved by using an additional measurement—or by using a measurement with a time-varying sensitivity matrix. Next, consider what can go wrong even with time-varying measurement sensitivities if the sampling rate is chosen badly. For that purpose, consider the problem of estimating unknown parameters (constant states) with both constant and sinusoidal measurement sensitivity matrix components:

$$H(t) = [1 \quad \cos(\omega t)],$$

as plotted in Figure 7.4. The equivalent model for use in the discrete Kalman filter is

$$x_k^1 = x_{k-1}^1, \qquad x_k^2 = x_{k-1}^2, \qquad H_k = H(kT), \qquad z_k = H_k \begin{bmatrix} x_k^1 \\ x_k^2 \end{bmatrix} + v_k,$$

where T is the intersample interval.

What happens when Murphy's law takes effect. With the choice of intersampling interval as $T = 2\pi/\omega$ and $t_k = kT$, the components of the measurement sensitivity matrix become equal and constant:

$$\begin{aligned} H_k &= [1 \quad \cos(\omega kT)] \\ &= [1 \quad \cos(2\pi k)] \\ &= [1 \quad 1], \end{aligned}$$

as shown in Figure 7.4 (This is the way many engineers discover "aliasing.") The states x^1 and x^2 are *unobservable* with this choice of sampling interval (see Figure

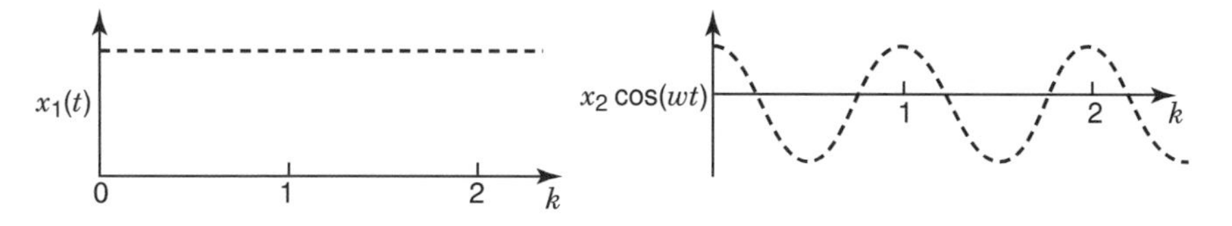

Fig. 7.4 Aliased measurement components.

7.4). With this choice of sampling times, the system and filter behave as in the previous example.

Methods for detecting and correcting unobservability include those given in Example 7.6 plus the more obvious remedy of changing the sampling interval T to obtain observability, for example,

$$T = \frac{\pi}{\omega} \tag{7.32}$$

is a better choice.

Causes of unpredicted nonconvergence. Unpredictable nonconvergence may be caused by

1. bad data,
2. numerical problems, or
3. mismodeling.

Example 7.8: Unpredicted Nonconvergence Due to Bad Data "Bad data" are caused by something going wrong, which is almost sure to happen in real-world applications of Kalman filtering. These verifications of Murphy's law occur principally in two forms:

- The initial estimate is badly chosen, for example,

$$|\hat{x}(0) - x|^2 = |\tilde{x}|^2 \gg \text{ trace } P_0. \tag{7.33}$$

- The measurement has an exogenous component (a mistake, not an error) that is excessively large, for example,

$$|v|^2 \gg \text{ trace } R. \tag{7.34}$$

Asymptotic recovery from bad data. In either case, if the system is truly linear, the Kalman filter will (in theory) recover in finite time as it continues to use measurements z_k to estimate the state x. (The best way is to prevent bad data from getting into the filter in the first place!) See Figure 7.5.

Practical limitations of recovery. Often, in practice, the recovery is not adequate in finite time. The interval $(0, T)$ of measurement availability is fixed and may be too

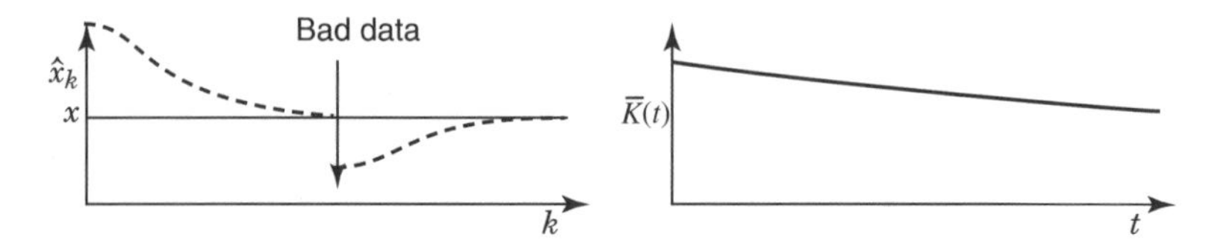

Fig. 7.5 Asymptotic recovery from bad data.

short to allow sufficient recovery (see Figure 7.6). The normal behavior of the gain matrix $\overline{K}$ may be too rapidly converging toward its steady-state value of $\overline{K} = 0$. (See Figure 7.7.)

Remedies for "heading off" bad data:

- Inspection of $P(t)$ and $\overline{K}(t)$ is useless, because they are not affected by data.
- Inspection of the state estimates $\hat{x}(t)$ for sudden jumps (*after* a bad measurement has already been used by the filter) is sometimes used, but it still leaves the problem of undoing the damage to the estimate after it has been done.
- Inspection of the "innovations" vector $[z - H\hat{x}]$ for sudden jumps or large entries (*before* bad measurement is processed by the filter) is much more useful, because the discrepancies can be interpreted probabilistically, and the data can be discarded before it has spoiled the estimate (see Section 7.3).

The best remedy for this problem is to implement a "bad data detector" to reject the bad data before it contaminates the estimate. If this is to be done in real time, it is sometimes useful to *save* the bad data for off-line examination by an "exception handler" (often a human, but sometimes a second-level data analysis program) to locate and remedy the causes of the bad data that are occurring.

Artificially increasing process noise covariance to improve bad data recovery. If bad data are detected after they have been used, one can keep the filter "alive" (to pay more attention to subsequent data) by increasing the process noise covariance Q in the system model assumed by the filter. Ideally, the new process noise covariance should reflect the actual measurement error covariance, including the bad data as well as other random noise.

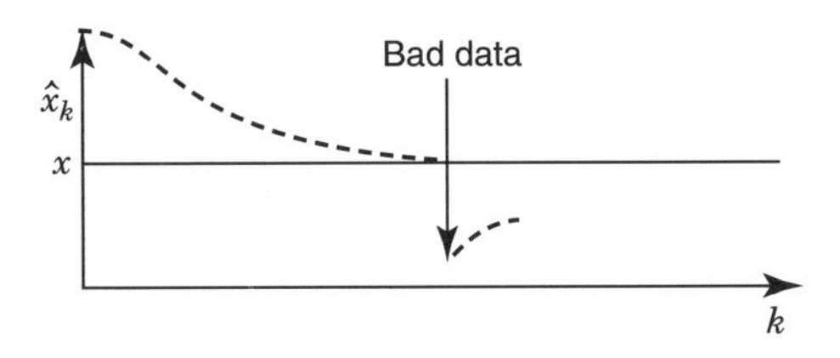

Fig. 7.6 *Failure to recover in short period.*

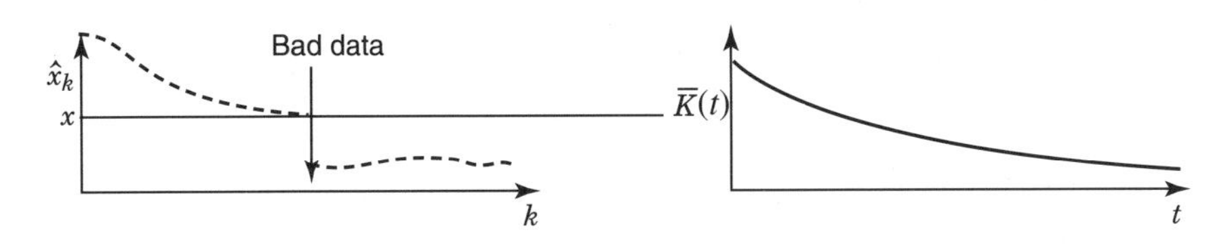

Fig. 7.7 *Failure to recover due to gain decay.*

Example 7.9: Nonconvergence Due to Numerical Problems This is sometimes detected by observing *impossible P_k behavior*. The terms on the main diagonal of P_k may become negative, or *larger*, immediately after a measurement is processed than immediately before, that is, $\sigma(+) > \sigma(-)$. A less obvious (but detectable) failure mode is for the *characteristic values* of P to become negative. This can be detected by eigenvalue–eigenvector decomposition of P. Other means of detection include using simulation (with known states) to compare estimation errors with their estimated covariances. One can also use double precision in place of single precision to detect differences due to precision. Causes of numerical problems can sometimes be traced to inadequate wordlength (precision) of the host computer. These problems tend to become worse with larger numbers of state variables.

Remedies for numerical problems. These problems have been treated by many "brute-force" methods, such as using higher precision (e.g., double instead of single). One can try reducing the number of states by merging or eliminating unobservable states, eliminating states representing "small effects," or using other suboptimal filter techniques such as decomposition into lower dimensional state spaces.

Possible remedies include the use of *more numerically stable methods* (to obtain more computational accuracy with the same computer precision) and the use of *higher precision*. The latter approach (higher precision) will increase the execution time but will generally require less reprogramming effort. One can sometimes use a better algorithm to improve the accuracy of matrix inverse $(HPH^{\mathrm{T}} + R)^{-1}$ (e.g., the Cholesky decomposition method shown in Chapter 6), or eliminate the inverse altogether by diagonalizing R and processing the measurements sequentially (also shown in Chapter 6).

7.2.4 Effects Due to Mismodeling

Lying to the Filter. The Kalman gain and the covariance matrix P are correct *if* the models used in computing them are correct. With mismodeling, the P matrix can be erroneous and of little use in detecting nonconvergence, or P can even converge to zero while the state estimation error $\tilde{x}$ is actually diverging. (It happens.)

The problems that result from bad initial estimates of x or P have already been addressed. There are four other types that will now be addressed:

1. unmodeled state variables of the dynamic system,
2. unmodeled process noise,
3. errors in the dynamic coefficients or state transition matrix, and
4. overlooked nonlinearities.

Example 7.10: Nonconvergence Caused by Unmodeled State Variables Consider the following example:

Real World (Creeping State)	Kalman Filter Model (Constant State)
$\dot{x}_1 = 0$	$\dot{x}_2 = 0$
$\dot{x}_2 = x_1$	$z = x_2 + v$
$z = x_2 + v$	
$\Rightarrow x_2(t) = x_2(0) + x_1(0)t$	$\Rightarrow x_2(t) = x_2(0)$

(7.35)

for which, in the filter model, the Kalman gain $\overline{K}(t) \to 0$ as $t \to \infty$. The filter is unable to provide feedback for the error in the estimate of $x_2(t)$ as it grows in time (even if it grows slowly). Eventually $\tilde{x}_2(t) = \hat{x}_2(t) - x_2(t)$ diverges as shown in Figure 7.8.

Detecting unmodeled state dynamics by Fourier analysis of the filter innovations. It is difficult to diagnose unmodeled state variables unless all other causes for nonconvergence have been ruled out. If there is high confidence in the model being used, then simulation can be used to rule out any of the other causes above. Once these other causes have been eliminated, Fourier analysis of the filter innovations (the prediction errors $\{z_k - H_k\hat{x}_k\}$) can be useful for spotting characteristic frequencies of the unmodeled state dynamics. If the filter is modeled properly, then the innovations should be uncorrelated, having a power spectral density (PSD) that is essentially flat. Persistent peaks in the PSD would indicate the characteristic frequencies of unmodeled effects. These peaks can be at zero frequency, which would indicate a bias error in the innovations.

Remedies for unmodeled state variables. The best cure for nonconvergence caused by unmodeled states is to correct the model, but this not always easy to do. As an ad hoc fix, additional "fictitious" process noise can be added to the system model assumed by the Kalman filter.

Example 7.11: Adding "Fictitious" Process Noise to the Kalman Filter Model Continuing with the continuous-time problem of Example 7.10, consider the alternative Kalman filter model

$$\dot{x}_2(t) = w(t), \qquad z(t) = x_2(t) + v(t).$$

"Type-1 servo" behavior. The behavior of this filter can be analyzed by applying the continuous Kalman filter equations from Chapter 4, with parameters

$$F = 0, \qquad H = 1, \qquad G = 1,$$

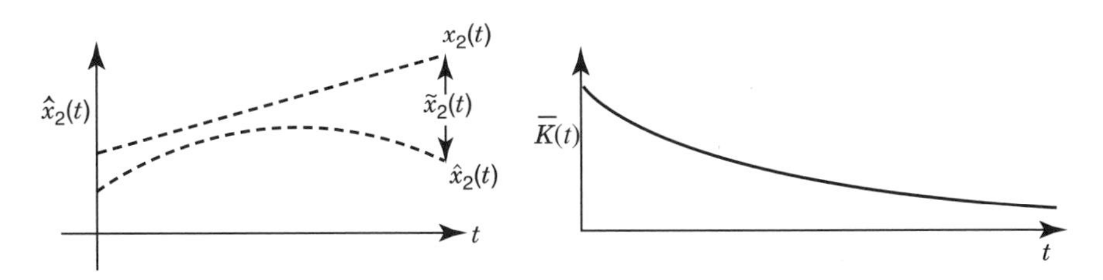

Fig. 7.8 Divergence due to mismodeling.

transforming the general Riccati differential equation

$$\begin{aligned}\dot{P} &= FP + PF^{\mathrm{T}} - PH^{\mathrm{T}}R^{-1}HP + GQG^{\mathrm{T}} \\ &= \frac{-P^2}{R} + Q\end{aligned}$$

to a scalar equation with steady-state solution (to $\dot{P} = 0$)

$$P(\infty) = \sqrt{RQ}.$$

The steady-state Kalman gain

$$\overline{K}(\infty) = \frac{P(\infty)}{R} = \sqrt{\frac{Q}{R}}. \tag{7.36}$$

The equivalent steady-state model of the Kalman filter can now be formulated as follows[2]:

$$\dot{\hat{x}}_2 = F\hat{x}_2 + \overline{K}[z - H\hat{x}_2]. \tag{7.37}$$

Here, $F = 0$, $H = 1$, $\overline{K} = \overline{K}(\infty)$, $\hat{x} = \hat{x}_2$, so that

$$\dot{\hat{x}}_2 + \overline{K}(\infty)\hat{x}_2 = \overline{K}(\infty)z. \tag{7.38}$$

The steady-state response of this estimator can be determined analytically by taking Laplace transforms:

$$[s + \overline{K}(\infty)]\hat{x}_2(s) = \overline{K}(\infty)z(s) \tag{7.39}$$

$$\Rightarrow \quad \frac{\hat{x}_2(s)}{z(s)} = \frac{\overline{K}(\infty)}{s + \overline{K}(\infty)}. \tag{7.40}$$

Figure 7.9*a* shows a “type 1 servo”. Its steady-state following error (even in the noise-free case) is not zero in the real-world case:

$$z(t) = x_2(t) = x_2(0) + x_1(0)t \quad \text{with } v = 0.$$

Taking the Laplace transform of the equation yields

$$z(s) = x_2(s) = \frac{x_2(0)}{s} + \frac{x_1(0)}{s^2}.$$

[2]This steady-state form is the Wiener filter discussed in Chapter 4.

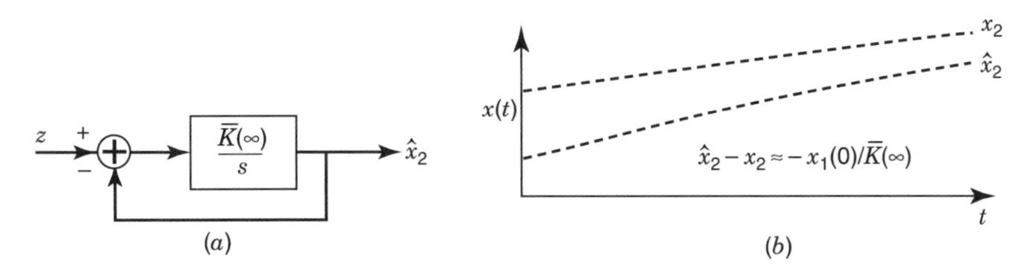

Fig. 7.9 *Type 1 servo (*a) and estimates (*b*).

The error in $x_2(t)$ is

$$\tilde{x}_2(t) = \hat{x}_2(t) - x_2(t).$$

Taking the Laplace of the equation and substituting the value of $\hat{x}_2(s)$ from Equation 7.40 give

$$\begin{aligned}\tilde{x}_2(s) &= \frac{\overline{K}(\infty)}{s + \overline{K}(\infty)}\, x_2(s) - x_2(s) \\ &= \left[-\frac{s}{s + \overline{K}(\infty)} \right] x_2(s).\end{aligned}$$

Applying the final-value theorem, one gets

$$\begin{aligned}\tilde{x}_2(\infty) = \left[\hat{x}_2(\infty) - x_2(\infty)\right] &= \lim_{s \to 0} s\left[\hat{x}_2(s) - x_2(s)\right] \\ &= \lim_{s \to 0} s\left[-\frac{s}{s + \overline{K}(\infty)} \right]\left[x_2(s)\right] \\ &= \lim_{s \to 0} s\left[-\frac{s}{s + \overline{K}(\infty)} \right]\left[\frac{x_2(0)}{s} + \frac{x_1(0)}{s^2}\right] \\ &= -\frac{x_1(0)}{\overline{K}(\infty)} \qquad \text{(a bias).}\end{aligned}$$

This type of behavior is shown in Figure 7.9*b*.

If the steady-state bias in the estimation error is unsatisfactory with the approach in Example 7.11, one can go one step further by adding another state variable and fictitious process noise to the system model assumed by the Kalman filter.

Example 7.12: Effects of Adding States and Process Noise to the Kalman Filter Model Suppose that the model of Example 7.11 was modified to the following form:

$$
\begin{array}{ll}
\underline{\text{Real World}} & \underline{\text{Kalman Filter Model}} \\
\dot{x}_1 = 0 & \dot{x}_1 = w \\
\dot{x}_2 = x_1 & \dot{x}_2 = x_1 \\
z = x_2 + v & z = x_2 + v
\end{array}
\tag{7.41}
$$

That is, $x_2(t)$ is now modeled as an "integrated random walk." In this case, the steady-state Kalman filter has an additional integrator and behaves like a "type 2 servo" (see Figure 7.10*a*). The type 2 servo has zero steady-state following error to a ramp. However, its transient response may become more sluggish and its steady-state error due to noise is not zero, the way it would be with the real world correctly modeled. Here,

$$
F = \begin{bmatrix} 0 & 0 \\ 1 & 0 \end{bmatrix}, \qquad G = \begin{bmatrix} 1 \\ 0 \end{bmatrix}, \qquad H = [0 \ \ 1], \qquad Q = \operatorname{cov}(w), \qquad R = \operatorname{cov}(v)
\tag{7.42}
$$

$$
\dot{P} = FP + PF^{\mathrm{T}} + GQG^{\mathrm{T}} - PH^{\mathrm{T}}R^{-1}HP \ \text{ and } \ \overline{K} = PH^{\mathrm{T}}R^{-1}
\tag{7.43}
$$

become in the steady state

$$
\left.
\begin{aligned}
\dot{p}_{11} &= Q - \frac{p_{12}^2}{R} = 0 \\
\dot{p}_{12} &= p_{11} - \frac{p_{12}p_{22}}{R} = 0 \\
\dot{p}_{22} &= 2p_{12} - \frac{p_{22}^2}{R} = 0
\end{aligned}
\right\}
\Rightarrow
\begin{aligned}
&p_{12}(\infty) = \sqrt{RQ}, \\
&p_{22}(\infty) = \sqrt{2}(R^3Q)^{1/4}, \\
&p_{11}(\infty) = \sqrt{2}(Q^3R)^{1/4}, \\
&\overline{K}(\infty) = \begin{bmatrix} \sqrt{\dfrac{Q}{R}} \\ \sqrt{2}\sqrt[4]{\dfrac{Q}{R}} \end{bmatrix} = \begin{bmatrix} \overline{K}_1(\infty) \\ \overline{K}_2(\infty) \end{bmatrix},
\end{aligned}
\tag{7.44}
$$

and these can be Laplace transformed to yield

$$
\hat{x}(s) = [sI - F + \overline{K}(\infty)H]^{-1}\overline{K}(\infty)z(s).
\tag{7.45}
$$

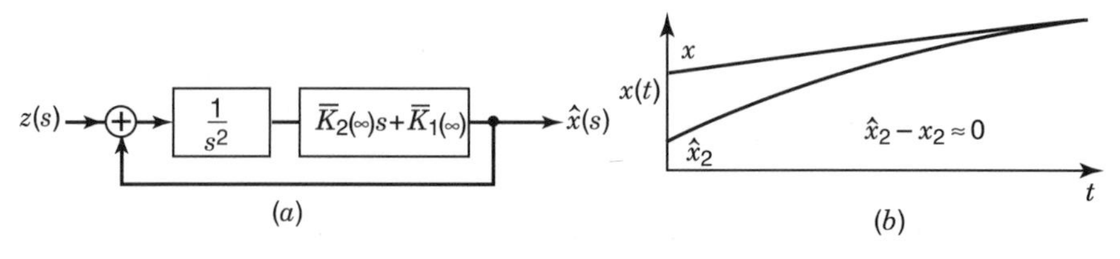

Fig. 7.10 *Type 2 servo (a) and servo estimates (b).*

In component form, this becomes:

$$\hat{x}_2(s) = \frac{(\overline{K}_2 s + \overline{K}_1)/s^2}{1 + (\overline{K}_2 s + \overline{K}_1)/s^2} z(s). \tag{7.46}$$

The resulting steady-state following error to a ramp (in the noise-free case) is easily determined:

$$z(t) = x_2(t) = x_2(0) + x_1(0)t, \qquad v = 0,$$

$$\tilde{x}_2(s) = \hat{x}_2(s) - x_2(s) = -\left(\frac{s^2}{s^2 + \overline{K}_2\, s + \overline{K}_1}\right) x_2(s),$$

$$\tilde{x}_2(\infty) = \lim_{s \to 0} s\left(\frac{-s^2}{s^2 + \overline{K}_2\, s + \overline{K}_1}\right)\left(\frac{x_2(0)}{s} + \frac{x_1(0)}{s^2}\right) = 0.$$

This type of behavior is shown in Figure 7.10*b*.

Example 7.13: Statistical Modeling Errors Due to Unmodeled System Driving Noise With the models

$$\begin{array}{ll} \underline{\text{Real World}} & \underline{\text{Kalman Filter Model}} \\ \dot{x}_1 = w & \dot{x}_1 = 0 \\ \dot{x}_2 = x_1 & \dot{x}_2 = x_1 \\ z = x_2 + v & z = x_2 + v \end{array} \tag{7.47}$$

the Kalman filter equations yield the following relationships:

$$\left.\begin{aligned} \dot{p}_{11} &= \frac{-1}{R} p_{12}^2 \\ \dot{p}_{12} &= p_{11} - \frac{p_{12} p_{22}}{R} \\ \dot{p}_{22} &= 2p_{12} - \frac{p_{22}^2}{R} \end{aligned}\right\} \Rightarrow \text{ in the steady state: } \begin{array}{ll} p_{12} = 0, & \overline{K} = PH^{\mathrm{T}}R^{-1} = 0, \\ p_{11} = 0, & \hat{x}_1 = \text{const}, \\ p_{22} = 0, & \hat{x}_2 = \text{ramp}. \end{array} \tag{7.48}$$

Since x_1 is not constant (due to the driving noise w), the state estimates will not converge to the states, as illustrated in the simulated case plotted in Figure 7.11. This figure shows the behavior of the Kalman gain $\overline{K}_1$, the estimated state component $\hat{x}_1$, and the "true" state component x_1 in a discrete-time simulation of the above model. Because the assumed process noise is zero, the Kalman filter gain $\overline{K}_1$ converges to zero. Because the gain converges to zero, the filter is unable to track the errant state vector component x_1, a random-walk process. Because the filter is unable to track the true state, the innovations (the difference between the predicted measurement and the actual measurement) continue to grow without bound. Even though the gains are

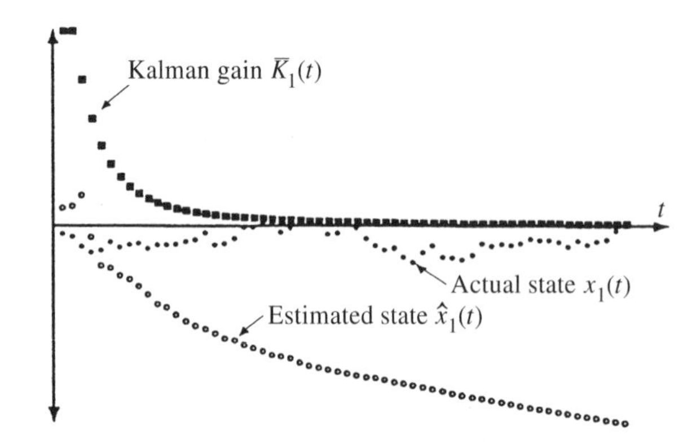

Fig. 7.11 *Diverging estimation error due to unmodeled process noise.*

converging to zero, the product of the gain and the innovations (used in updating the state estimate) can be significant.

In this particular example, $\sigma^2_{x_1}(t) = \sigma^2_{x_1}(0) + \sigma^2_w t$. That is, the variance of the system state itself diverges. As in the case of unmodeled states, the innovations vector $[z - H\hat{x}]$ *will* show effects of the "missing" system driving noise.

Example 7.14: Parametric Modeling Errors Having the wrong parameters in the system dynamic coefficients F, state transition matrix Φ, or output matrix H can and does bring about nonconvergence of the filter. This type of nonconvergence can be demonstrated by an example with the wrong period of a sinusoid in the output matrix:

Real World	Kalman Filter Model
$\begin{bmatrix} \dot{x}_1 \\ \dot{x}_2 \end{bmatrix} = 0$	$\begin{bmatrix} \dot{x}_1 \\ \dot{x}_2 \end{bmatrix} = 0$
$z = [\sin \Omega t \vert \cos \Omega t] \begin{bmatrix} x_1 \\ x_2 \end{bmatrix} + v$	$z = [\sin \omega t \vert \cos \omega t] \begin{bmatrix} x_1 \\ x_2 \end{bmatrix} + v$
v = white noise	v = white noise
No a priori information exists	No a priori information exists

(7.49)

In this case, the optimum filter is the "least-squares" estimator acting over the measurement interval $(0, T)$. Since this is a continuous case,

$$J = \int_0^T (z - H\hat{x})^T (z - H\hat{x})\, dt \tag{7.50}$$

is the performance index being minimized.[3] Its gradient

$$\frac{\partial J}{\partial \hat{x}} = 0 \quad \Rightarrow \quad 0 = 2\int_0^{\mathrm{T}} H^{\mathrm{T}} z\,dt - 2\int_0^{\mathrm{T}} (H^{\mathrm{T}}H)\hat{x}\,dt, \tag{7.51}$$

where $\hat{x}$ is unknown but constant. Therefore,

$$\hat{x} = \left[\int_0^{\mathrm{T}} H^{\mathrm{T}} H\, dt\right]^{-1} \left[\int_0^{\mathrm{T}} H^{\mathrm{T}} z\, dt\right], \tag{7.52}$$

where

$$H = [\sin \omega t | \cos \omega t]; \qquad z = [\sin \Omega t | \cos \Omega t] \begin{bmatrix} x_1 \\ x_2 \end{bmatrix} + v \tag{7.53}$$

and

$$\omega = 2\pi f = \frac{2\pi}{p}, \tag{7.54}$$

where p is the period of the sinusoid. For simplicity, let us choose the sampling time as $T = Np$, an integer multiple of the period, so that

$$\hat{x} = \begin{bmatrix} \frac{Np}{2} & 0 \\ 0 & \frac{Np}{2} \end{bmatrix}^{-1} \begin{bmatrix} \int_0^{Np} \sin(\omega t) z(t)\, dt \\ \int_0^{Np} \cos(\omega t) z(t)\, dt \end{bmatrix} = \begin{bmatrix} \frac{2}{Np}\int_0^{Np} z(t) \sin \omega t\, dt \\ \frac{2}{Np}\int_0^{Np} z(t) \cos \omega t\, dt \end{bmatrix}. \tag{7.55}$$

Concentrating on the first component $\hat{x}_1$, one can obtain its solution as

$$\begin{aligned} \hat{x}_1 = & \left\{ \frac{2}{Np} \left[\frac{\sin(\omega - \Omega)t}{2(\omega - \Omega)} - \frac{\sin(\omega + \Omega)t}{2(\omega + \Omega)} \right] \Bigg|_{t=0}^{t=Np} \right\} x_1 \\ & + \left\{ \frac{2}{Np} \left[\frac{-\cos(\omega - \Omega)t}{2(\omega - \Omega)} - \frac{\cos(\omega + \Omega)t}{2(\omega + \Omega)} \right] \Bigg|_{t=0}^{t=Np} \right\} x_2 \\ & + \frac{2}{Np} \int_0^{Np} v(t) \sin \omega t\, dt. \end{aligned}$$

[3]See Chapter 3.

By setting $v = 0$ (ignoring the estimation error due to measurement noise), one obtains the result

$$\hat{x}_1 = \frac{2}{Np}\left[\frac{\sin(\omega - \Omega)Np}{2(\omega - \Omega)} - \frac{\sin(\omega + \Omega)Np}{2(\omega + \Omega)}\right] x_1$$
$$+ \frac{2}{Np}\left[\frac{1 - \cos(\omega - \Omega)Np}{2(\omega - \Omega)} + \frac{1 - \cos(\omega + \Omega)t}{2(\omega + \Omega)}\right] x_2.$$

For the case that $\omega \to \Omega$,

$$\left.\begin{aligned}
\frac{\sin(\omega - \Omega)Np}{2(\omega - \Omega)} &= \frac{Np}{2}\frac{\sin x}{x}\bigg|_{x\to 0} &&= \frac{Np}{2},\\
\frac{\sin(\omega + \Omega)Np}{2(\omega + \Omega)} &= \frac{\sin[(4\pi/p)Np]}{2\Omega} &&= 0,\\
\frac{1 - \cos(\omega - \Omega)Np}{2(\omega - \Omega)} &= \frac{1 - \cos x}{x}\bigg|_{x\to 0} &&= 0,\\
\frac{1 - \cos(\omega + \Omega)Np}{2(\omega + \Omega)} &= \frac{1 - \cos[(4\pi/p)Np]}{2\Omega} &&= 0,
\end{aligned}\right\} \tag{7.56}$$

and $\hat{x}_1 = x_1$. In any other case, $\hat{x}_1$ would be a biased estimate of the form

$$\hat{x}_1 = \Upsilon_1 x_1 + \Upsilon_2 x_2, \qquad \text{where } \Upsilon_1 \neq 1, \quad \Upsilon_2 \neq 0. \tag{7.57}$$

Similar behavior occurs with $\hat{x}_2$.

Wrong parameters in the system and/or output matrices may not cause the filter covariance matrix or state vector to look unusual. However, the innovations vector $[z - H\hat{x}]$ will generally show detectable effects of nonconvergence.

This can only be cured by making sure that the right parameter values are used in the filter model. In the real world, this is often impossible to do precisely, since the "right" values are not known and can only be estimated. If the amount of degradation obtained is unacceptable, consider letting the questionable parameters become state variables to be estimated by extended (linearized nonlinear) filtering.

Example 7.15: Parameter Estimation This reformulation provides an example of a nonlinear estimation problem:

$$\begin{bmatrix} \dot{x}_1 \\ \dot{x}_2 \\ \dot{x}_3 \end{bmatrix} = 0, \qquad x_3 = \Omega. \tag{7.58}$$

Here, *something* is known about Ω, but it is not known precisely enough. One must choose

$$\hat{x}_3(0) = \text{"best guess" value of } \Omega,$$
$$P_{33}(0) = \sigma^2_{\tilde{x}_3}(0) = \text{a measure of uncertainty in } \tilde{x}_3(0).$$

Nonlinearities in the real-world system also cause nonconvergence or even divergence of the Kalman estimates.

7.2.5 Analysis and Repair of Covariance Matrices

Covariance matrices must be *nonnegative definite*. By definition, their characteristic values must be nonnegative. However, if any of them are *theoretically* zero—or even close to zero—then there is always the risk that roundoff will cause some roots to become negative. If roundoff errors should produce an *indefinite* covariance matrix (i.e., one with both positive and negative characteristic values), then there is a way to replace it with a "nearby" nonnegative-definite matrix.

Testing for Positive Definiteness. Checks that can be made for the definiteness of a symmetric matrix P include the following:

- If a *diagonal element* $a_{ii} < 0$, then the matrix is *not* positive definite, *but the matrix can have all positive diagonal elements and still fail to be positive definite.*
- If *Cholesky decomposition* $P = CC^{\mathrm{T}}$ fails due to a negative argument in a square root, the matrix is indefinite, or at least close enough to being indefinite that roundoff errors cause the test to fail.
- If *modified Cholesky decomposition* $P = UDU^{\mathrm{T}}$ produces an element $d_{ii} \leq 0$ in the diagonal factor D, then the matrix is not positive definite.
- *Symmetric eigenvalue–eigenvector decomposition* yields all the characteristic values and vectors of a symmetric matrix.

The first of these tests is not very reliable unless the dimension of the matrix is 1.

Example 7.16: The following two 3×3 matrices have positive diagonal values and consistent correlation coefficients, yet neither of them is *positive definite* and the first is actually indefinite:

Matrix	Correlation Matrix	Singular Values
$\begin{bmatrix} 343.341 & 248.836 & 320.379 \\ 248.836 & 336.83 & 370.217 \\ 320.379 & 370.217 & 418.829 \end{bmatrix}$	$\begin{bmatrix} 1. & 0.73172 & 0.844857 \\ 0.73172 & 1. & 0.985672 \\ 0.844857 & 0.985672 & 1. \end{bmatrix}$	$\{1000, 100, -1\}$
$\begin{bmatrix} 343.388 & 248.976 & 320.22 \\ 248.976 & 337.245 & 369.744 \\ 320.22 & 369.744 & 419.367 \end{bmatrix}$	$\begin{bmatrix} 1. & 0.731631 & 0.843837 \\ 0.731631 & 1. & 0.983178 \\ 0.843837 & 0.983178 & 1. \end{bmatrix}$	$\{1000, 100, 0\}$

Repair of indefinite covariance matrices. The symmetric eigenvalue–eigenvector decomposition is the more informative of the test methods, because it yields the actual eigenvalues (characteristic values) and their associated eigenvectors (characteristic vectors). The characteristic vectors tell the combinations of states with equivalent negative variance. This information allows one to compose a matrix with the identical characteristic vectors but with the offending characteristic values "floored" at zero:

$$P = TDT^{\mathrm{T}} \quad \text{(symmetric eigenvalue–eigenvector decomposition)}, \tag{7.59}$$

$$D = \mathrm{diag}_i\{d_i\}, \tag{7.60}$$

$$d_1 \geq d_2 \geq d_3 \geq \cdots \geq d_n. \tag{7.61}$$

If

$$d_n < 0. \tag{7.62}$$

then replace P with

$$P^* = TD^*T^{\mathrm{T}}, \tag{7.63}$$

$$D^* = \mathrm{diag}_i\{d_i^*\}, \tag{7.64}$$

$$d_i^* = \begin{cases} d_i & \text{if } d_i \geq 0, \\ 0 & \text{if } d_i < 0. \end{cases} \tag{7.65}$$

7.3 PREFILTERING AND DATA REJECTION METHODS

Prefilters perform data compression of the inputs to Kalman filters. They can be linear continuous, linear discrete, or nonlinear. They are beneficial for several purposes:

1. They allow a discrete Kalman filter to be used on a continuous system without "throwing away" information. For example, the integrate-and-hold prefilter shown in Figure 7.12 integrates over a period T, where T is a sampling time chosen sufficiently small that the dynamic states cannot change significantly between estimates.
2. They attenuate some states to the point that they can be safely ignored (in a suboptimal filter).
3. They can reduce the required iteration rate in a discrete filter, thereby saving computer time [170].
4. They tend to reduce the range of the dynamic variables due to added noise, so that the Kalman filter estimate is less degraded by nonlinearities.

7.3.1 Continuous (Analog) Linear Filters

Example 7.17: Continuous linear filters are usually used for the first three purposes and must be inserted before the sampling process. An example of the continuous linear prefilter is shown in Figure 7.13. An example of such a prefilter is a digital voltmeter (DVM). A DVM is essentially a time-gated averager with sampler and quantizer.

Thus the input signal is continuous and the output discrete. A functional representation is given in Figures 7.14–7.16, where

$$\begin{aligned} \Delta T &= \text{sampling interval} \\ \varepsilon &= \text{dead time for re-zeroing the integrator} \\ \Delta T - \varepsilon &= \text{averaging time} \end{aligned}$$

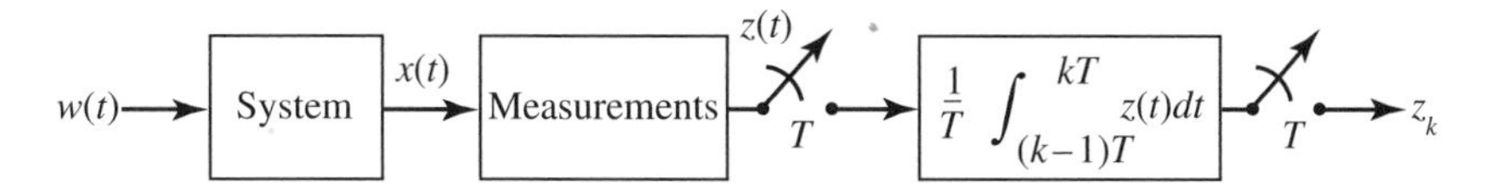

Fig. 7.12 Integrate-and-hold prefilter.

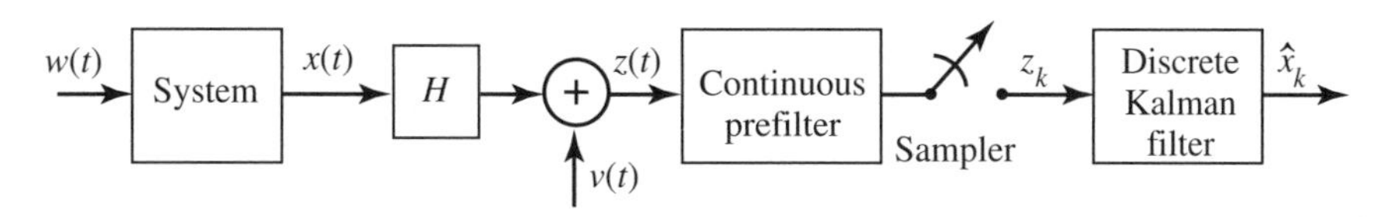

Fig. 7.13 Continuous linear prefiltering and sampling.

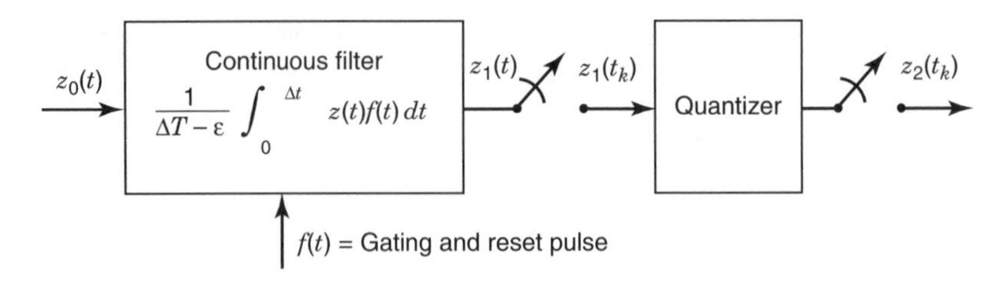

Fig. 7.14 *Block diagram of DVM.*

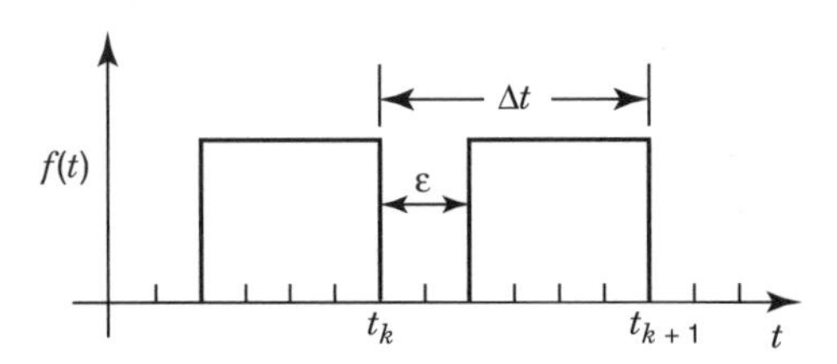

Fig. 7.15 *DVM gating waveform.*

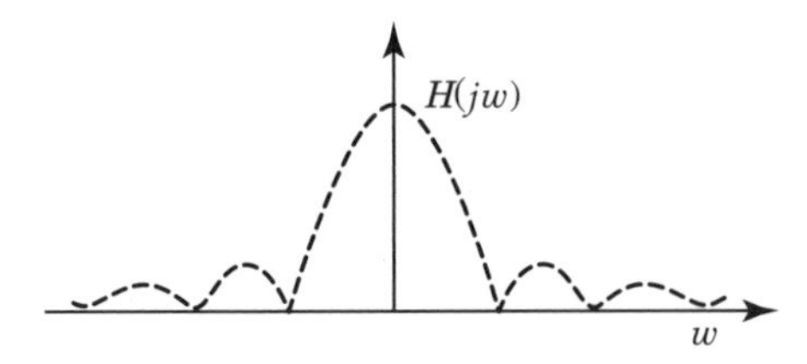

Fig. 7.16 *DVM frequency response.*

and the output

$$z^1(t_i) = \frac{1}{\Delta T - \varepsilon} \int_{t_i - \Delta T + \varepsilon}^{t_i} z(t)\, dt. \tag{7.66}$$

It can be shown that the frequency response of the DVM is:

$$|H(j\omega| = \left| \frac{\sin \omega[(\Delta T - \varepsilon)/2]}{\omega[(\Delta T - \varepsilon)/2]} \right| \quad \text{and} \quad \theta(j\omega) = -\omega\left(\frac{\Delta T - \varepsilon}{2}\right). \tag{7.67}$$

With *white-noise* continuous input, the output is a white-noise sequence (since the averaging intervals are non overlapping). With an *exponentially correlated* random continuous input with correlation time τ_c and autocovariance

$$\psi_z(\tau) = \sigma_z^2 e^{-(|\tau|/\tau_c)}, \tag{7.68}$$

the variance and autocorrelation function of the output random sequence can be shown to be

$$\left.\begin{aligned} \sigma_{z^1}^2 &= \psi_{z^1 z^1}(0) = f(u)\sigma_z^2, \\ \psi(j-i) &= g(u)\sigma_z^2 e^{-(j-i)\frac{\Delta T}{\tau_c}}, \end{aligned} \qquad \begin{aligned} u &= \frac{\Delta T - \varepsilon}{T_c}, \\ f(u) &= \frac{2}{u^2}(e^{-u} + u - 1) \\ g(u) &= \frac{e^u + e^{-u} - 2}{u^2} \end{aligned} \right\} \tag{7.69}$$

7.3.2 Discrete Linear Filters

These can be used effectively for the attenuation of the effects of some of the state variables to the point that they can be safely ignored. (However, the sampling rate input to the filter must be sufficiently high to avoid aliasing.)

Note that discrete filters can be used for the third purpose to reduce the discrete filter iteration rate. The input sampling period can be chosen shorter than the output sampling period. This can greatly reduce the computer (time) load (see Figure 7.17).

Example 7.18: For the simple digital averager shown in Figure 7.18, let

$$z^1(t_i^1) \equiv \frac{1}{N} \sum_{j=i-N+1}^{i} z(t_j), \qquad t_i^1 = iT^1, \tag{7.70}$$

which is the average of N adjacent samples of $z(t_j)$. Note that $z^1(t_i^1)$ and $z^1(t_{i+1}^1)$ use nonoverlapping samples of $z(t_j)$. Then it can be shown that the frequency response is

$$|H(j\omega)| = \frac{|\sin(N\omega T/2)|}{N|\sin(\omega T/2)|}, \qquad \theta(j\omega) = -\left(\frac{N-1}{2}\right)\omega T. \tag{7.71}$$

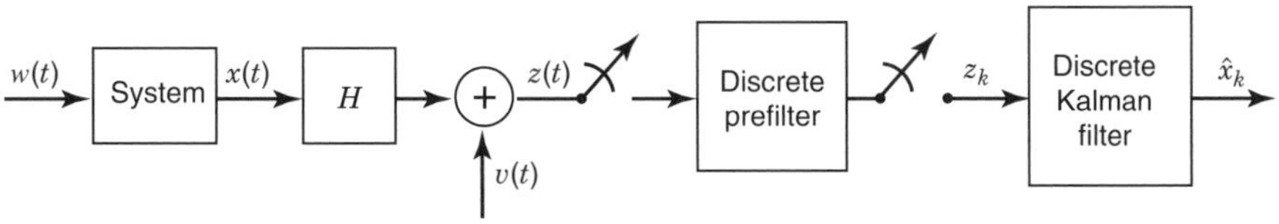

$w(t)$ System $x(t)$ H + $z(t)$ Discrete prefilter z_k Discrete Kalman filter $\hat{x}_k$ $v(t)$

Fig. 7.17 Discrete linear prefilter.

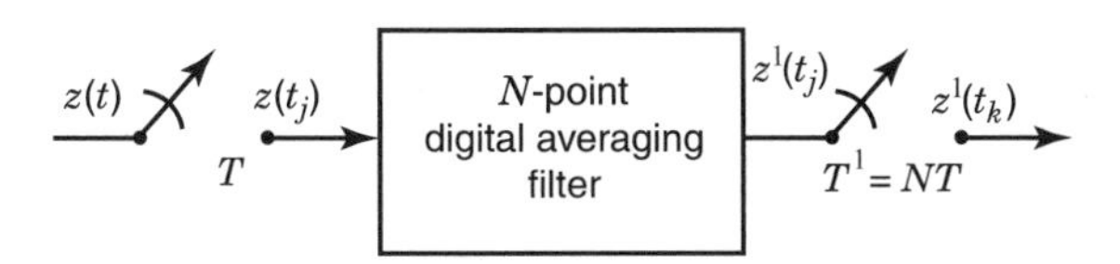

Fig. 7.18 Discrete linear averager.

If adequate knowledge of the innovations vector $[z - H\hat{x}]$ exists, nonlinear "data rejection filters" can be implemented. Some simple examples are cited below.

Example 7.19: Data Rejection Filters For excess amplitude:

$$\text{If } |(z - H\hat{x})_i| > A_{\max}, \text{ then reject data.} \tag{7.72}$$

For excess rate (or change):

$$\text{If } |(z - H\hat{x})_{i+1} - (z - H\hat{x})_i| > \delta A_{\max}, \text{ then reject data.} \tag{7.73}$$

Many other ingenious techniques have been used, but they usually depend on the specifics of the problem.

7.4 STABILITY OF KALMAN FILTERS

The dynamic stability of a system usually refers to the behavior of the state variables, not the estimation errors. This applies as well to the behavior of the homogeneous part of the filter equations. However, the mean-squared estimation errors may remain bounded, even if the system is unstable.[4]

If the actual measurement processing in the Kalman filter state equations is neglected, then the resulting equations characterize the stability of the filter itself. In the continuous case, these equations are

$$\dot{\hat{x}}(t) = [F(t) - \overline{K}(t)H(t)]\,\hat{x}(t), \tag{7.74}$$

and in the discrete case,

$$\begin{aligned}\hat{x}_k(+) &= \Phi_{k-1}\hat{x}_{k-1}(+) - \overline{K}_k H_k\, \Phi_{k-1}\hat{x}_{k-1}(+) \\ &= [I - \overline{K}_k H_k]\, \Phi_{k-1}\hat{x}_{k-1}(+).\end{aligned} \tag{7.75}$$

The solution of the filter equation 7.74 or 7.75 is uniformly asymptotically stable, which means bounded input-bounded output (BIBO) stability. Mathematically, it implies that

$$\lim_{t\to\infty} \|\hat{x}(t)\| = 0 \tag{7.76}$$

or

$$\lim_{k\to\infty} \|\hat{x}_k(+)\| = 0, \tag{7.77}$$

[4] See, for example, Gelb et al. [21], pp. 22, 31, 36, 53, 72, or Maybeck [30], p. 278.

no matter what the initial conditions are. In other words, the filter is uniformly asymptotically stable if the system model is stochastically controllable and observable. See Chapter 4 for the solution of the matrix Riccati equation $P(t)$ or $P_k(+)$ uniformly bounded from above for large t or $\overline{K}$ independent of $P(0)$. Bounded Q, R (above and below) and bounded F (or Φ) will guarantee stochastic controllability and observability.

The most important issues relating to stability are described in the sections on unmodeled effects, finite wordlength, and other errors (Section 7.2).

7.5 SUBOPTIMAL AND REDUCED-ORDER FILTERS

Suboptimal Filters. The Kalman filter has a reputation for being robust against certain types of modeling errors, such as those in the assumed values of the statistical parameters R and Q. This reputation is sometimes tested by deliberate simplification of the known (or, at least, "believed") system model. The motive for these actions is usually to reduce implementation complexity by sacrificing some optimality. The result is called a suboptimal filter.

7.5.1 Rationale for Suboptimal Filtering

It is often the case that real hardware is nonlinear but, in the filter model, approximated by a linear system. The algorithms developed in Chapters 4–6 will provide suboptimal estimates. These are

1. Kalman filters (linear optimal estimate),
2. linearized Kalman filters, and
3. extended Kalman filters.

Even if there is good reason to believe that the real hardware is truly linear, there may still be reasons to consider suboptimal filters. Where there is doubt about the absolute certainty of the model, there is always a motivation to meddle with it, especially if meddling can decrease the implementation requirements. Optimal filters are generally demanding on computer throughput, and optimality is unachievable if the required computer capacity is not available. Suboptimal filtering can reduce the requirements for computer memory, throughput, and cost. A suboptimal filter design may be "best" if factors other than theoretical filter performance are considered in the trade-offs.

7.5.2 Techniques for Suboptimal Filtering

These techniques can be divided into three categories:

1. modifying the optimal gain $\overline{K}_k$ or $\overline{K}(t)$,

2. modifying the filter model, and
3. other techniques.

Techniques for Evaluating Suboptimal Filters. The covariance matrix P may not represent the actual estimation error and the estimates may be biased. In the following section, the dual-state technique for evaluating performance of linear suboptimal filters will be discussed.

Modification of $\overline{K}(t)$ or $\overline{K}_k$ Consider the system

$$\begin{aligned} \dot{x} &= Fx + Gw, & Q &= \text{cov}(w), \\ z &= Hx + v, & R &= \text{cov}(v), \end{aligned} \tag{7.78}$$

with state transition matrix Φ.

The state estimation portion of the optimal linear estimation algorithm is then, for the continuous case,

$$\dot{\hat{x}} = F\hat{x} + \overline{K}(t)[z - H\hat{x}] \quad \text{with} \quad \hat{x}(0) = \hat{x}_0, \tag{7.79}$$

and for the discrete case,

$$\hat{x}_k(+) = \Phi_{k-1}\hat{x}_{k-1}(+) + \overline{K}_k[z_k - H_k(\Phi_{k-1}\hat{x}_{k-1}(+))] \tag{7.80}$$

with initial conditions $\hat{x}_0$.

Schemes for retaining the structure of these algorithms using modified gains include the Wiener filter and approximating functions.

First, consider the *Wiener filter*, which is a useful suboptimal filtering method when the gain vector $\overline{K}(t)$ is time varying but quickly reaches a constant nonzero steady-state value. Typical settling behaviors are shown in Figure 7.19.

A Wiener filter results from the approximation

$$\overline{K}(t) \approx \overline{K}(\infty). \tag{7.81}$$

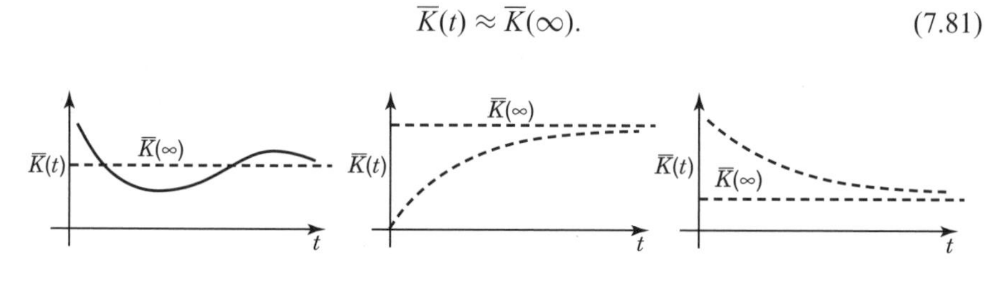

Fig. 7.19 *Settling of Kalman gains.*

If, in addition, the matrices F and H are time invariant, the matrix of transfer functions characterizing the Wiener filter can easily be computed:

$$\dot{\hat{x}} = F\hat{x} + \overline{K}(\infty)[z - H\hat{x}] \tag{7.82}$$

$$\Rightarrow \frac{\hat{x}(s)}{z(s)} = [sI - F + \overline{K}(\infty)H]^{-1}. \tag{7.83}$$

The corresponding steady-state sinusoidal frequency response matrix is

$$\frac{|\hat{x}(j\omega)|}{|z(j\omega)|} = \left|[(j\omega)I - F + \overline{K}(\infty)H]^{-1}\right|. \tag{7.84}$$

Among the advantages of the Wiener filter are that—its structure being identical to conventional filters—all the tools of pole-zero, frequency response, and transient response analysis using Laplace transforms can be employed to gain "engineering insight" into the behavior of the filter. Among the disadvantages of this approach is that it cannot be used if $\overline{K}(\infty) \neq$ constant or $\overline{K}(\infty) = 0$. The typical penalty is poorer transient response (slower convergence) than with optimal gains.

The second scheme for retaining the structure of the algorithms using modified gains is that of *approximating functions*. The optimal time-varying gains $\overline{K}_{\mathrm{OP}}(t)$ are often approximated by simple functions.

For example, one can use piecewise constant approximation, $\overline{K}_{\mathrm{pwc}}$, a piecewise linear approximation, $\overline{K}_{\mathrm{pwl}}$, or a curve fit using smooth functions $\overline{K}_{\mathrm{CF}}$, as shown in Figure 7.20:

$$\overline{K}_{\mathrm{CF}}(t) = C_1 e^{-a_1 t} + C_2(1 - e^{-a_2 t}). \tag{7.85}$$

The advantages of approximating functions over the Wiener filter are that this can handle cases where $\overline{K}(\infty)$ is not constant or $\overline{K}(\infty) = 0$. A result closer to optimal performance is obtained.

Modification of the Filter Model. Let the real-world model (actual system S) be linear,

$$\dot{x}^S = F_S x_S + G_S w^S, \qquad z^S = H_S x_S + v^S.$$

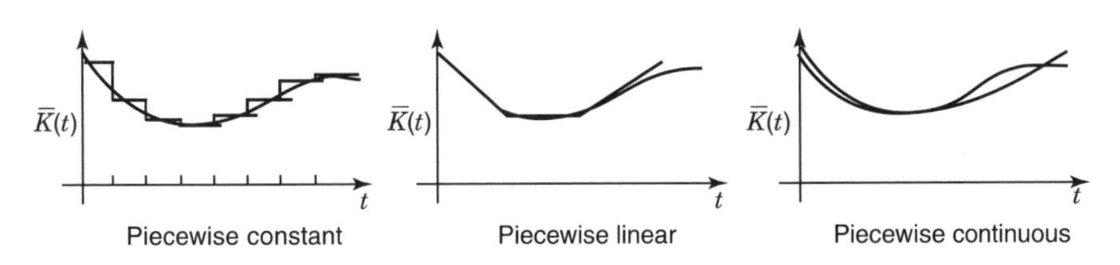

Fig. 7.20 *Approximating time-varying Kalman gains.*

The filter model of the system will, in general, be (intentionally or unintentionally) different:

$$\dot{x}_F = F_F x_F + G_F w_F, \qquad z_F = H_F x_F + v_F.$$

Usually, the intent is to make the filter model less complex than the actual system. This can be done by ignoring some states, prefiltering to attenuate some states, decoupling states, or with frequency domain approximations. Ignoring some states reduces model complexity and provides a suboptimal design. Often, however, little performance degradation occurs.

Example 7.20: In this example, one combines two nonobservable states with identical propagation into z.

$$\begin{aligned} &\text{From:} \quad \dot{x}_1 = -ax_1, \\ &\dot{x}_2 = -ax_2, \qquad\qquad \text{To:} \quad \dot{x}^1 = -ax^1, \\ &z = [2 \;\; 3]\begin{bmatrix} x_1 \\ x_2 \end{bmatrix} + v, \qquad\qquad z = x^1 + v, \end{aligned} \tag{7.86}$$

Of course, $x^1 \equiv 2x_1 + 3x_2$ and the a priori information must be represented as

$$\begin{aligned} \hat{x}^1(0) &= 2\hat{x}_1(0) + 3\hat{x}_2(0), \\ P_{x^1x^1}(0) &= 4P_{x_1x_1}(0) + 12P_{x_1x_2}(0) + 9P_{x_2x_2}(0). \end{aligned}$$

Example 7.21: Continuing where Example 7.101 left off, one can combine two states if they are "close functions" over the entire measurement interval:

$$\begin{aligned} &\text{From:} \quad \dot{x}_1 = 0, \\ &\dot{x}_2 = 0, \qquad\qquad \text{To:} \quad \dot{x}^1 = 0, \\ &z = [t \mid \sin t]\begin{bmatrix} x_1 \\ x_2 \end{bmatrix} + v, \qquad\qquad z = tx^1 + v, \end{aligned} \tag{7.87}$$

where the a priori information on x^1 must be formulated and where z is available on the interval $(0, \;\; \pi/20.)$

Example 7.22: Ignoring Small Effects

$$\begin{aligned} &\text{From:} \quad \dot{x}_1 = -ax_1, \qquad\qquad \text{To:} \quad \dot{x}_2 = -bx_2, \\ &\dot{x}_2 = -bx_2, \qquad\qquad z = x_2 + v, \\ &z = x_1 + x_2 + v \quad \text{on } (0, \;\; T), \end{aligned} \tag{7.88}$$

with

$$E(x_1^2) = 0.1, \qquad E(x_2^2) = 10.0, \qquad \text{desired} \quad P_{22}(T) = 1.0. \tag{7.89}$$

Example 7.23: Relying on Time Orthogonality of States

$$\dot{x}_1 = 0,$$
$$\dot{x}_2 = 0,$$

From: $$z = [1 | \sin t]\begin{bmatrix} x_1 \\ x_2 \end{bmatrix} + v,$$ To: $$\dot{x}_2 = 0, \quad z = (\sin t)x_2 + v, \tag{7.90}$$

z available on$(0,\ T)$ with T large

$P_{22}(0) = \text{large},$

Prefiltering to Simplify the Model. A second technique for modifying the filter model of the real world is to ignore states after prefiltering to provide attenuation. This is illustrated in Figure 7.21.

Of course, prefiltering has deleterious side effects, too. It may require other changes to compensate for the measurement noise v becoming time correlated after passing through the prefilter and to account for some "distortion" of those states in the passband of the prefilter (e.g., amplitude and phase of sinusoidals and wave shape of signals). The filter may be modified as shown in Figure 7.22 to compensate for such effects. Hopefully, the net result is a smaller filter than before.

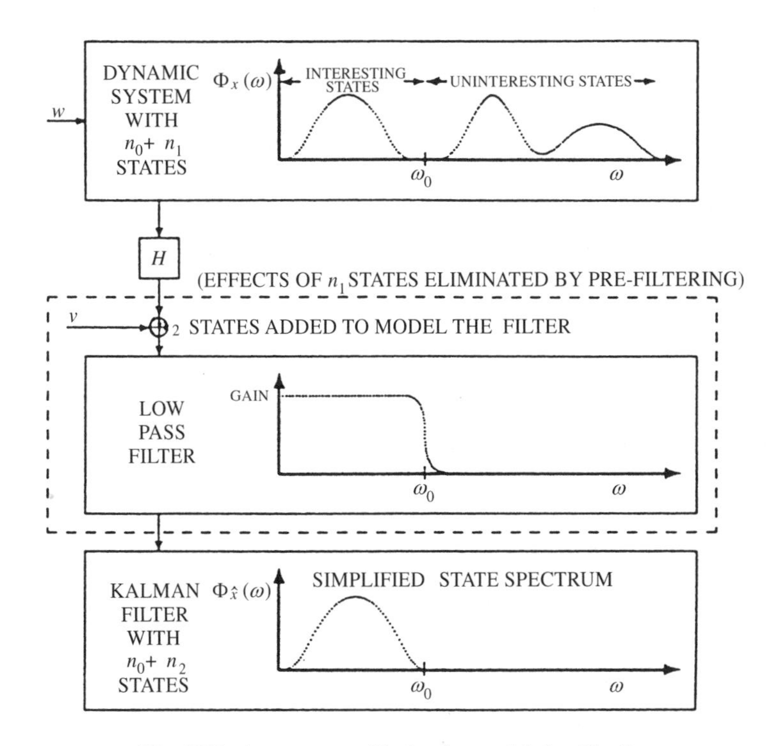

Fig. 7.21 *Low-pass prefiltering for model simplification.*

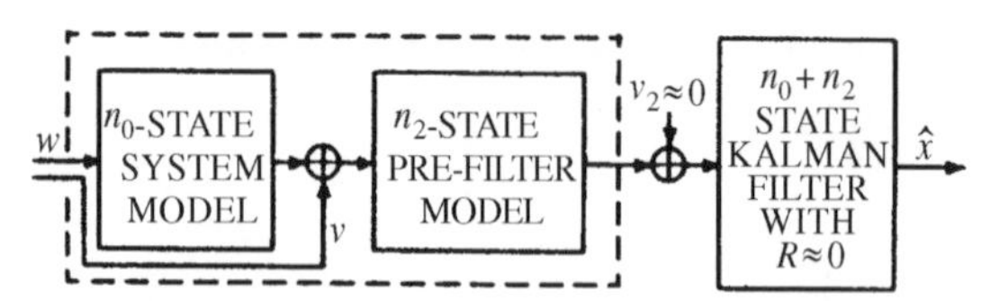

Fig. 7.22 Modified system model for low-pass prefiltering.

Example 7.24: Consider the second-order system with coupling coefficient τ, represented by the equations

$$\begin{bmatrix} \dot{x}_1 \\ \dot{x}_2 \end{bmatrix} = \begin{bmatrix} -\alpha_1 & 0 \\ \tau & -\alpha_2 \end{bmatrix} \begin{bmatrix} x_1 \\ x_2 \end{bmatrix} + \begin{bmatrix} w_1 \\ w_2 \end{bmatrix},$$

$$\begin{bmatrix} z_1 \\ z_2 \end{bmatrix} = \begin{bmatrix} 1 & 0 \\ 0 & 1 \end{bmatrix} \begin{bmatrix} x_1 \\ x_2 \end{bmatrix} + \begin{bmatrix} v_1 \\ v_2 \end{bmatrix}.$$

If τ is sufficiently small, one can treat the system as two separate uncoupled systems with two separate corresponding Kalman filters:

$$\begin{aligned} \dot{x}_1 &= -\alpha_1 x_1 + w_1, & \dot{x}_2 &= -\alpha_2 x_2 + w_2, \\ z_1 &= x_1 + v_1, & z_2 &= x_2 + v_2. \end{aligned} \tag{7.91}$$

The advantage of this decoupling method is a large reduction in computer load. However, the disadvantage is that the estimation error has increased variance and can be biased.

Frequency-domain approximations. These can be used to formulate a suboptimal filter for a system of the sort

$$\dot{x} = Fx + Gw, \qquad z = Hx + v.$$

Example 7.25: Often, some of the states are stationary random processes whose power spectral densities can be approximated as shown in Figure 7.23.

A filter designed for the approximate spectrum may well use fewer states.

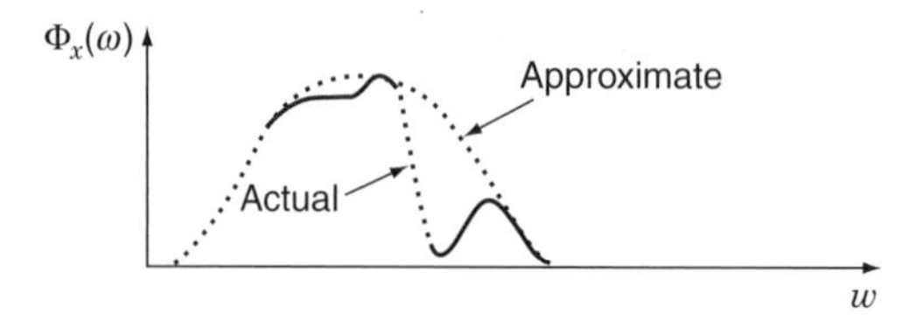

Fig. 7.23 Frequency-domain approximation.

Example 7.26: Sometimes the general structure of the random process model is known, but the parameters are not known precisely:

$$\dot{x} = -\alpha x + w \quad \text{and} \quad \sigma_w \text{ are "uncertain."} \tag{7.92}$$

Replacing this model by a random walk

$$\dot{x} = w \tag{7.93}$$

in conjunction with "sensitivity studies" will often allow a judicious choice for σ_w with small sensitivity to α uncertainties and small degradation in filter performance.

Example 7.27: White-Noise Approximation of Broadband Noise If the system-driving noise and measurement noise are "broadband" but with sufficiently flat PSDs, they can be replaced by "white-noise" approximations, as illustrated in Figure 7.24.

Least-Squares Filters. Among the other techniques used in practice to intentionally suboptimize linear filters is least-squares estimation. It is equivalent to Kalman filtering if there are no state dynamics ($F = 0$ and $Q = 0$) and is often considered as a candidate for a suboptimal filter if the influence of the actual values of Q and F (or Φ) on the values of the Kalman gain is small.

Observer Methods. These simpler filters can be designed by choosing the eigenvalues of a filter of special structure. The design of suboptimal filters using engineering insight is often possible. Ingenuity based on physical insights with regard to design of suboptimal filters is considered an art, not a science, by some practitioners.

7.5.3 Dual-State Evaluation of Suboptimal Filters

Dual-State Analysis. This form of analysis takes its name from the existence of two views of reality:

1. The so-called *system model* (or "truth model") of the actual system under study. This model is used to generate the observations input to the suboptimal filter. It should be a reasonably complete model of the actual system under consideration, including all known phenomena that are likely to influence the performance of the estimator.

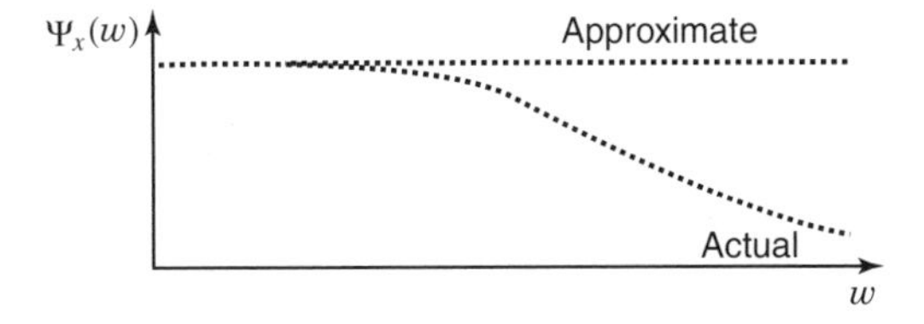

Fig. 7.24 *White-noise approximation of broadband noise.*

2. The *filter model*, which is a reduced-order version of the system model, it is usually constrained to contain all the states in the *domain* of the measurement sensitivity matrix (i.e., the states that contribute to the measurement) and possibly other state components of the system model as well. The filter model is to be used by a proposed filter implementation with significantly reduced computational complexity, but (hopefully) not greatly reduced fidelity. The performance of the reduced-order filter implementation is usually measured by how well its estimates agree with those of the actual system state. It may not estimate *all* of the components of the state vector of the system model, however. In that case, the evaluation of its estimation accuracy is restricted to just the common state vector components.

Performance analysis of suboptimal filter errors requires the simultaneous consideration of both (system and filter) state vectors, which are combined into one *dual-state vector.*

Notation. Let us use a superscript notation to distinguish between these models. A superscript S will denote the *system model*, and a superscript F will denote the filter model, as illustrated later in Figure 7.30.[5] There are two commonly used definitions for the dual-state vector:

1. $x_k^{\text{dual}} = \begin{bmatrix} x_k^S \\ \hat{x}_k^F(+) \end{bmatrix}$ (concatenation of the two vectors) and

2. $\tilde{x}_k^{\text{dual}} = \begin{bmatrix} \tilde{x}_k^S(+) \\ x_k^S \end{bmatrix}, \quad \tilde{x}_k^S(+) = \hat{x}_k^F(+) - x_k^S.$

In the second definition, the filter state vector x^F may have lower dimension than the system state vector x^S—due to neglected state components of the system in the suboptimal filter model. In that case, the missing components in x^F can be padded with zeros to make the dimensions match.

In dual-state analysis for evaluating suboptimal filters, let the following definitions apply:

$$\begin{array}{ll} \underline{\text{Actual System}} & \underline{\text{Filter Model}} \\ \dot{x}^S = F_S x^S + G_S w^S & \dot{x}^F = F_F x^F + G_F w^F \\ z^S = H_S x^S + v^S & z^F = H_F x^F + v^F \end{array} \tag{7.94}$$

$$z^F = z^S, \qquad H_F = H_S, \qquad v^F = v^S, \tag{7.95}$$

$$\left.\begin{aligned} \eta_k^S &\equiv \int_{t_{k-1}}^{t_k} \Phi_S(t_k - \tau) G_S(\tau) w^S(\tau)\, d\tau, \\ \hat{x}_k^F(+) &= \Phi_F \hat{x}_{k-1}^F(+) + \overline{K}_k [z_k^F - H_F \Phi_F \hat{x}_{k-1}^F(+)], \\ \Phi_S &\equiv \text{System state transition matrix}, \\ \Phi_F &\equiv \text{State transition matrix of filter model}, \\ Q_S &\equiv \text{cov}(w^S), \\ Q_F &\equiv \text{cov}(w^F), \end{aligned}\right] \tag{7.96}$$

[5] The "system" model is called the "truth" model in the figure. That nomenclature here would lead us to use a superscript T, however, and that notation is already in use to denote transposition.

Let $\Gamma_k \equiv (I - \overline{K}_k H_F)$, and let

$$A \equiv \begin{bmatrix} \Phi_F & \Phi_F - \Phi_S \\ 0 & \Phi_S \end{bmatrix}, \tag{7.97}$$

$$B \equiv \begin{bmatrix} \Gamma_k & 0 \\ 0 & I \end{bmatrix}. \tag{7.98}$$

Let the "prediction" estimation error be

$$\tilde{x}_k^S(+) \equiv \hat{x}_k^F(+) - x_k^S \tag{7.99}$$

and let the "filtered" estimation error be

$$\tilde{x}_{k-1}^S(+) \equiv \hat{x}_{k-1}^F(+) - x_{k-1}^S. \tag{7.100}$$

Then the prediction error equation is

$$\begin{bmatrix} \tilde{x}_k^S(-) \\ x_k^S \end{bmatrix} = A \begin{bmatrix} \tilde{x}_{k-1}^S(+) \\ x_{k-1}^S \end{bmatrix} + \begin{bmatrix} -\eta_{k-1}^S \\ \eta_{k-1}^S \end{bmatrix} \tag{7.101}$$

and the filtered error equation is

$$\begin{bmatrix} \tilde{x}_{k-1}^S(+) \\ x_{k-1}^S \end{bmatrix} = B \begin{bmatrix} \tilde{x}_{k-1}^S(-) \\ x_{k-1}^S \end{bmatrix} + \begin{bmatrix} \overline{K}_{k-1} v_{k-1}^S \\ 0 \end{bmatrix}. \tag{7.102}$$

Taking the expected value E and the covariance of Equations 7.101 and 7.102 yields the following recursive relationships:

$$\left.\begin{aligned} E_{\text{cov}} &= A \cdot P_{\text{cov}} \cdot A^{\mathrm{T}} + \operatorname{cov} L \\ P_{\text{cov}} &= B \cdot E_{\text{cov}} \cdot B^{\mathrm{T}} + \operatorname{cov} P \end{aligned}\right\} \quad \text{(covariance propagation)}, \tag{7.103}$$

$$\left.\begin{aligned} E_{\text{DX}} &= A \cdot P_{\text{DX}} \\ P_{\text{DX}} &= B \cdot E_{\text{DX}} \end{aligned}\right\} \quad \text{(bias propagation)}, \tag{7.104}$$

where the newly introduced symbols are defined as follows:

$$
\begin{aligned}
E_{\text{cov}} &\equiv \operatorname{cov}\begin{bmatrix} \tilde{x}_k^S(-) \\ x_k^S \end{bmatrix} \quad \text{(predicted dual-state vector)}, \\
\operatorname{cov} L &\equiv \operatorname{cov}\begin{bmatrix} -\eta_{k-1}^S \\ \eta_{k-1}^S \end{bmatrix}, \\
P_{\text{cov}} &\equiv \operatorname{cov}\begin{bmatrix} \tilde{x}_{k-1}^S(+) \\ x_{k-1}^S \end{bmatrix} \quad \text{(filtered covariance)}, \\
\operatorname{cov} P &\equiv \operatorname{cov}\begin{bmatrix} \overline{K}_{k-1} v_{k-1}^S \\ 0 \end{bmatrix}, \\
&= \begin{bmatrix} \overline{K}_{k-1} \operatorname{cov}(v_{k-1}^S) \overline{K}_{k-1}^{\mathrm{T}} & 0 \\ 0 & 0 \end{bmatrix}, \qquad (7.105) \\
P_{\text{DX}} &= E\begin{bmatrix} \tilde{x}_{k-1}^S(+) \\ x_{k-1}^S \end{bmatrix} \quad \text{(expected value of filtered dual-state vector)}, \\
E_{\text{DX}} &= E\begin{bmatrix} \tilde{x}_k^S(-) \\ x_k^S \end{bmatrix} \quad \text{(expected value of predicted dual-state vector)}.
\end{aligned}
$$

The suboptimal estimate $\hat{x}$ can be biased. The estimation error covariance can depend on the system state covariance. To show this, one can use the dual-state equations. The bias propagation equations

$$
E_{\text{DX}} = A \cdot P_{\text{DX}}, \qquad P_{\text{DX}} = B \cdot E_{\text{DX}}
$$

become

$$
E\begin{bmatrix} \tilde{x}_k^S(-) \\ x_k^S \end{bmatrix} = AE\begin{bmatrix} \tilde{x}_{k-1}^S(+) \\ x_{k-1}^S \end{bmatrix} \qquad (7.106)
$$

and

$$
E\begin{bmatrix} \tilde{x}_{k-1}^S(+) \\ x_{k-1}^S \end{bmatrix} = BE\begin{bmatrix} \tilde{x}_{k-1}^S(-) \\ x_{k-1}^S \end{bmatrix}. \qquad (7.107)
$$

Clearly, if $E\left[\tilde{x}_k^S(+)\right] \neq 0$, then the estimate becomes biased. If

$$\Phi_F \neq \Phi_S \tag{7.108}$$

and

$$E(x^S) \neq 0, \tag{7.109}$$

which often is the case (for example, x^S may be a deterministic variable, so that $E(x^S) = x^S \neq 0$, x^S may be a random variable with nonzero mean). Similarly, an examination of the covariance propagation equations

$$\begin{aligned} E_{\text{cov}} &= A(P_{\text{cov}})A^{\text{T}} + \text{cov}\, L \\ P_{\text{cov}} &= B(E_{\text{cov}})B^{\text{T}} + \text{cov}\, P \end{aligned}$$

show that $\text{cov}\left[\tilde{x}_k^S(-)\right]$ depends on $\text{cov}\,(x_k^S)$. (See Problem 7.5.)

If

$$\Phi_F \neq \Phi_S \tag{7.110}$$

and

$$\text{cov}(x^S) \neq 0, \tag{7.111}$$

which often is the case with the suboptimal filter, the estimate $\hat{x}$ is unbiased and the estimation error covariance is independent of the system state.

7.6 SCHMIDT–KALMAN FILTERING

7.6.1 Historical Background

Stanley F. Schmidt was an early and successful advocate of Kalman filtering. He was working at NASA Ames Laboratory in Mountain View, California, when Kalman presented his results there in 1959. Schmidt immediately began applying it to a problem then under study at Ames, which was the space navigation problem (i.e., trajectory estimation) for the upcoming Apollo project for manned exploration of the moon. (In the process, Schmidt discovered what is now called *extended* Kalman filtering.) Schmidt was so impressed with his results that he set about proselytizing his professional colleagues and peers to try Kalman filtering.

Schmidt also derived and evaluated many practical methods for improving the numerical stability of the procedures used and for reducing the computational requirements of Kalman filtering. Many of these results were published in journal articles, technical reports, and books. In [45] Schmidt presents an approach (now called *Schmidt–Kalman filtering*) for reducing the computational complexity of Kalman filters by eliminating some of the computation for "nuisance variables," which are state variables that are of no interest for the problem at hand—except that they are part of the system state vector.

Schmidt's approach is suboptimal, in that it sacrifices estimation performance for computational performance. It enabled Kalman filters to be approximated so that they could be implemented in real time on the computers of that era (the mid-1960s). However, it still finds useful application today for implementing Kalman filters on small embedded microprocessors.

The types of nuisance variables that find their way into the Kalman filter state vector include those used for modeling correlated measurement noise (e.g., colored, pastel, or random-walk noise). We generally have no interest in the memory state of such noise. We just want to filter it out.

Because the dynamics of measurement noise are generally *not* linked to the other system state variables, these added state variables are not dynamically coupled to the other state variables. That is, the elements in the dynamic coefficient matrix linking the two state variable types (states related to correlated measurement noise and states not related to correlated measurement noise) are zero. In other words, if the ith state variable is of one type and the jth state variable is of the other type, then the element f_{ij} in the ith row and jth column of the dynamic coefficient matrix F will always be zero.

Schmidt was able to take advantage of this, because it means that the state variables could be reordered in the state vector such that the nuisance variables appear last. The resulting dynamic equation then has the form

$$\frac{d}{dt}\mathbf{x}(t) = \begin{bmatrix} F_\varepsilon(t) & 0 \\ 0 & F_\nu(t) \end{bmatrix}\mathbf{x}(t) + \mathbf{w}(t) \tag{7.112}$$

such that F_ν represents the dynamics of the nuisance variables and F_ε represents the dynamics of the other state variables.

It is this partitioning of the state vector that leads to the reduced-order, suboptimal filter called the Schmidt–Kalman filter.

7.6.2 Derivation

Partitioning the Model Equations. Let[6]

$n = n_\varepsilon + n_\nu$ be the total number of state variables,

n_ε be the number of essential variables, whose values are of interest for the application, and

n_ν be the number of *nuisance* variables, whose values are of no intrinsic interest and whose dynamics are not coupled with those of the essential state variables.

[6]This derivation follows that in [12].

Then the state variables can be reordered in the state vector such that the essential variables precede the nuisance variables:

$$\mathbf{x} = \left[\begin{array}{ll} \left.\begin{array}{c} x_1 \\ x_2 \\ x_3 \\ \vdots \\ x_{n_\varepsilon} \end{array}\right\} & \text{essential variables} \\ \left.\begin{array}{c} x_{n_\varepsilon+1} \\ x_{n_\varepsilon+2} \\ x_{n_\varepsilon+3} \\ \vdots \\ x_{n_\varepsilon+n_\nu} \end{array}\right\} & \text{nuisance variables} \end{array}\right] \tag{7.113}$$

$$= \left[\begin{array}{c} \mathbf{x}_\varepsilon \\ \hline \mathbf{x}_\nu \end{array}\right], \tag{7.114}$$

where the state vector has been partitioned into a subvector $\mathbf{x}_\varepsilon$ of the essential state variables and a subvector $\mathbf{x}_\nu$ of nuisance variables.

Partitioning of State Dynamic Models. We know that these two state variable types are not linked dynamically, so that the system dynamic model has the form

$$\frac{d}{dt}\left[\begin{array}{c} \mathbf{x}_\varepsilon(t) \\ \hline \mathbf{x}_\nu(t) \end{array}\right] = \left[\begin{array}{c|c} F_\varepsilon(t) & 0 \\ \hline 0 & F_\nu(t) \end{array}\right]\left[\begin{array}{c} \mathbf{x}_\varepsilon(t) \\ \hline \mathbf{x}_\nu(t) \end{array}\right] + \left[\begin{array}{c} \mathbf{w}_\varepsilon(t) \\ \hline \mathbf{w}_\nu(t) \end{array}\right] \tag{7.115}$$

in continuous time, where the process noise vectors $\mathbf{w}_\varepsilon$ and $\mathbf{w}_\nu$ are uncorrelated. That is, the covariance matrix of process noise

$$Q = \left[\begin{array}{c|c} Q_{\varepsilon\varepsilon} & 0 \\ \hline 0 & Q_{\nu\nu} \end{array}\right] \tag{7.116}$$

for the continuous-time model (as well as for the discrete-time model). That is, the cross-covariance block $Q_{\varepsilon\nu} = 0$.

Partitioned Covariance Matrix. The covariance matrix of estimation uncertainty (the dependent variable of the Riccati equation) can also be partitioned as

$$P = \left[\begin{array}{c|c} P_{\varepsilon\varepsilon} & P_{\varepsilon\nu} \\ \hline P_{\nu\varepsilon} & P_{\nu\nu} \end{array}\right], \tag{7.117}$$

where

the block $P_{\varepsilon\varepsilon}$ is dimensioned $n_\varepsilon \times n_\varepsilon$,
the block $P_{\varepsilon\nu}$ is dimensioned $n_\varepsilon \times n_\nu$,
the block $P_{\nu\varepsilon}$ is dimensioned $n_\nu \times n_\varepsilon$, and
the block $P_{\nu\nu}$ is dimensioned $n_\nu \times n_\nu$.

Temporal Covariance Update in Discrete Time. The corresponding state transition matrix for the discrete-time model will then be of the form

$$\Phi_k = \left[\begin{array}{c|c} \Phi_{\varepsilon k} & 0 \\ \hline 0 & \Phi_{\nu k} \end{array}\right], \tag{7.118}$$

$$\Phi_{\varepsilon k} = \exp\left(\int_{t_k}^{t_{k+1}} F_\varepsilon(t)\, dt\right), \tag{7.119}$$

$$\Phi_{\nu k} = \exp\left(\int_{t_k}^{t_{k+1}} F_\nu(t)\, dt\right), \tag{7.120}$$

and the temporal update of P will have the partitioned form

$$\begin{aligned} &\left[\begin{array}{c|c} P_{\varepsilon\varepsilon\, k+1-} & P_{\varepsilon\nu\, k+1-} \\ \hline P_{\nu\varepsilon\, k+1-} & P_{\nu\nu\, k+1-} \end{array}\right] \\ &\quad = \left[\begin{array}{c|c} \Phi_{\varepsilon k} & 0 \\ \hline 0 & \Phi_\nu\, k \end{array}\right] \left[\begin{array}{c|c} P_{\varepsilon\varepsilon\, k+} & P_{\varepsilon\nu\, k+} \\ \hline P_{\nu\varepsilon\, k+} & P_{\nu\nu\, k+} \end{array}\right] \left[\begin{array}{c|c} \Phi_{\varepsilon k}^{\mathrm{T}} & 0 \\ \hline 0 & \Phi_\nu\, k^{\mathrm{T}} \end{array}\right] \\ &\qquad + \left[\begin{array}{c|c} Q_{\varepsilon\varepsilon} & 0 \\ \hline 0 & Q_{\nu\nu} \end{array}\right], \end{aligned} \tag{7.121}$$

or, in terms of the individual blocks,

$$P_{\epsilon\epsilon k+1-} = \Phi_{\epsilon k} P_{\epsilon\epsilon k+} \Phi_{\epsilon k}^{\mathrm{T}} + Q_{\epsilon\epsilon}, \tag{7.122}$$

$$P_{\epsilon v k+1-} = \Phi_{\epsilon k} P_{\epsilon v k+} \Phi_{v} k^{\mathrm{T}}, \tag{7.123}$$

$$P_{v\epsilon k+1-} = \Phi_{v} k P_{v\epsilon k+} \Phi_{\epsilon k}^{\mathrm{T}}, \tag{7.124}$$

$$P_{vvk+1-} = \Phi_{v} k P_{vvk+} \Phi_{v} k^{\mathrm{T}} + Q_{vv}. \tag{7.125}$$

Partitioned Measurement Sensitivity Matrix. With this partitioning of the state vector, the measurement model will have the form

$$z = \left[H_{\varepsilon} \mid H_{v} \right] \left[\frac{\mathbf{x}_{\varepsilon}(t)}{\mathbf{x}_{v}(t)} \right] + \mathbf{v} \tag{7.126}$$

$$= \underbrace{H_{\varepsilon}\mathbf{x}_{\varepsilon}}_{\substack{\text{essential}\\ \text{state}\\ \text{dependence}}} + \underbrace{H_{v}\mathbf{x}_{v}}_{\substack{\text{correlated}\\ \text{noise}}} + \underbrace{\mathbf{v}}_{\substack{\text{uncorrelated}\\ \text{noise}}} . \tag{7.127}$$

7.6.2.1 Schmidt–Kalman Gain

Kalman Gain. The Schmidt–Kalman filter does not use the Kalman gain matrix. However, we need to write out its definition in partitioned form to show how its modification results in the Schmidt–Kalman gain.

The Kalman gain matrix would be partitionable conformably, such that

$$\overline{K} = \left[\frac{K_{\epsilon}}{K_{v}} \right] \tag{7.128}$$

$$= \left[\begin{array}{c|c} P_{\varepsilon\epsilon} & P_{\epsilon v} \\ \hline P_{v\varepsilon} & P_{vv} \end{array} \right] \left[\frac{H_{\epsilon}^{\mathrm{T}}}{H_{v}^{\mathrm{T}}} \right]$$

$$\left\{ \left[H_{\epsilon} \mid H_{v} \right] \left[\begin{array}{c|c} P_{\epsilon\epsilon} & P_{\epsilon v} \\ \hline P_{v\epsilon} & P_{vv} \end{array} \right] \left[\frac{H_{\epsilon}^{\mathrm{T}}}{H_{v}^{\mathrm{T}}} \right] + R \right\}^{-1} \tag{7.129}$$

and the individual blocks

$$K_{\epsilon} = \{ P_{\epsilon\epsilon} H_{\epsilon}^{\mathrm{T}} + P_{\epsilon v} H_{v}^{\mathrm{T}} \} \mathcal{C}, \tag{7.130}$$

$$K_{v} = \{ P_{v\epsilon} H_{\epsilon}^{\mathrm{T}} + P_{vv} H_{v}^{\mathrm{T}} \} \mathcal{C} \tag{7.131}$$

where the common factor

$$\mathcal{C} = \left\{ \left[H_\epsilon \mid H_\nu \right] \left[\begin{array}{c|c} P_{\epsilon\epsilon} & P_{\epsilon\nu} \\ \hline P_{\nu\epsilon} & P_{\nu\nu} \end{array} \right] \left[\begin{array}{c} H_\epsilon^{\mathrm{T}} \\ \hline H_\nu^{\mathrm{T}} \end{array} \right] + R \right\}^{-1} \tag{7.132}$$

$$= \{ H_\epsilon P_{\epsilon\epsilon} H_\epsilon^{\mathrm{T}} + H_\epsilon P_{\epsilon\nu} H_\nu^{\mathrm{T}} + H_\nu P_{\nu\epsilon} H_\epsilon^{\mathrm{T}} + H_\nu P_{\nu\nu} H_\nu^{\mathrm{T}} + R \}^{-1}. \tag{7.133}$$

However, the Schmidt–Kalman filter will, in effect, force K_ν to be zero and redefine the upper block (no longer the K_ϵ of the Kalman filter) to be optimal under that constraint.

Suboptimal Approach. The approach will be to define a suboptimal filter that does not estimate the nuisance state variables but does keep track of the influence they will have on the gains applied to the other state variables.

The suboptimal gain matrix for the Schmidt–Kalman filter has the form

$$\overline{K}_{\text{suboptimal}} = \begin{bmatrix} K_{\text{SK}} \\ 0 \end{bmatrix}, \tag{7.134}$$

where K_{SK} is the $n_\varepsilon \times \ell$ Schmidt–Kalman gain matrix.

This suboptimal filter effectively ignores the nuisance states.

However, the calculation of the covariance matrix P used in defining the gain K_{SK} must still take into account the effect that this constraint has on the state estimation uncertainties and must optimize K_{SK} for that purpose. Here, K_{SK} will effectively be optimal for the constraint that the nuisance states are not estimated. However, the filter will still be *sub*optimal in the sense that filter performance using both Kalman gain blocks (K_ϵ and K_ν) would be superior to that with K_{SK} alone.

The approach still propagates the full covariance matrix P, but the observational update equations are changed to reflect the fact that (in effect) $K_\nu = 0$.

Suboptimal Observational Update. The observational update equation for arbitrary gain K_k can be represented in the form

$$P_k(+) = (I_n - K_k H_k) P_k(-) (I_n - K_k H_k)^{\mathrm{T}} + KRK^{\mathrm{T}}, \tag{7.135}$$

where n is the dimension of the state vector, I_n is the $n \times n$ identity matrix, ℓ is the dimension of the measurement, H_k is the $\ell \times n$ measurement sensitivity matrix, and R_k is the $\ell \times \ell$ covariance matrix of uncorrelated measurement noise.

In the case that the suboptimal gain K_k has the partitioned form shown in Equation 7.134, the partitioned observational update equation for P will be

$$\begin{aligned}\left[\begin{array}{c|c} P_{\epsilon\epsilon,k+} & P_{\epsilon\nu,k+} \\ \hline P_{\nu\epsilon,k+} & P_{\nu\nu,k+} \end{array}\right] &= \left(\left[\begin{array}{c|c} I_{n_\epsilon} & 0 \\ \hline 0 & I_{n_\nu} \end{array}\right] - \left[\begin{array}{c} K_{\mathrm{SK},k} \\ \hline 0 \end{array}\right] \left[H_{\epsilon,k} \mid H_{\nu,k} \right]\right) \\ &\quad \times \left[\begin{array}{c|c} P_{\epsilon\epsilon,k-} & P_{\epsilon\nu,k-} \\ \hline P_{\nu\epsilon,k-} & P_{\nu\nu,k-} \end{array}\right] \\ &\quad \times \left(\left[\begin{array}{c|c} I_{n_\epsilon} & 0 \\ \hline 0 & I_{n_\nu} \end{array}\right] - \left[\begin{array}{c} K_{\mathrm{SK},k} \\ \hline 0 \end{array}\right] \left[H_{\epsilon,k} \quad H_{\nu,k} \right]\right)^{\mathrm{T}} \\ &\quad + \left[\begin{array}{c} K_{\mathrm{SK},k} \\ \hline 0 \end{array}\right] R_k \left[K_{\mathrm{SK},k}^{\mathrm{T}} \quad 0 \right]. \end{aligned} \tag{7.136}$$

The summed terms in parentheses can be combined into the following form for expansion:

$$\begin{aligned} &= \left[\begin{array}{c|c} I_{n_\epsilon} - K_{\mathrm{SK},k} H_{\epsilon,k} & -K_{\mathrm{SK},k} H_{\nu,k} \\ \hline 0 & I_{n_\nu} \end{array}\right] \times \left[\begin{array}{c|c} P_{\epsilon\epsilon,k-} & P_{\epsilon\nu,k-} \\ \hline P_{\nu\epsilon,k-} & P_{\nu\nu,k-} \end{array}\right] \\ &\quad \times \left[\begin{array}{c|c} I_{n_\epsilon} - H_{\epsilon,k}^{\mathrm{T}} K_{\mathrm{SK},k}^{\mathrm{T}} & 0 \\ \hline -H_{\nu,k}^{\mathrm{T}} K_{\mathrm{SK},k}^{\mathrm{T}} & I_{n_\nu} \end{array}\right] + \left[\begin{array}{c|c} K_{\mathrm{SK},k} R_k K_{\mathrm{SK},k}^{\mathrm{T}} & 0 \\ \hline 0 & 0 \end{array}\right], \end{aligned} \tag{7.137}$$

which can then be expanded to yield the following formulas for the blocks of P (with annotation showing intermediate results that can be reused to reduce the computation):

$$\begin{aligned} P_{\epsilon\epsilon,k+} &= \underbrace{(I_{n_\epsilon} - K_{\mathrm{SK},k} H_{\epsilon,k})}_{\mathcal{A}} P_{\epsilon\epsilon,k-} \underbrace{(I_{n_\epsilon} - K_{\mathrm{SK},k} H_{\epsilon,k})^{\mathrm{T}}}_{\mathcal{A}^{\mathrm{T}}} \\ &\quad - \underbrace{\overbrace{(I_{n_\epsilon} - K_{\mathrm{SK},k} H_{\epsilon,k})}^{\mathcal{A}} P_{\nu\epsilon,k-} H_{\nu,k}^{\mathrm{T}} K_{\mathrm{SK},k}^{\mathrm{T}}}_{\mathcal{B}} \\ &\quad - \overbrace{K_{\mathrm{SK},k} H_{\nu,k} P_{\epsilon\nu,k-} (I_{n_\epsilon} - K_{\mathrm{SK},k} H_{\epsilon,k})^{\mathrm{T}}}^{\mathcal{B}^{\mathrm{T}}} \\ &\quad + K_{\mathrm{SK},k} R_k K_{\mathrm{SK},k}^{\mathrm{T}}, \end{aligned} \tag{7.138}$$

$$P_{\epsilon\nu,k+} = \overbrace{\left(I_{n_\epsilon} - K_{\mathrm{SK},k} H_{\epsilon,k}\right)}^{\mathcal{A}} P_{\epsilon\nu,k-} - K_{\mathrm{SK},k} H_{\nu,k} P_{\nu\nu,k-}, \tag{7.139}$$

$$P_{\nu\epsilon,k+} = P_{\epsilon\nu,k+}^{\mathrm{T}}, \tag{7.140}$$

$$P_{\nu\nu,k+} = P_{\nu\nu,k-}. \tag{7.141}$$

Note that $P_{\nu\nu}$ is unchanged by the observational update because $\mathbf{x}_\nu$ is not updated.

This completes the derivation of the Schmidt–Kalman filter. The temporal update of P in the Schmidt–Kalman filter will be the same as for the Kalman filter. This happens because the temporal update only models the propagation of the state variables, and the propagation model is the same in both cases.

7.6.3 Implementation Equations

We can now summarize just the essential equations from the derivation above, as listed in Table 7.1. These have been rearranged slightly to reuse intermediate results.

7.6.4 Computational Complexity

The purpose of the Schmidt-Kalman filter was to reduce the computational requirements over those required for the full Kalman filter. Although the equations appear to be more complicated, the dimensions of the matrices involved are smaller than the matrices in the Kalman filter.

We will now do a rough operations count of those implementation equations, just to be sure that they do, indeed, decrease computational requirements.

Table 7.2 is a breakdown of the operations counts for implementing the equations in Table 7.1. The formulas (in angular brackets) above the matrix formulas give the rough operations counts for implementing those formulas. An "operation" in this accounting is roughly equivalent to a multiply-and-accumulate. The operations counts are expressed in terms of the number of measurements (ℓ, the dimension of the measurement vector), the number of essential state variables (n_ϵ), and the number of nuisance state variables (n_ν).

These complexity formulas are based on the matrix dimensions listed in Table 7.3.

A MATLAB implementation of the Schmidt-Kalman filter is in the m-file KFvsSKF.m on the accompanying diskette.

7.7 MEMORY, THROUGHPUT, AND WORDLENGTH REQUIREMENTS

These may not be important issues for off-line implementations of Kalman filters on mainframe scientific computers, but they can become critical issues for real-time implementations in embedded processors, especially as the dimensions of the state

TABLE 7.1 Implementation Equations of Schmidt–Kalman Filter

Observational update:

$$\mathcal{C} = \{H_{\epsilon k}(P_{\epsilon\epsilon k-}H_{\epsilon k}^{\mathrm{T}} + P_{\epsilon\nu k-}H_{\nu k}^{\mathrm{T}}) + H_{\nu k}(P_{\nu\epsilon k-}H_{\epsilon k}^{\mathrm{T}} + P_{\nu\nu k-}H_{\nu k}^{\mathrm{T}}) + R_k\}^{-1}$$

$$K_{\mathrm{SK},k} = \{P_{\epsilon\epsilon k-}H_{\epsilon k}^{\mathrm{T}} + P_{\epsilon\nu k-}H_{\nu k}^{\mathrm{T}}\}\mathcal{C}$$

$$\mathbf{x}_{\epsilon,k+} = \mathbf{x}_{\epsilon,k-} + K_{\mathrm{S-K},k}\{\mathbf{z}_k - H_{\epsilon k}\mathbf{x}_{\epsilon,k-}\}$$

$$\mathcal{A} = I_{n_\epsilon} - K_{\mathrm{SK},k}H_{\epsilon,k}$$

$$\mathcal{B} = \mathcal{A}P_{\nu\epsilon,k-}H_{\nu,k}^{\mathrm{T}}K_{S-K,k}^{\mathrm{T}}$$

$$P_{\epsilon\epsilon,k+} = \mathcal{A}P_{\epsilon\epsilon,k-}\mathcal{A}^{\mathrm{T}} - \mathcal{B} - \mathcal{B}^{\mathrm{T}} + K_{\mathrm{SK},k}R_kK_{\mathrm{SK},k}^{\mathrm{T}}$$

$$P_{\epsilon\nu,k+} = \mathcal{A}P_{\epsilon\nu,k-} - K_{\mathrm{SK},k}H_{\nu,k}P_{\nu\nu,k-}$$

$$P_{\nu\epsilon,k+} = P_{\epsilon\nu,k+}^{\mathrm{T}}$$

$$P_{\nu\nu,k+} = P_{\nu\nu,k-}$$

Temporal update:

$$\mathbf{x}_{\epsilon,k+1-} = \Phi_{\epsilon k}\mathbf{x}_{\epsilon,k+}$$

$$P_{\epsilon\epsilon k+1-} = \Phi_{\epsilon k}P_{\epsilon\epsilon k+}\Phi_{\epsilon k}^{\mathrm{T}} + Q_{\epsilon\epsilon}$$

$$P_{\epsilon\nu k+1-} = \Phi_{\epsilon k}P_{\epsilon\nu k+}\Phi_{\nu k}^{\mathrm{T}}$$

$$P_{\nu\epsilon k+1-} = P_{\epsilon\nu k+1-}^{\mathrm{T}}$$

$$P_{\nu\nu k+1-} = \Phi_{\nu k}P_{\nu\nu k+}\Phi_{\nu k}^{\mathrm{T}} + Q_{\nu\nu}$$

vector or measurement become larger. We present here some methods for assessing these requirements for a given application and for improving feasibility in marginal cases. These include order-of-magnitude plots of memory requirements and computational complexity as functions of the dimensions of the state vector and measurement vector. These plots cover the ranges from 1 to 1000 for these dimensions, which should include most problems of interest.

7.7.1 Wordlength Problems

Precision Problems. Wordlength issues include precision problems (related to the number of significant bits in the mantissa field) and dynamic range problems (related to the number of bits in the exponent field). The issues and remedies related to precision are addressed in Chapter 6.

TABLE 7.2 Operations Counts for Schmidt–Kalman Filter

Scalar Operation Counts for Matrix Operations	Totals by Rows
$\mathcal{C} = \left\{ \overbrace{H_{\epsilon k}\times}^{\langle n_\epsilon \ell^2\rangle} \left(\underbrace{\overbrace{P_{\epsilon\epsilon k-}H^{\mathrm{T}}_{\epsilon k}}^{\langle n_\epsilon^2\ell\rangle} + \overbrace{P_{\epsilon v k-}H^{\mathrm{T}}_{vk}}^{\langle n_\epsilon n_v \ell\rangle}}_{\langle\text{used again below}\rangle}\right)\right.$	$n_\epsilon\ell^2 + n_\epsilon^2\ell + n_\epsilon n_v\ell$
$+ \overbrace{H_{vk}\times}^{\langle n_v\ell^2\rangle}\left(\overbrace{P_{v\epsilon k-}H^{\mathrm{T}}_{\epsilon k}}^{\langle n_\epsilon n_v\ell\rangle} + \overbrace{P_{vvk-}H^{\mathrm{T}}_{vk}}^{\langle n_v^2\ell\rangle}\right)$	$n_v\ell^2 + n_\epsilon n_v\ell + n_v^2\ell$
$\left. + R_k\right\}^{-1\langle\ell^3\rangle}$ (matrix inverse)	ℓ^3
$K_{SK,k} = \left\{\overbrace{P_{\epsilon\epsilon k-}H^{\mathrm{T}}_{\epsilon k} + P_{\epsilon v k-}H^{\mathrm{T}}_{vk}}^{\langle\text{already computed above}\rangle}\right\}\overbrace{\times\mathcal{C}}^{\langle n_\epsilon\ell^2\rangle}$	$n_\epsilon\ell^2$
$\mathbf{x}_{\epsilon,k+} = \mathbf{x}_{\epsilon,k-} + \overbrace{K_{\mathrm{SK},k}}^{\langle n_\epsilon\ell\rangle} \times \overbrace{\{\mathbf{z}_k - H_{\epsilon k}\mathbf{x}_{\epsilon,k-}\}}^{\langle n_\epsilon\ell\rangle}$	$2n_\epsilon\ell$
$\mathcal{A} = I_{n_\epsilon} - \overbrace{K_{\mathrm{SK},k}}^{\langle n_\epsilon^2\ell\rangle} H_{\epsilon,k}$	$n_\epsilon^2\ell$
$\mathcal{B} = \underbrace{\overbrace{\mathcal{A}\times P_{\epsilon v,k-}}^{\langle n_\epsilon^2 n_v\rangle} \times \overbrace{H^{\mathrm{T}}_{v,k}\times K^{\mathrm{T}}_{\mathrm{SK},k}}^{\langle n_\epsilon n_v\ell\rangle}}_{\langle n_\epsilon^2 n_v\rangle}$	$2n_\epsilon^2 n_v + n_\epsilon n_v\ell$
$P_{\epsilon\epsilon,k+} = \mathcal{A}P_{\epsilon\epsilon,k-}\mathcal{A}^{\mathrm{T}} - \mathcal{B} - \mathcal{B}^{\mathrm{T}}$	$\frac{3}{2}n_\epsilon^3 + \frac{1}{2}n_\epsilon^2$
$+ \underbrace{K_{\mathrm{SK},k}\overbrace{[H_{vk}, P_{vv,k-}H^{\mathrm{T}}_{v,k} + R_k]}^{\langle n_v^2\ell + \frac{1}{2}n_v\ell^2 + \frac{1}{2}n_v\ell\rangle}K^{\mathrm{T}}_{\mathrm{SK},k}}_{\langle\frac{1}{2}n_\epsilon^2\ell + n_\epsilon\ell^2 + \frac{1}{2}n_\epsilon\ell\rangle}$	$\frac{3}{2}n_v^3 + n_v^2\ell + \frac{1}{2}n_v\ell^2 + \frac{1}{2}n_v^2 + \frac{1}{2}n_v\ell$
$P_{\epsilon v,k+} = \overbrace{\mathcal{A}\times P_{\epsilon v,k-}}^{\langle n_\epsilon^2 n_v\rangle} - \overbrace{K_{\mathrm{SK},k}\times H_{v,k-}}^{\langle n_\epsilon n_v\ell\rangle}\overbrace{\times P_{vv,k-}}^{\langle n_\epsilon n_v^2\rangle}$	$n_\epsilon^2 n_v + n_\epsilon n_v\ell + n_\epsilon n_v^2$
$P_{v\epsilon,k+} = P^{\mathrm{T}}_{\epsilon v,k+}$	0
$P_{vv,k+} = P_{vv,k-}$	0
Total for Observational Update	$3n_\epsilon\ell^2 + \frac{5}{2}n_\epsilon^2\ell + 4n_\epsilon n_v\ell$ $+\frac{3}{2}n_v\ell^2 + 2n_v^2\ell + \ell^3$ $+\frac{5}{2}n_\epsilon\ell + 3n_\epsilon^2 n_v$ $+\frac{1}{2}n_v\ell + n_\epsilon n_v^2$

TABLE 7.2 (continued)

Scalar Operation Counts for Matrix Operations	Totals by Rows
Temporal Update	
$\hat{\mathbf{x}}_{\epsilon,k+1-} = \Phi_{\epsilon k}\hat{\mathbf{x}}_{\epsilon,k+}$	n_ϵ^2
$P_{\epsilon\epsilon\, k+1-} = \Phi_{\epsilon k} P_{\epsilon\epsilon\, k+} \Phi_{\epsilon k}^{\mathrm{T}} + Q_{\epsilon\epsilon}$	$\frac{3}{2}n_\epsilon^3 + \frac{1}{2}n_\epsilon^2$
$P_{\epsilon\nu\, k+1-} = \Phi_{\epsilon k} P_{\epsilon\nu\, k+} \Phi_{\nu k}^{\mathrm{T}}$	$n_\epsilon n_\nu^2 + n_\nu n_\epsilon^2$
$P_{\nu\epsilon\, k+1-} = P_{\epsilon\nu k+1-}^{\mathrm{T}}$	0
$P_{\nu\nu\, k+1-} = \Phi_{\nu k} P_{\nu\nu\, k+} \Phi_{\nu k}^{\mathrm{T}} + Q_{\nu\nu}$	$\frac{3}{2}n_\nu^3 + \frac{1}{2}n_\nu^2$
Total for Temporal Update	$\frac{3}{2}n_\epsilon^2 + \frac{3}{2}n_\epsilon^3 + n_\epsilon n_\nu^2 + n_\epsilon^2 n_\nu + \frac{3}{2}n_\nu^3 + \frac{1}{2}n_\nu^2$
Total for Schmidt–Kalman Filter	$3n_\epsilon\ell^2 + \frac{5}{2}n_\epsilon^2\ell + 4n_\epsilon n_\nu \ell + \frac{3}{2}n_\nu\ell^2 + 2n_\nu^2\ell + \ell^3 + \frac{5}{2}n_\epsilon\ell + 4n_\epsilon^2 n_\nu + \frac{3}{2}n_\epsilon^3 + \frac{3}{2}n_\epsilon^2 + \frac{1}{2}n_\nu\ell + \frac{3}{2}n_\nu^3 + \frac{1}{2}n_\nu^2 + \frac{1}{2}n_\nu\ell + 2n_\epsilon n_\nu^2$

Scaling Problems. Underflows and overflows are symptoms of dynamic range problems. These can often be corrected by rescaling the variables involved. This is equivalent to changing the units of measure, such as using kilometers in place of centimeters to represent length. In some cases, but not all cases, the condition number of a matrix can be improved by rescaling. For example, the two covariance matrices

$$\begin{bmatrix} 1 & 0 \\ 0 & \epsilon^2 \end{bmatrix} \quad \text{and} \quad \frac{1}{2}\begin{bmatrix} 1+\epsilon^2 & 1-\epsilon^2 \\ 1-\epsilon^2 & 1+\epsilon^2 \end{bmatrix}$$

have the same condition number ($1/\epsilon^2$), which can be troublesome for very small values of ϵ. The condition number of the matrix on the left can be made equal to 1 by simply rescaling the second component by $1/\epsilon$.

7.7.2 Memory Requirements

In the early years of Kalman filter implementation, a byte of memory cost about as much as a labor hour at minimum wage. With these economic constraints, programmers developed many techniques for reducing the memory requirements of Kalman filters. A few of these techniques have been mentioned in Chapter 6, although they are not as important as they once were. Memory costs have dropped

TABLE 7.3 Array Dimensions

Symbol	Rows	Columns
$\mathcal{A}$	n_ϵ	n_ϵ
$\mathcal{B}$	n_ϵ	n_ϵ
$\mathcal{C}$	ℓ	ℓ
H_ϵ	ℓ	n_ϵ
H_v	ℓ	n_v
$K_{\rm SK}$	n_ϵ	ℓ
$P_{\epsilon\epsilon}$	n_ϵ	n_ϵ
$P_{\epsilon v}$	n_ϵ	n_v
$P_{v\epsilon}$	n_v	n_ϵ
P_{vv}	n_v	n_v
$Q_{\epsilon\epsilon}$	n_ϵ	n_ϵ
Q_{vv}	n_v	n_v
R	ℓ	ℓ
Φ_ϵ	n_ϵ	n_ϵ
Φ_v	n_v	n_v
$\hat{\mathbf{x}}_\epsilon$	n_ϵ	1
$\mathbf{z}$	ℓ	1

dramatically since these methods were developed. The principal reason for paying attention to memory requirements nowadays is to determine the limits on problem size with a fixed memory allocation. Memory is cheap, but still finite.

Program Memory versus Data Memory. In the "von Neumann architecture" for processing systems, there is no distinction between the memory containing the algorithms and that containing the data used by the algorithms. In specific applications, the program may include formulas for calculating the elements of arrays such as Φ or H. Other than that, the memory requirements for the algorithms tend to be independent of application and the "problem size." For the Kalman filter, the problem size is specified by the dimensions of the state (n), measurement (ℓ), and process noise (p). The data memory requirements for storing the arrays with these dimensions is very much dependent on the problem size. We present here some general formulas for this dependence.

Data Memory and Wordlength. The data memory requirements will depend upon the data wordlengths (in bits) as well as the size of the data structures. The data requirements are quoted in "floating-point words." These are either 4- or 8-byte words (in IEEE floating-point standard formats) for the examples presented in this book.

Data Memory Requirements. These are also influenced somewhat by programming style, particularly by the ways that data structures containing partial results are reused.

The data memory requirements for a more-or-less "conventional" implementation of the Kalman filter are listed in Table 7.4 and plotted in Figure 7.25. This is the Kalman filter implementation diagrammed in Figure 6.2, which reuses some partial results. The array dimensions are associated with the results of matrix subexpressions that they contain. These are divided into three groups:

1. those arrays common to both the Riccati equation (for covariance and gain computations) and the linear estimation computations;
2. the additional array expressions required for solving the Riccati equation, which provides the Kalman gain as a partial result; and
3. the additional array expressions required for linear estimation of the state variables, given the Kalman gains.

Expressions grouped together by curly braces are assumed to be kept in the same data structures. This implementation assumes that

- the product GQG^{T} is input and not computed (which eliminates any dependence on p, the dimension of Q);
- given ΦP, Φ, and GQG^{T}, the operation $P \leftarrow \Phi P \Phi^{\mathrm{T}} + GQG^{\mathrm{T}}$ is performed in place;

TABLE 7.4 Array Requirements for "Conventional" Kalman Filter Implementation

Functional Grouping	Matrix Expression	Array Dimensions	Total Memory[a]
Riccati equation	P	$n \times n$	
	ΦP	$n \times n$	
	GQG^{T}	$n \times n$	$3n^2$
	$\left.\begin{matrix} HP \\ PH^{\mathrm{T}} \end{matrix}\right\}$	$\ell \times n$	$+\ell n$
	R	$\ell \times \ell$	
	$\left.\begin{matrix} HPH^{\mathrm{T}} + R \\ [HPH^{\mathrm{T}} + R]^{-1} \end{matrix}\right\}$	$\ell \times \ell$	$+2\ell^2$
Common	Φ	$n \times n$	n^2
	H	$\ell \times n$	
	$\overline{K}$	$n \times \ell$	$+2\ell n$
Linear Estimation	$\left.\begin{matrix} z \\ z - Hx \end{matrix}\right\}$	ℓ	ℓ
	x	n	
	Φx	n	$+2n$

[a]In units of floating-point data words.

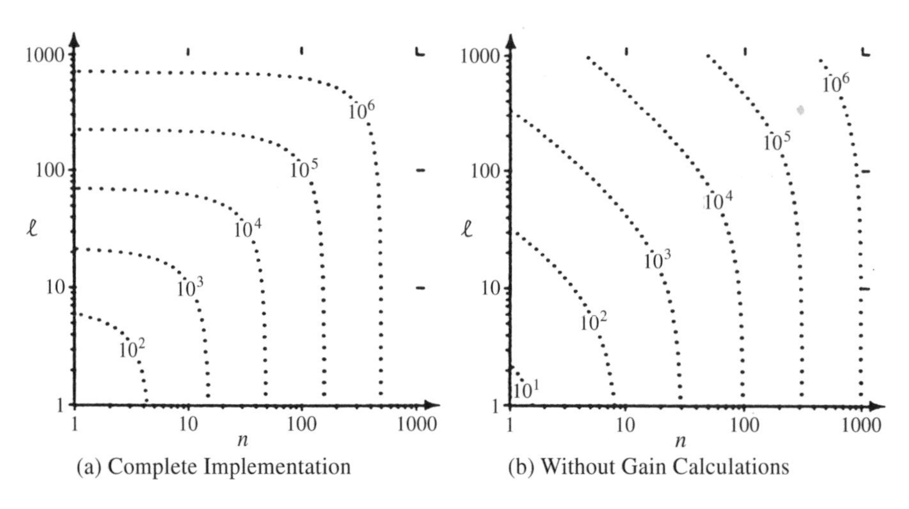

Fig. 7.25 *Conventional filter memory requirements (in words) versus state dimension (n) and measurement dimension (ℓ).*

- computations involving the subexpression PH^{T} can be implemented with HP by changing the indexing;
- $HPH^{\mathrm{T}} + R$ can be computed in place (from HP, H, and R) and inverted in place;
- $z - Hx$ can be computed in place (in the z array); and
- the state update computation $x \leftarrow (\Phi x) + \overline{K}[z - H(\Phi x)]$ requires additional memory only for the intermediate result (Φx).

Figure 7.25 illustrates the numerical advantage of precomputing and storing the Kalman gains in bulk memory. It saves about a factor of 4 in data memory requirements for small dimensions of the measurement relative to the state and even more for larger measurement dimensions.

Eliminating Data Redundancy. Reuse of temporary arrays is not the only way to save memory requirements. It is also possible to save data memory by eliminating redundancies in data structures. The symmetry of covariance matrices is an example of this type of redundancy. The methods discussed here depend on such constraints on matrices that can be exploited in designing their data structures. They do require additional programming effort, however, and the resulting run time program code may require slightly more memory and more processing time. The difference will be primarily from index computations, which are not the standard ones used by optimizing compilers. Table 7.5 lists some common constraints on square matrices, the minimum memory requirements (as multiples of the memory required for a scalar variable), and corresponding indexing schemes for packing the matrices in singly subscripted arrays. The indexing schemes are given as formulas for the single

TABLE 7.5 Minimum Memory Requirements for $n \times n$ Matrices[a]

Matrix Type	Minimum Memory[b]	Indexing $k(i,j)$
Symmetric	$\frac{n(n+1)}{2}$	$i+\frac{j(j-1)}{2}$
		$\frac{(2n-i)(i-1)}{2}+j$
Upper triangular	$\frac{n(n+1)}{2}$	$i+\frac{j(j-1)}{2}$
Unit upper triangular	$\frac{n(n+1)}{2}$	$i+\frac{(j-1)(j-2)}{2}$
Strictly upper triangular	$\frac{n(n-1)}{2}$	$i+\frac{(j-1)(j-2)}{2}$
Diagonal	n	i
Toeplitz	n	$i+j-1$

[a]Note: n is the dimension of the matrix; i and j are the indices of a two-dimensional array; k is the corresponding index of a one-dimensional array;
[b]In units of data words.

subscript (k) corresponding to the row (i) and column (j) indices of a two-dimensional array. The two formulas given for symmetric matrices correspond to the two alternative indexing schemes:

$$
\begin{bmatrix}
1 & 2 & 4 & \dots & \frac{1}{2}n(n-1)+1 \\
 & 3 & 5 & \dots & \frac{1}{2}n(n-1)+2 \\
 & & 6 & \dots & \frac{1}{2}n(n-1)+3 \\
 & & & \ddots & \vdots \\
 & & & \dots & \frac{1}{2}n(n+1)
\end{bmatrix}
\quad \text{or} \quad
\begin{bmatrix}
1 & 2 & 3 & \dots & n \\
 & n+1 & n+2 & \dots & 2n-1 \\
 & & 2n & \dots & 3n-3 \\
\vdots & \vdots & \vdots & \ddots & \vdots \\
 & & & \dots & \frac{1}{2}n(n+1)
\end{bmatrix},
$$

where the element in the ith row and jth column is $k(i,j)$.

Just by exploiting symmetry or triangularity, these methods can save about a factor of 2 in memory with a fixed state vector dimension, or allow about a factor of $\sqrt{2}$ increase in the state vector dimension (about 40% increase in the dimension) with the same amount of memory.

Arrays Can Be Replaced by Algorithms. In special cases, data arrays can be eliminated entirely by using an algorithm to compute the matrix elements "on the fly." For example, the companion form coefficient matrix (F) and the corresponding state transition matrix (Φ) for the differential operator d^n/dt^n are

$$F = \begin{bmatrix} 0 & 1 & 0 & \cdots & 0 \\ 0 & 0 & 1 & \cdots & 0 \\ \vdots & \vdots & \vdots & \ddots & \vdots \\ 0 & 0 & 0 & \cdots & 1 \\ 0 & 0 & 0 & \cdots & 0 \end{bmatrix}, \tag{7.142}$$

$$\Phi(t) = e^{Ft} \tag{7.143}$$

$$= \begin{bmatrix} 1 & t & \frac{1}{2}t^2 & \cdots & \frac{1}{(n-1)!}t^{n-1} \\ 0 & 1 & t & \cdots & \frac{1}{(n-2)!}t^{n-2} \\ 0 & 0 & 1 & \cdots & \frac{1}{(n-3)!}t^{n-3} \\ \vdots & \vdots & \vdots & \ddots & \vdots \\ 0 & 0 & 0 & \cdots & 1 \end{bmatrix}, \tag{7.144}$$

where t is the discrete-time interval. The following algorithm computes the product $M = \Phi P$, with Φ as given in Equation 7.144, using only P and t:

```
for i=1:n,
    for j=1:n,
        s=P(n,j);
        m=n-1;
        for k=n-1:1:i,
            s=P(k,j)+s*t/m;
            m=m-1;
        end;
        M(i,j)=s;
    end;
end;
```

It requires about half as many arithmetic operations as a general matrix multiply and requires no memory allocation for one of its matrix factors.

7.7.3 Throughput, Processor Speed, and Computational Complexity

Influence of Computational Complexity on Throughput. The "throughput" of a Kalman filter implementation is related to how many updates it can perform in a unit of time. This depends upon the speed of the host processor, in floating-point operations (flops) per second, and the computational complexity of the application, in flops per filter update:

$$\text{throughput}\left(\frac{\text{updates}}{\text{s}}\right) = \frac{\text{processor speed (flops/s)}}{\text{computational complexity (flops/update)}}.$$

The numerator of the expression on the right-hand side depends upon the host processor. Formulas for the denominator, as functions of the application problem size, are derived in Chapter 6. These are the *maximum* computational complexities of the implementation methods listed in Chapter 6. The computational complexity of an application can be made smaller if it includes sparse matrix operations that can be implemented more efficiently.

Conventional Kalman Filter. The maximum computational complexity of the conventional Kalman filter implementation is plotted versus problem size in Figure 7.26. This implementation uses all the shortcuts listed in Section 7.7.2, and also eliminates redundant computations in symmetric matrix products. The right hand plot assumes that the matrix Riccati equation for the Kalman gain computations has been solved off-line, as in a gain-scheduled or steady-state implementation.

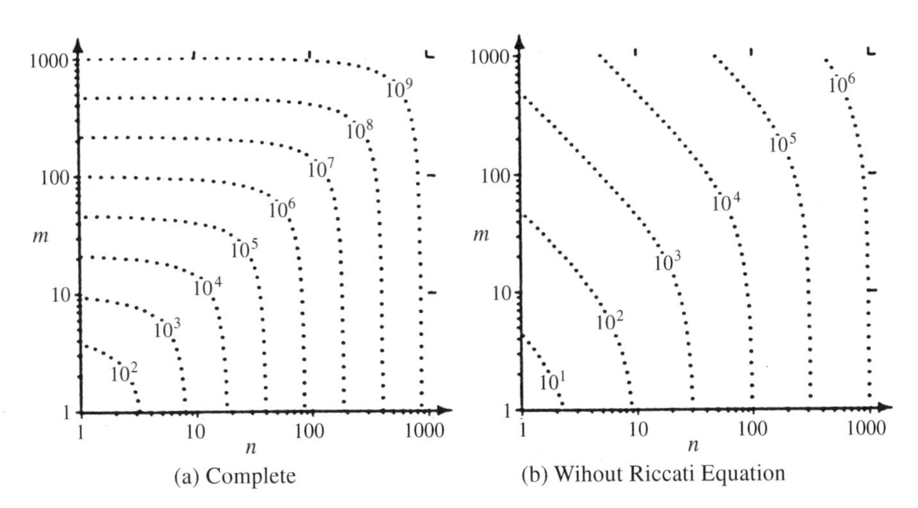

Fig. 7.26 Contour plots of computational complexity (in flops per measurement) of the Kalman filter as a function of state dimension (n) and measurement dimension (m).

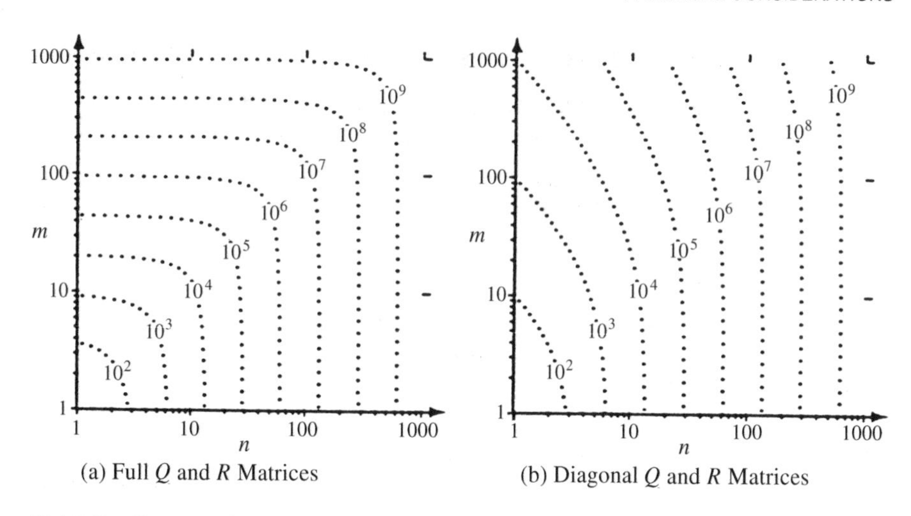

Fig. 7.27 Contour plots of computational complexity (in flops per measurement) of the Bierman–Thornton implementation as a function of state dimension (n) and measurement dimension (m).

Bierman-Thornton Square Root Implementation. The corresponding dependence of computational complexity on problem size for the *UD* filter implementation is plotted in Figure 7.27*a*. These data include the computational cost of diagonalizing Q and R on each temporal and observational update, respectively. The corresponding results for the case that Q and R are already diagonal are displayed in Figure 7.27*b*.

7.7.4 Programming Cost versus Run Time Cost

The computational complexity issue in Kalman filtering is usually driven by the need to execute in real time. The computational complexity grows so fast with the problem size that it will overwhelm even the fastest processors for sufficiently large system models. For that reason, the issue of computational complexity is one that must be addressed early on in the filter design cycle.

Another trade-off in the design of Kalman filters is between the one-time cost of programming the implementation and the recurring cost of running it on a computer. As computers grow less expensive compared to programmers, this trade-off tends to favor the most straightforward methods, even those that cause numerical analysts to wince. Keep in mind, however, that this is a low-cost/high-risk approach. Remember that the reason for the development of better implementation methods was the failure of the straightforward programming solutions to produce acceptable results.

7.8 WAYS TO REDUCE COMPUTATIONAL REQUIREMENTS

7.8.1 Reducing Complexities of Matrix Products

Implementing Products of Two Matrices. The number of flops required to compute the product of general $\ell \times m$ and $m \times n$ matrices is $\ell m^2 n$. This figure can be reduced substantially for matrices with predictable patterns of zeros or symmetry properties. These tricks can be used to advantage in computing matrix products that are symmetric and for products involving diagonal or triangular factors. They should always be exploited whenever H or Φ is a sparse matrix.

Implementing Products of Three Matrices. It is of considerable practical importance that *associativity of matrix multiplication does not imply invariance of computational complexity.* The associativity of matrix multiplication is the property that

$$M_1 \times \left(M_2 \times M_3\right) = (M_1 \times M_2) \times M_3 \tag{7.145}$$

for conformably dimensioned matrices M_1, M_2, and M_3. That is, the *result* is guaranteed to be independent of the *order* in which the two matrix multiplications are performed. However, the *effort* required to obtain the result is *not* always independent of the order of multiplication. This distinction is evident if one assigns conformable dimensions to the matrices involved and evaluates the number of scalar multiplications required to compute the result, as shown in Table 7.6. The number of flops depends on the order of multiplication, being $n_2\left(n_3^3 + n_1 n_2\right)n_4$ in one case and $n_1\left(n_2^2 + n_3 n_4\right)n_3$ in the other case. The implementation $M_1 \times \left(M_2 \times M_3\right)$ is favored if $n_1 n_2^2\left(n_4 - n_3\right) < (n_1 - n_2)n_3^2 n_4$, and the implementation $(M_1 \times M_2) \times M_3$ is favored if the inequality is reversed. The correct selection is used to advantage in the more practical implementations of the Kalman filter, such as the De Vries implementation (see Section 6.6.1.4).

7.8.2 Off-Line versus On-Line Computational Requirements

The Kalman filter is a "real-time" algorithm, in the sense that it calculates an estimate of the current state of a system given measurements obtained in real time. In order that the filter be implementable in real time, however, it must be possible to execute the algorithm in real time with the available computational resources. In this

TABLE 7.6 Computational Complexities of Triple Matrix Product

Attribute	Value	
Implementation	$\underbrace{M_1}_{n_1\times n_2} \times \left(\underbrace{M_2}_{n_2\times n_3} \times \underbrace{M_3}_{n_3\times n_4}\right)$	$\left(\underbrace{M_1}_{n_1\times n_2} \times \underbrace{M_2}_{n_2\times n_3}\right) \times \underbrace{M_3}_{n_3\times n_4}$
Number of flops		
First multiply	$n_2 n_3^2 n_4$	$n_1 n_2^2 n_3$
Second multiply	$n_1 n_2^2 n_4$	$n_1 n_3^2 n_4$
Total	$n_2(n_3^3 + n_1 n_2)n_4$	$n_1(n_2^2 + n_3 n_4)n_3$

assessment, it is important to distinguish between those parts of the filter algorithm that must be performed "on-line" and those that can be performed "off-line" (i.e., carried out beforehand, with the results stored in memory, including bulk media, such as magnetic tape or CDROM, and read back in real time[7]). The on-line computations are those that depend upon the measurements of the real-time system. Those calculations cannot be made until their input data become available.

It is noted in Chapter 4 that the computations required for calculating the Kalman gains do not depend upon the real-time data, and for that reason they can be executed off-line. It is repeated here for emphasis and to formalize some of the practical methods used for implementation.

The most straightforward method is to precompute the gains and store them for retrieval in real time. This is also the method with the most general applicability. Some methods of greater efficiency (but less generality) are discussed in the following subsections. Methods for performance analysis of these suboptimal estimation methods are discussed in Section 7.5.

7.8.3 Gain Scheduling

This is an approximation method for estimation problems in which the rate of change of the Kalman gains is very slow compared to the sampling rate. Typically, the relative change in the Kalman gain between observation times may be a few percent or less. In that case, one value of the Kalman gain may be used for several observation times. Each gain value is used for a "stage" of the filtering process.

This approach is typically used for problems with constant coefficients. The gains in this case have an asymptotic constant value but go through an initial transient due to larger or smaller initial uncertainties than the steady-state uncertainties. A few "staged" values of the gains during that transient phase may be sufficient to achieve adequate performance. The values used may be sampled values in the interior of the stage in which they are used or weighted averages of all the exact values over the range.

The performance trade-off between the decreased storage requirements (for using fewer values of the gains) and the increased approximation error (due to differences between the optimal gains and the scheduled gains) can be analyzed by simulation.

7.8.4 Steady-State Gains for Time-Invariant Systems

This is the limiting case of gain scheduling—with only one stage—and it is one of the more common uses of the algebraic Riccati equation. In this case, only the asymptotic values of the gains are used. This requires the solution of the algebraic (steady-state) matrix Riccati equation.

[7]In assessing the real-time implementation requirements, one must trade off the time to read these prestored values versus the time required to compute them. In some cases, the read times may exceed the computation times.

There are several methods for solving the steady-state matrix Riccati equation in the following subsections. One of these (the doubling method) is based on the linearization method for the Riccati equation presented in Chapter 4. Theoretically, it converges exponentially faster than the serial iteration method. In practice, however, convergence can stall (due to numerical problems) before an accurate solution is attained. However, it can still be used to obtain a good starting estimate for the Newton–Raphson method (described in Chapter 4).

7.8.4.1 Doubling Method for Time-Invariant Systems. This is an iterative method for approximating the asymptotic solution to the *time-invariant* Riccati equation, based on the formula given in Lemma 2 in Chapter 4. As in the continuous case, the asymptotic solution should equal the solution of the *steady-state equation*:

$$P_\infty = \Phi\left[P_\infty - P_\infty H^{\mathrm{T}}\left(HP_\infty H^{\mathrm{T}} + R\right)^{-1} HP_\infty\right]\Phi^{\mathrm{T}} + Q, \tag{7.146}$$

although this is not the form of the equation that is used. Doubling methods generate the sequence of solutions

$$P_1(-), P_2(-), P_4(-), P_8(-), \ldots, P_{2^k}(-), P_{2^{k+1}}(-), \ldots$$

of the nonalgebraic matrix Riccati equation as an initial-value problem—by doubling the time interval between successive solutions. The doubling speedup is achieved by successive squaring of the equivalent state transition matrix for the time-invariant Hamiltonian matrix

$$\Psi = \begin{bmatrix} \left(\Phi + Q\Phi^{-\mathrm{T}} HR^{-1}H^{\mathrm{T}}\right) & Q\Phi^{-\mathrm{T}} \\ \Phi^{-\mathrm{T}}R^{-1} & \Phi^{-\mathrm{T}} \end{bmatrix}. \tag{7.147}$$

The pth squaring of Ψ will then yield Ψ^{2^p} and the solution

$$P_{2^p}(-) = A_{2^p} B_{2^p}^{-1} \tag{7.148}$$

for

$$\begin{bmatrix} A_{2^p} \\ B_{2^p} \end{bmatrix} = \Psi^{2^p} \begin{bmatrix} A_0 \\ B_0 \end{bmatrix}. \tag{7.149}$$

Davison–Maki–Friedlander–Kailath Squaring Algorithm. Note that if one expresses Ψ^{2^N} in symbolic form as

$$\Psi^{2^N} = \begin{bmatrix} \mathcal{A}_N^{\mathrm{T}} + \mathcal{C}_N \mathcal{A}_N^{-1} \mathcal{B}_N & \mathcal{C}_N \mathcal{A}_N^{-1} \\ \mathcal{A}_N^{-1} \mathcal{B}_N & \mathcal{A}_N^{-1} \end{bmatrix}, \tag{7.150}$$

then its square can be put in the form

$$\Psi^{2^{N+1}} = \begin{bmatrix} \mathcal{A}_N^{\mathrm{T}} + \mathcal{C}_N\mathcal{A}_N^{-1}\mathcal{B}_N & \mathcal{C}_N\mathcal{A}_N^{-1} \\ \mathcal{A}_N^{-1}\mathcal{B}_N & \mathcal{A}_N^{-1} \end{bmatrix}^2 \tag{7.151}$$

$$= \begin{bmatrix} \mathcal{A}_{N+1}^{\mathrm{T}} + \mathcal{C}_{N+1}\mathcal{A}_{N+1}^{-1}\mathcal{B}_{N+1} & \mathcal{C}_{N+1}\mathcal{A}_{N+1}^{-1} \\ \mathcal{A}_{N+1}^{-1}\mathcal{B}_{N+1} & \mathcal{A}_{N+1}^{-1} \end{bmatrix}, \tag{7.152}$$

$$\mathcal{A}_{N+1} = \mathcal{A}_N(I + \mathcal{B}_N\mathcal{C}_N)^{-1}\mathcal{A}_N, \tag{7.153}$$

$$\mathcal{B}_{N+1} = \mathcal{B}_N + \mathcal{A}_N(I + \mathcal{B}_N\mathcal{C}_N)^{-1}\mathcal{B}_N\mathcal{A}_N^{\mathrm{T}}, \tag{7.154}$$

$$\mathcal{C}_{N+1} = \mathcal{C}_N + \mathcal{A}_N^{\mathrm{T}}\mathcal{C}_N(I + \mathcal{B}_N\mathcal{C}_N)^{-1}\mathcal{A}_N. \tag{7.155}$$

The last three equations define an algorithm for squaring Ψ^{2^N}, starting with the values of $\mathcal{A}_N$, $\mathcal{B}_N$, and $\mathcal{C}_N$ for $N = 0$, given by Equation 7.147:

$$\mathcal{A}_0 = \Phi^{\mathrm{T}}, \tag{7.156}$$

$$\mathcal{B}_0 = H^{\mathrm{T}}R^{-1}H, \tag{7.157}$$

$$\mathcal{C}_0 = Q. \tag{7.158}$$

Initial Conditions. If the initial value of the Riccati equation is with $P_0 = 0$, the zero matrix, it can be represented by $P_0 = A_0B_0^{-1}$ for $A_0 = 0$ and any nonsingular B_0. Then the Nth iterate of the doubling algorithm will yield

$$\begin{bmatrix} A_{2^N} \\ B_{2^N} \end{bmatrix} = \Psi^{2^N}\begin{bmatrix} A_0 \\ B_0 \end{bmatrix} \tag{7.159}$$

$$= \begin{bmatrix} \mathcal{A}_N^{\mathrm{T}} + \mathcal{C}_N\mathcal{A}_N^{-1}\mathcal{B}_N & \mathcal{C}_N\mathcal{A}_N^{-1} \\ \mathcal{A}_N^{-1}\mathcal{B}_N & \mathcal{A}_N^{-1} \end{bmatrix}\begin{bmatrix} 0 \\ B_1 \end{bmatrix} \tag{7.160}$$

$$= \begin{bmatrix} \mathcal{C}_N\mathcal{A}_N^{-1}B_1 \\ \mathcal{A}_N^{-1}B_1 \end{bmatrix}, \tag{7.161}$$

$$P_{2^N} = A_{2^N}B_{2^N}^{-1} \tag{7.162}$$

$$= \mathcal{C}_N\mathcal{A}_N^{-1}B_1(\mathcal{A}_N^{-1}B_1)^{-1} \tag{7.163}$$

$$= \mathcal{C}_N. \tag{7.164}$$

That is, after the Nth squaring step, the submatrix

$$\mathcal{C}_N = P_{2^N}. \tag{7.165}$$

TABLE 7.7 Davison–Maki–Friedlander–Kailath Squaring Algorithm

Initialization	Iteration (N times)
$\mathcal{A} = \Phi^T$	$\mathcal{A} \leftarrow \mathcal{A}(I + \mathcal{BC})^{-1}\mathcal{A}^T$
$\mathcal{B} = H^T R^{-1} H$	$\mathcal{B} \leftarrow \mathcal{B} + \mathcal{A}(I + \mathcal{BC})^{-1}\mathcal{BA}^T$
$\mathcal{C} = Q\ (= P_1)$	$\mathcal{C} \leftarrow \mathcal{C} + \mathcal{A}^T\mathcal{C}(I + \mathcal{BC})^{-1}\mathcal{A}$
Termination	
$P_{2^N} = \mathcal{C}$	

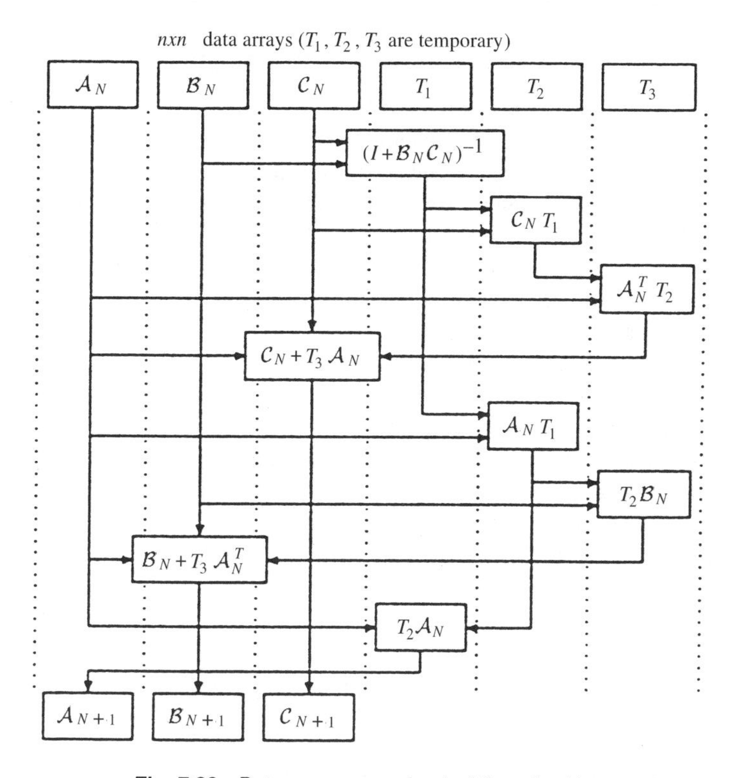

Fig. 7.28 Data array usage for doubling algorithm.

The resulting algorithm is summarized in Table 7.7. It has complexity $\mathcal{O}(n^3 \log k)$ flops for computing P_k, requiring one $n \times n$ matrix inverse and 8 $n \times n$ matrix products per iteration[8]. An array allocation scheme for performing the squaring algorithm using only 6 $n \times n$ arrays is shown in Figure 7.28.

[8]The matrices $\mathcal{B}_N$ and $\mathcal{C}_N$ and the matrix products $(I + \mathcal{B}_N\mathcal{C}_N)^{-1}B_N$ and $\mathcal{C}_N(I + \mathcal{B}_N\mathcal{C}_N)^{-1}$ are symmetric. That fact can be exploited to eliminate $2n^2(n-1)$ flops per iteration. With these savings, this algorithm requires slightly fewer flops per iteration than the straightforward squaring method. It requires about one-fourth less memory than straightforward squaring, also.

Numerical Convergence Problems. Convergence can be stalled by precision limitations before it is complete. The problem is that the matrix $\mathcal{A}$ is effectively squared on each iteration and appears quadratically in the update equations for $\mathcal{B}$ and $\mathcal{C}$. Consequently, if $\|\mathcal{A}_N\| \ll 1$, then the computed values of $\mathcal{B}_N$ and $\mathcal{C}_N$ may become stalled numerically as $\|\mathcal{A}_N\| \to 0$ exponentially. The value of $\mathcal{A}_N$ can be monitored to test for this stall condition. Even in those stall situations, the doubling algorithm is still an efficient method for getting an approximate nonnegative definite solution.

7.9 ERROR BUDGETS AND SENSITIVITY ANALYSIS

7.9.1 Design Problem for Statistical Performance Requirements

This is the problem of estimating the statistical performance of a sensor system that will make measurements of some dynamic and stochastic process and estimate its state. Statistical performance is defined by mean-squared errors at the "system level"; these depend on mean-squared errors at the subsystem level; and so on down to the level of individual sensors and components. The objective of this activity is to be able to justify the apportionment of these lower level performance requirements.

This type of performance analysis is typically performed during the preliminary design of estimation systems. The objective of the analysis is to evaluate the feasibility of an estimation system design for meeting some prespecified acceptable level of uncertainty in the estimates that will be obtained.

The Kalman filter does not design sensor systems, but it provides the tool for doing it defensibly. That tool is the model for estimation uncertainty. The covariance propagation equations derived from the model can be used in characterizing estimation uncertainty as a function of the "parameters" of the design. Some of these parameters are statistical, such as the noise models of the sensors under consideration. Others are deterministic. The deterministic parameters may also be discrete valued—such as the sensor type—or continuous—such as the sensor location.

One of the major uses of Kalman filtering theory is in the design of sensor systems:

1. Vehicle navigation systems containing some combination of sensors, such as:
   - (a) Attitude and attitude rate sensors
     - i. Magnetic compass (field sensor)
     - ii. Displacement gyroscopes
     - iii. Star trackers or sextants
     - iv. Rate gyroscopes
     - v. Electric field sensors (for earth potential field)
   - (b) Acceleration sensors (accelerometers)
   - (c) Velocity sensors (e.g., onboard Doppler radar)

   (d) Position sensors
      i. Global Positioning System (GPS)
      ii. Terrain-mapping radar
      iii. Long-range navigation (LORAN)
      iv. Instrument Landing System (ILS)
2. Surface-based, airborne, or spaceborne tracking systems
   (a) Range and Doppler radar
   (b) Imaging sensors (e.g., visible or infrared cameras)

In the design of these systems, it is assumed that a Kalman filter will be used in estimating the dynamic state (position and velocity) of the vehicle. Therefore, the associated covariance equations can be used to estimate the performance in terms of the covariance of estimation uncertainty.

7.9.2 Error Budgeting

Large systems such as spacecraft and aircraft contain many sensors of many types, and the Kalman filter provides a methodology for the integrated design of such systems. Error budgeting is a specialized form of sensitivity analysis. It uses the error covariance equations of the Kalman filter to formalize the dependence of system accuracy on the component accuracies of its individual sensors. This form of covariance analysis is significantly more efficient than Monte Carlo analysis for this purpose, although it does depend upon linearity of the underlying dynamic processes.

Error budgeting is a process for trading off performance requirements among sensors and subsystems of a larger system for meeting a diverse set of overall performance constraints imposed at the "system level." The discussion here is limited to the system-level requirements related to *accuracy*, although most system requirements include other factors related to cost, weight, size, and power.

The error budget is an allocation of accuracy requirements down through the hierarchy of subsystems to individual sensors, and even to their component parts. It is used for a number of purposes, such as:

1. Assessing theoretical or technological performance limits by determining whether the performance requirements for a given application of a given system are *achievable* within the performance capabilities of available, planned, or theoretically attainable sensor subsystems.
2. Determining the extent of *feasible design space*, which is the range of possible sensor types and their design parameters (e.g., placement, orientation, sensitivity, and accuracy) for meeting the system performance requirements.

3. Finding a *feasible apportionment* of individual subsystem or sensor accuracies for meeting overall system accuracy requirements.
4. Identifying the *critical* subsystems, that is, those for which slight degradation or derating of performance would most severely affect system performance. These are sometimes called "the long poles in the tent," because they tend to "stick out" in this type of assessment.
5. Finding feasible *upgrades and redesigns* of existing systems for meeting new performance requirements. This may include relaxation of some requirements and tightening of others.
6. *Trading off* requirements among subsystems. This is done for a number of reasons:
   (a) Reapportionment of error budgets to meet a new set of requirements (item 5, above).
   (b) Relaxing accuracy requirements where they are difficult (or expensive) to attain and compensating by tightening requirements where they are easier to attain. This approach can sometimes be used to overcome sensor problems uncovered in concurrent development and testing.
   (c) Reducing other system-level performance attributes, such as cost, size, weight, and power. This also includes such practices as suboptimal filtering methods to reduce computational requirements.

7.9.3 Error Budget

Multiple Performance Requirements. System-level performance requirements can include constraints on the mean-squared values of several error types at several different times. For example, the navigational errors of a space-based imaging system may be constrained at several points corresponding to photographic missions or planetary encounters. These constraints may include errors in pointing (attitude), position, and velocity. The error budget must then consider how each component, component group, or subsystem contributes to each of these performance requirements. The budget will then have a two-dimensional breakout—like a spreadsheet—as shown in Figure 7.29. The rows represent the contributions of major sensor subsystems, and the columns represent their contributions to each of the multiple system-level error constraints. The formulas determining how each error source contributes to each of the system-level error categories are more complex than those of the usual spreadsheet, however.

7.9.4 Error Sensitivity Analysis and Budgeting

Nonlinear Programming Problem. The dependence of mean-squared system-level errors on mean-squared subsystem-level errors is nonlinear, and the budgeting process seeks a satisfactory apportionment or the subsystem-level error covariances by a gradientlike method. This includes sensitivity analysis to determine the

ERROR BUDGET					
ERROR SOURCE GROUP	SYSTEM ERRORS				
	E_1	E_2	E_3	...	E_n
G_1				...	
G_2				...	
G_3				...	
⋮	⋮	⋮	⋮	...	⋮
G_m				...	
TOTAL				...	

Fig. 7.29 *Error budget breakdown.*

gradients of the various mean-squared system-level errors with respect to the mean-squared subsystem-level errors.

Dual-State System Model. Errors considered in the error budgeting process may include known "modeling errors" due to simplifying assumptions or other measures to reduce the computational burden of the filter. For determining the effects that errors of this type will have on system performance, it is necessary to carry both models in the analysis: the "truth model" and the "filter model." The budgeting model used in this analysis is diagrammed in Figure 7.30. In sensitivity analysis, equivalent variations of some parameters must be made in both models. The resulting variations in the projected performance characteristics of the system are then used to establish the sensitivities to the corresponding variations in the subsystems. These sensitivities are then used to plan how one can modify the

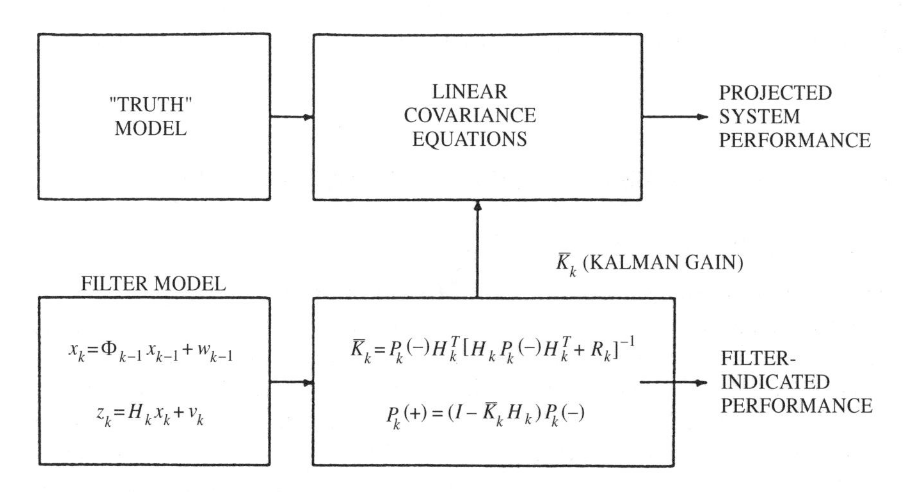

Fig. 7.30 *Error budgeting model.*

current “protobudget” to arrive at an error budget allocation that will meet all performance requirements. Often, this operation must be repeated many times, because the sensitivities estimated from variations are only accurate for small changes in the budget entries.

There Are Two Stages of the Budgeting Process. The first stage results in a “sufficing” error budget. It should meet system-level performance requirements and be reasonably close to attainable subsystem-level performance capabilities. The second stage includes “finessing” these subsystem-level error allocations to arrive at a more reasonable distribution.

7.9.5 Budget Validation by Monte Carlo Analysis

It is possible to validate some of the assumptions used in the error budgeting process by analytical and empirical methods. Although covariance analysis is more efficient for developing the error budget, Monte Carlo analysis is useful for assessing the effects of nonlinearities that have been approximated by variational models. This is typically done after the error budget is deemed satisfactory by linear methods. Monte Carlo analysis can then be performed on a dispersion of actual trajectories about some nominal trajectory to test the validity of the results estimated from the nominal trajectory. This is the only way to test the influence of nonlinearities, but it can be computationally expensive. Typically, very many Monte Carlo runs must be made to obtain reasonable confidence in the results.

Monte Carlo analysis has certain advantages over covariance analysis, however. The Monte Carlo simulations can be integrated with actual hardware, for example, to test the system performance in various stages of development. This is especially useful for testing filter performance in onboard computer implementations using actual system hardware as it becomes available. Sign errors in the filter algorithms that may be unimportant in covariance analysis will tend to show up under these test conditions.

7.10 OPTIMIZING MEASUREMENT SELECTION POLICIES

7.10.1 Measurement Selection Problem

Relation to Kalman Filtering and Error Budgeting. You have seen how Kalman filtering solves the optimization problem related to the *use* of data obtained from a measurement and how error budgeting is used to quantify the relative merits of alternative sensor designs. However, there is an even more fundamental optimization problem related to the *selection* of those measurements. This is not an estimation problem, strictly speaking, but a *decision problem*. It is usually considered to be a problem in the general theory of optimal control, because the decision to make a measurement is considered to be a generalized control action. The problem can also be ill-posed, in the sense that there may be no unique optimal solution [131].

Optimization with Respect to a Quadratic Loss Function. The Kalman filter is optimal with respect to all quadratic loss functions defining performance as a function of estimation error, but the measurement selection problem does not have that property. It depends very much on the particular loss function defining performance.

We present here a solution method based on what is called "maximum marginal benefit." It is computationally efficient but suboptimal with respect to a given quadratic loss function of the resulting estimation errors $\hat{x}-x$:

$$\mathcal{L} = \sum_{\ell=1}^{N} \left\| A_\ell \left(\hat{x}_\ell(+) - x_\ell \right) \right\|^2, \tag{7.166}$$

where the given matrices A_ℓ transform the estimation errors to other "variables of interest," as illustrated by the following examples:

1. If only the *final values* of the estimation errors are of interest, then $A_N = I$ (the identity matrix) and $A_\ell = 0$ (a matrix of zeros) for $\ell < N$.
2. If only a *subset of the state vector components* are of interest, then the A_ℓ will all equal the projection onto those components that are of interest.
3. If *any linear transformation* of the estimation errors is of interest, then the A_ℓ will be defined by that transformation.
4. If any *temporally weighted combination* of linear transformations of the estimation errors is of interest, then the corresponding A_ℓ will be the weighted matrices of those linear transformations. That is, $A_\ell = f_\ell B_\ell$, where $0 \leq f_\ell$ is the temporal weighting and the B_ℓ are the matrices of the linear transformations.

7.10.2 Marginal Optimization

The loss function is defined above as a function of the a posteriori estimation errors following measurements. The next problem will be to represent the dependence of the associated risk[9] function on the selection of measurements.

Parameterizing the Possible Measurements. As far as the Kalman filter is concerned, a measurement is characterized by H (its measurement sensitivity matrix)

[9]The term "risk" is here used to mean the expected loss.

and R (its covariance matrix of measurement uncertainty). A sequence of measurements is then characterized by the sequence

$$\{\{H_1, R_1\}, \{H_2, R_2\}, \{H_3, R_3\}, \ldots, \{H_N, R_N\}\}$$

of pairs of these parameters. This sequence will be called *marginally optimal* with respect to the above risk function if, for each k, the kth measurement is chosen to minimize the risk of the subsequence

$$\{\{H_1, R_1\}, \{H_2, R_2\}, \{H_3, R_3\}, \ldots, \{H_k, R_k\}\}.$$

That is, marginal optimization assumes that:

1. The previous selections of measurements have already been decided.
2. No further measurements will be made after the current one is selected.

Admittedly, a marginally optimal solution is not necessarily a globally optimal solution. However, it does yield an efficient suboptimal solution method.

Marginal Risk. Risk is the expected value of loss. The *marginal risk function* represents the functional dependence of risk on the selection of the kth measurement, *assuming that it is the last*. Marginal risk will depend only on the a posteriori estimation errors after the decision has been made. It can be expressed as an implicit function of the decision in the form

$$\mathcal{R}_k(P_k(+)) = \mathrm{E}\left\{\sum_{\ell=k}^{N} \|A_\ell(\hat{x}_\ell(+) - x_\ell)\|^2\right\}, \tag{7.167}$$

where $P_k(+)$ will depend on the choice for the kth measurement and, for $k < \ell \le N$,

$$\hat{x}_{\ell+1}(+) = \hat{x}_{\ell+1}(-), \tag{7.168}$$

$$\hat{x}_{\ell+1}(+) - x_{\ell+1} = \Phi_\ell(\hat{x}_\ell(+) - x_\ell) - w_\ell, \tag{7.169}$$

so long as no additional measurements are used.

Marginal Risk Function. Before proceeding further with the development of a solution method, it will be necessary to derive an explicit representation of the marginal risk as a function of the measurement used. For that purpose, one can use a *trace formulation* of the risk function, as presented in the following lemma.

LEMMA 4 For $0 \le k \le N$, the risk function defined by Equation 7.167 can be represented in the form

$$\mathcal{R}_k(P_k) = \text{trace}\,\{P_k W_k + V_k\}, \tag{7.170}$$

where

$$W_N = A_N^{\mathrm{T}} A_N, \tag{7.171}$$

$$V_N = 0, \tag{7.172}$$

and, for $\ell < N$,

$$W_\ell = \Phi_\ell^{\mathrm{T}} W_{\ell+1} \Phi_\ell + A_\ell^{\mathrm{T}} A_\ell, \tag{7.173}$$

$$V_\ell = Q_\ell W_{\ell+1} + V_{\ell+1}. \tag{7.174}$$

Proof: A formal proof of the equivalence of the two equations requires that each be entailed by (derivable from) the other. We give a proof here as a reversible chain of equalities, starting with one form and ending with the other form. This proof is by backward induction, starting with $k = N$ and proceeding by induction back to any $k \leq N$. The property that the trace of a matrix product is invariant under cyclical permutations of the order of multiplication is used extensively.

Initial step: The initial step of a proof by induction requires that the statement of the lemma hold for $k = N$. By substituting from Equations 7.171 and 7.172 into Equation 7.170, and substituting N for k, one can obtain the following sequence of equalities:

$$\begin{aligned}
\mathcal{R}_N(P_N) &= \text{trace}\{P_N W_N + V_N\} \\
&= \text{trace}\{P_N A_N^{\mathrm{T}} A_N + 0_{n\times n}\} \\
&= \text{trace}\{A_N P_N A_N^{\mathrm{T}}\} \\
&= \text{trace}\{A_N \mathrm{E}\langle(\hat{x}_N - x_N)(\hat{x}_N - x_N)^{\mathrm{T}}\rangle A_N^{\mathrm{T}}\} \\
&= \text{trace}\{\mathrm{E}\langle A_N(\hat{x}_N - x_N)(\hat{x}_N - x_N)^{\mathrm{T}} A_N^{\mathrm{T}}\rangle\} \\
&= \text{trace}\{\mathrm{E}\langle[A_N(\hat{x}_N - x_N)][A_N(\hat{x}_N - x_N)]^{\mathrm{T}}\rangle\} \\
&= \text{trace}\{\mathrm{E}\langle[A_N(\hat{x}_N - x_N)]^{\mathrm{T}}[A_N(\hat{x}_N - x_N)]\rangle\} \\
&= \mathrm{E}\langle\|A_N(\hat{x}_N - x_N)\|^2\rangle.
\end{aligned}$$

The first of these is Equation 7.170 for $k = N$, and the last is Equation 7.167 for $k = N$. That is, the statement of the lemma is true for $k = N$. This completes the initial step of the induction proof.

Induction step: One can suppose that Equation 7.170 is equivalent to Equation 7.167 for $k = \ell + 1$ and seek to prove from that it must also be the case for $k = \ell$. Then start with Equation 7.167, noting that it can be written in the form

$$\begin{aligned}
\mathcal{R}_\ell(P_\ell) &= \mathcal{R}_{\ell+1}(P_{\ell+1}) + \mathrm{E}\langle \|A_\ell(\hat{x}_\ell - x_\ell)\|^2\rangle \\
&= \mathcal{R}_{\ell+1}(P_{\ell+1}) + \mathrm{trace}\{\mathrm{E}\langle \|A_\ell(\hat{x}_\ell - x_\ell)\|^2\rangle\} \\
&= \mathcal{R}_{\ell+1}(P_{\ell+1}) + \mathrm{trace}\{\mathrm{E}\langle [A_\ell(\hat{x}_\ell - x_\ell)]^{\mathrm{T}}[A_\ell(\hat{x}_\ell - x_\ell)]\rangle\} \\
&= \mathcal{R}_{\ell+1}(P_{\ell+1}) + \mathrm{trace}\{\mathrm{E}\langle [A_\ell(\hat{x}_\ell - x_\ell)][A_\ell(\hat{x}_\ell - x_\ell)]^{\mathrm{T}}\rangle\} \\
&= \mathcal{R}_{\ell+1}(P_{\ell+1}) + \mathrm{trace}\{A_\ell \mathrm{E}\langle (\hat{x}_\ell - x_\ell)(\hat{x}_\ell - x_\ell)^{\mathrm{T}}\rangle A_\ell^{\mathrm{T}}\} \\
&= \mathcal{R}_{\ell+1}(P_{\ell+1}) + \mathrm{trace}\{A_\ell P_\ell A_\ell^{\mathrm{T}}\} \\
&= \mathcal{R}_{\ell+1}(P_{\ell+1}) + \mathrm{trace}\{P_\ell A_\ell^{\mathrm{T}} A_\ell\}.
\end{aligned}$$

Now one can use the assumption that Equation 7.170 is true for $k = \ell + 1$ and substitute the resulting value for $\mathcal{R}_{\ell+1}$ into the last equation above. The result will be the following chain of equalities:

$$\begin{aligned}
\mathcal{R}_\ell(P_\ell) &= \mathrm{trace}\{P_{\ell+1} W_{\ell+1} + V_{\ell+1}\} + \mathrm{trace}\{P_\ell A_\ell^{\mathrm{T}} A_\ell\} \\
&= \mathrm{trace}\{P_{\ell+1} W_{\ell+1} + V_{\ell+1} + P_\ell A_\ell^{\mathrm{T}} A_\ell\} \\
&= \mathrm{trace}\{[\Phi_\ell P_\ell \Phi_\ell^{\mathrm{T}} + Q_\ell] W_{\ell+1} + V_{\ell+1} + P_\ell A_\ell^{\mathrm{T}} A_\ell\} \\
&= \mathrm{trace}\{\Phi_\ell P_\ell \Phi_\ell^{\mathrm{T}} W_{\ell+1} + Q_\ell W_{\ell+1} + V_{\ell+1} + P_\ell A_\ell^{\mathrm{T}} A_\ell\} \\
&= \mathrm{trace}\{P_\ell \Phi_\ell^{\mathrm{T}} W_{\ell+1} \Phi_\ell + Q_\ell W_{\ell+1} + V_{\ell+1} + P_\ell A_\ell^{\mathrm{T}} A_\ell\} \\
&= \mathrm{trace}\{P_\ell [\Phi_\ell^{\mathrm{T}} W_{\ell+1} \Phi_\ell + A_\ell^{\mathrm{T}} A_\ell] + [Q_\ell W_{\ell+1} + V_{\ell+1}]\} \\
&= \mathrm{trace}\{P_\ell [W_\ell] + [V_\ell]\},
\end{aligned}$$

where the Equations 7.173 and 7.174 were used in the last substitution. The last equation is Equation 7.170 with $k = \ell$, which was to be proved for the induction step. Therefore, by induction, the equations defining the marginal risk function are equivalent for $k \le N$, which was to be proved.

Implementation note: The last formula separates the marginal risk as the sum of two parts. The first part depends only upon the choice of the measurement and the deterministic state dynamics. The second part depends only upon the stochastic state dynamics and is unaffected by the choice of measurements. As a consequence of this separation, the decision process will use only the first part. However, an assessment of the marginal risk performance of the decision process itself would require the evaluation of the complete marginal risk function.

Marginal Benefit from Using a Measurement. The *marginal benefit* resulting from the use of a measurement will be defined as the associated *decrease* in the marginal risk. By this definition, the marginal benefit resulting from using a measurement with sensitivity matrix H and measurement uncertainty covariance R at time t_k will be the difference between the a priori and a posteriori marginal risks:

$$\mathcal{B}(H,R) = \mathcal{R}_k(P_k(-)) - \mathcal{R}_k(P_k(+)) \tag{7.175}$$

$$= \text{trace}\{[P_k(-) - P_k(+)]W_k\} \tag{7.176}$$

$$= \text{trace}\{[P_k(-)H^{\text{T}}(HP_k(-)H^{\text{T}} + R)^{-1}HP_k(-)]W_k\} \tag{7.177}$$

$$= \text{trace}\{(HP_k(-)H^{\text{T}} + R)^{-1}HP_k(-)W_kP_k(-)H^{\text{T}}\} \tag{7.178}$$

This last formula is in a form useful for implementation.

7.10.3 Solution Algorithm for Maximum Marginal Benefit

1. Compute the matrices W_ℓ using the formulas given by Equations 7.171 and 7.173.
2. Select the measurements in temporal order: for $k = 0, 1, 2, 3, \ldots, N$:
   (a) For each possible measurement, using Equation 7.178, evaluate the marginal benefit that would result from the use of that measurement.
   (b) Select the measurement that yields the *maximum* marginal benefit.

Again, note that this algorithm does *not* use the matrices V_ℓ in the "trace formulation" of the risk function. It is necessary to compute the V_ℓ only if the specific value of the associated risk is of sufficient interest to warrant the added computational expense.

7.10.3.1 Computational Complexity

Complexity of Computing the W_ℓ. Complexity will depend upon the dimensions of the matrices A_ℓ. If each matrix A_ℓ is $p \times n$, then the products $A_\ell^{\text{T}} A_\ell$ require $\mathcal{O}(pn^2)$ operations. The complexity of computing $\mathcal{O}(N)$ of the W_ℓ will then be $\mathcal{O}(Nn^2(p+n))$.

Complexity of Measurement Selection. The computational complexity of making a single determination of the marginal benefit of a measurement of dimension m is summarized in Table 7.8. On each line, the complexity figure is based on reuse of partial results from computations listed on lines above. If all possible measurements have the same dimension ℓ and the number of such measurements to be evaluated is μ, then the complexity of evaluating all of them[10] will be $\mathcal{O}(\mu\ell(\ell^2 + n^2))$. If this is repeated for each of $\mathcal{O}(N)$ measurement selections, then the total complexity will be $\mathcal{O}(N\mu\ell(\ell^2 + n^2))$.

[10]Although the intermediate product $P_k(-)W_kP_k(-)$ [of complexity $\mathcal{O}(n^3)$] does not depend on the choice of the measurement, no reduction in complexity would be realized even if it were computed only once and reused for all measurements.

TABLE 7.8 Complexity of Determining the Marginal Benefit of a Measurement

Operation	Complexity
$HP_k(-)$	$\mathcal{O}(\ell n^2)$
$HP_k(-)H^{\mathrm{T}} + R$	$\mathcal{O}(\ell^2 n)$
$[HP_k(-)H^{\mathrm{T}} + R]^{-1}$	$\mathcal{O}(\ell^3)$
$HP_k(-)W_k$	$\mathcal{O}(\ell n^2)$
$HP_k(-)W_k P_k(-)H^{\mathrm{T}}$	$\mathcal{O}(\ell^2 n)$
$\mathrm{trace}\{(HP_k(-)H^{\mathrm{T}} + R)^{-1} HP_k(-)W_k P_k(-)H^{\mathrm{T}}\}$	$\mathcal{O}(\ell^2)$
Total	$\mathcal{O}(\ell(\ell^2 + n^2))$

Note: ℓ is the dimension of the measurement vector; n is the dimension of the state vector.

7.11 APPLICATION TO AIDED INERTIAL NAVIGATION

This section will demonstrate the use of the *UD*-formulated extended Kalman filter for a full-scale example of aiding an inertial system with data provided by the Global Positioning System (GPS) of navigation satellites. For more examples and discussion, see reference [22]. There are two general approaches to this application:

1. INS-aided GPS and
2. GPS-aided INS (Inertial Navigation System).

In the first approach, the INS is being aided by GPS. That is, additional data to aid the INS implementation will be provided by GPS. These independent data may be used to correct inertial sensor scale factor and/or bias errors, for example. It may be robust against loss of GPS data, however. The aided system may even lose the GPS data for some periods of time, but the INS will continue to provide the position and velocity information.

The second approach provides a more conservative and robust design from the standpoint of dependence on inertial sensor performance. It essentially uses the inertial system to estimate otherwise undetectable perturbations in the propagation delays of GPS signals or to smooth over short-term zero-mean perturbations. It may also use an INS model with a minimum of Kalman filter states and use an inertial system of lowest allowable quality (which may not always be available). In this case, the GPS continues to provide the position and velocity information.

We will discuss the first in detail with models (process and measurement). If the user has an INS, its position indication and the satellite ephemeris data can be used to compute an INS–indicated range to the satellite. The difference of these two range indicators, called the *pseudorange*, serves as an input to a Kalman filter to yield an integrated GPS-aided INS. Another measurement is called *delta pseudorange* measurement and is in error by an amount proportional to the relative frequency error between the transmitter and receiver clocks.

7.11.1 Dynamic Process Model

The basic nine-state error model has three position errors, three velocity errors and three platform tilt errors—all specified by a 9×9 dynamic coefficient matrix, shown below (for values, see Table 7.9 later):

$$F(t) = \begin{bmatrix} -k_1 ee^{\mathrm{T}} & I_{3\times3} & 0_{3\times3} \\ \mathcal{A} & 2\mathcal{B} & \mathcal{C} \\ 0_{3\times3} & 0_{3\times3} & \mathcal{B} \end{bmatrix}, \tag{7.179}$$

$$\mathcal{A} = \left(3\omega_s^2 - k_2\right)ee^{\mathrm{T}} - \omega_s^2 I_{3\times3}, \tag{7.180}$$

$$\mathcal{B} = \begin{bmatrix} 0 & \Omega & 0 \\ -\Omega & 0 & 0 \\ 0 & 0 & 0 \end{bmatrix}, \tag{7.181}$$

$$\mathcal{C} = \begin{bmatrix} 0 & -f_3 & f_2 \\ f_3 & 0 & -f_1 \\ -f_2 & f_1 & 0 \end{bmatrix}, \tag{7.182}$$

where

e = unit vector in vertical direction
f = specific force vector
ω_s = Schuler frequency
Ω = earth spin rate
k_1 = vertical-channel position loop gain
k_2 = vertical-channel velocity loop gain

7.11.2 Measurement Model

As given in reference [184], the GPS pseudorange (PR) from a satellite is defined as

$$\mathrm{PR} = [(X_S - X_R)^2 + (Y_S - Y_R)^2 + (Z_S - Z_R)^2]^{1/2} + bc, \tag{7.183}$$

where

$$
\begin{aligned}
(X_S, Y_S, Z_S) &= \text{satellite position coordinates at the time of transmission} \\
(X_R, Y_R, Z_R) &= \text{receiver position coordinates at the time of reception} \\
b &= \text{receiver clock bias error} \\
c &= \text{carrier speed (speed of light)}
\end{aligned}
$$

The linearized observation equation implemented in the extended Kalman filter is

$$\delta PR = H_{\text{PR}} X + V_{\text{PR}}, \tag{7.184}$$

where X is the state vector with its states (three position errors, three velocity errors, and three platform tilt errors); V_{PR} is the additive measurement noise; and H_{PR}, the pseudorange observation matrix, is obtained by linearizing the pseudorange equation with respect to the filter states (Jacobian matrix):

$$
\begin{aligned}
H_{\text{PR}} &= \left.\frac{\partial \text{PR}}{\partial X}\right|_{X=\hat{X}} \\
&= [-U_x, \ -U_y, \ -U_z, \ 0, \ 0, \ 0, \ 0, \ 0, \ 0],
\end{aligned} \tag{7.185}
$$

where (U_x, U_y, U_z) is the user-to-satellite line-of-sight unit vector.

The GPS delta pseudorange is defined as the difference between two pseudoranges separated in time,

$$\text{DR} = \text{PR}(t_2) - \text{PR}(t_1), \quad t_2 > t_1. \tag{7.186}$$

Since the delta pseudorange represents the Doppler integrated over a finite time interval, any point within the integration interval can be chosen as the reference time at which the measurement is valid. In practice, either the beginning or end of the interval is selected as the reference time.

If the interval stop time is chosen as the reference, the linearized measurement model can be written as

$$\delta \text{DR} = H_{\text{DR}} X + V_{\text{DR}}, \tag{7.187}$$

where the measurement noise V_{DR}, not only accounts for the very small additive tracking error in the highly accurate carrier loop but also includes the integrated dynamics effects representing unmodeled jerk and higher order terms over the integration interval and

$$H_{\text{DR}} = \left.\frac{\partial \text{DR}}{\partial X}\right|_{X=\hat{X}} \tag{7.188}$$

$$\begin{aligned} &= -[\Delta U_x,\ \Delta U_y,\ \Delta U_z,\ \ \Delta t U_{x1},\ \Delta t U_{y1},\ \Delta t U_{z1}, \\ &\tfrac{1}{2}\Delta t^2(f_2 U_{z1} - f_3 U_{y1}),\ \tfrac{1}{2}\Delta t^2(f_3 U_{x1} - f_1 U_{z1}), \\ &\tfrac{1}{2}\Delta t^2(f_1 U_{y1} - f_2 U_{x1})] \end{aligned} \tag{7.189}$$

with

Δt = delta pseudorange integration interval

$U_{x1},\ U_{y1},\ U_{z1}$ = user-to-satellite line-of-sight vector at delta pseudorange start time

$\Delta U_x,\ \Delta U_y,\ \Delta U_z$ = line-of-sight vector change over delta pseudorange integration interval [184]

7.11.3 Kalman Filter State Configuration

Figure 7.31 shows a block diagram representation of an integrated navigation system using inertial and satellite information. The integrated GPS-aided inertial system provides the estimated position and velocity during GPS signal availability and extends the period of acceptable operation subsequent to the loss of GPS signals. As proposed by Bletzacker et al. [142], an important part of the filter design process involves the selection of a state configuration that can satisfy the performance requirements within existing throughput constraints. Other than the basic nine states

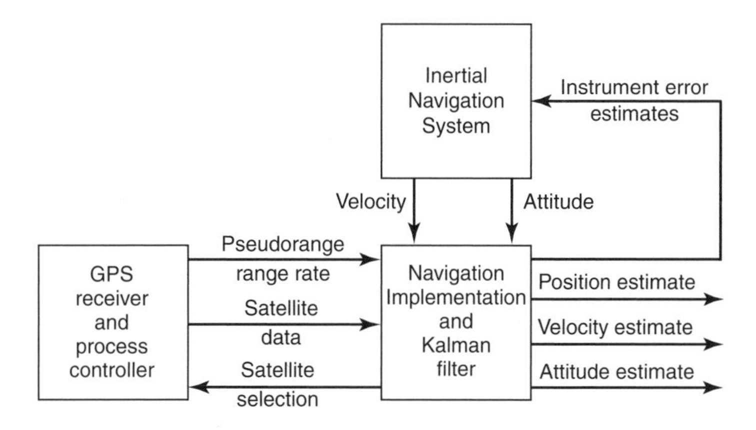

Fig. 7.31 Integrated GPS/INS navigation system.

(position, velocity, and platform error angles) required in any aided mechanization, the remaining states are chosen from the suite of inertial sensor error parameters. The effect that these parameters have on system performance depends critically upon whether the INS is gimbaled or strapdown. For example, errors in gyroscope scale factor, scale factor asymmetry, and nonorthogonality will be more important in a strapdown system, where the gyroscopes are exposed to a much higher level of dynamics, and their effect on system performance is reduced significantly. For this reason, all of the inertial sensor errors considered for inclusion as states, except for gyroscope drift, are related to the accelerometers. The selection process included a trade-off study involving options ranging from 9 (position, velocity, and platform tilts) to 24 (position, velocity, platform tilts, accelerometer bias, gyroscope drifts, accelerometer scale factor and scale factor asymmetry, and accelerometer nonorthogonality) states. The ultimate decision was based upon considerations of throughput, performance requirements, sensor characteristics, and mission applications. The result of this study was the selection of a 15-state Kalman filter with states of position, velocity, platform error angles, gyroscope drift, and accelerometer bias [142].[11]

The INS is a 0.5 nautical mile/hour (CEP rate) system. The INS vertical channel is controlled by an ideal barometric altimeter. The GPS pseudorange and delta pseudorange measurement errors are 0.6 meters and 2.0 centimeters, respectively. The INS vertical channel is controlled by an ideal barometric altimeter, with position, velocity, and acceleration loop gains of 0.03, 0.0003, and 0.000001, respectively [126]. The GPS control and space segments are assumed to have biased type errors of 5 meters. The GPS receiver clock has no G-sensitivity. All lever-arm effects have been omitted. The 18 satellite constellation is assumed to be operational with a GDOP between 3 and 4 continuously available. The flight profile includes a take-off and climb to 7 km with an acceleration (5 m/s/s) to a speed of 300 m/s. The aircraft then flies a race track with 180 km straight legs and 7 m/s/s turns [142]. The GPS is assumed to be available for the first 5000 seconds.

Table 7.9 gives typical error source characteristics for this application. A typical set of results is shown in Figure 7.32. The error growth in the receiver position and velocity estimates is caused by the inertial reference tilt errors while GPS data are lost (after 5000 s). The improved tilt estimation provided by the Kalman filter implementation may provide an order-of-magnitude improvement in the resulting (integrated) navigation solution.

7.12 SUMMARY

This chapter discussed methods for the design and evaluation of estimation systems using Kalman filters. Specific topics addressed include the following:

[11]Other investigators have evaluated filters with 39, 12, and 14 states. Maybeck [31] has mentioned a 96-state error state vector. In this example, we give the results of a 15-state filter.

TABLE 7.9 Inertial Sensor Error Sources (1σ)

Accelerometer	
G-insensitive	
Bias stability	40 μG
Scale factor stability	100 ppm
Scale factor asymmetry	100 ppm
Nonorthogonality	1.1_{10}-6 arc-sec
White noise	5 μG/Hz$^{1/2}$
Correlated noise	4 μG
Correlation time	20 min
G-sensitive	
Nonlinearity	5 μG/G^2
Cross axis coupling	5 μG/G^2
Gyroscope	
G-Insensitive	
Bias stability	0.001 deg/hr
Scale factor stability	100 ppm
Scale factor asymmetry	100 ppm
Nonorthogonality	1.1_{10}-6 arc-sec
White noise	0.002 deg/hr/Hz$^{1/2}$
Correlated noise	0.004 deg/hr
Correlation time	20 min
G-Sensitive	
Mass unbalance	0.008 deg/hr/G
Quadrature	0.008 deg/hr/G
Anisoelastic	0.001 deg/hr/G^2

1. methods for detecting and correcting anomalous behavior of estimators,
2. predicting and detecting the effects of mismodeling and poor unobservability,
3. evaluation of suboptimal filters (using dual-state filters) and sensitivity analysis methods,
4. comparison of memory, throughput, and worldlength requirements for alternative implementation methods,
5. methods for decreasing computational requirements,
6. methods for assessing the influence on estimator performance of sensor location and type and the number of sensors,
7. methods for top-down hierarchical system-level error budgeting, and
8. demonstration of the application of square-root filtering techniques to an INS-aided GPS navigator.

PROBLEMS

7.1 Show that the final value of the risk obtained by the marginal optimization technique of Section 7.10 will equal the initial risk minus the sum of the marginal benefits of the measurements selected.

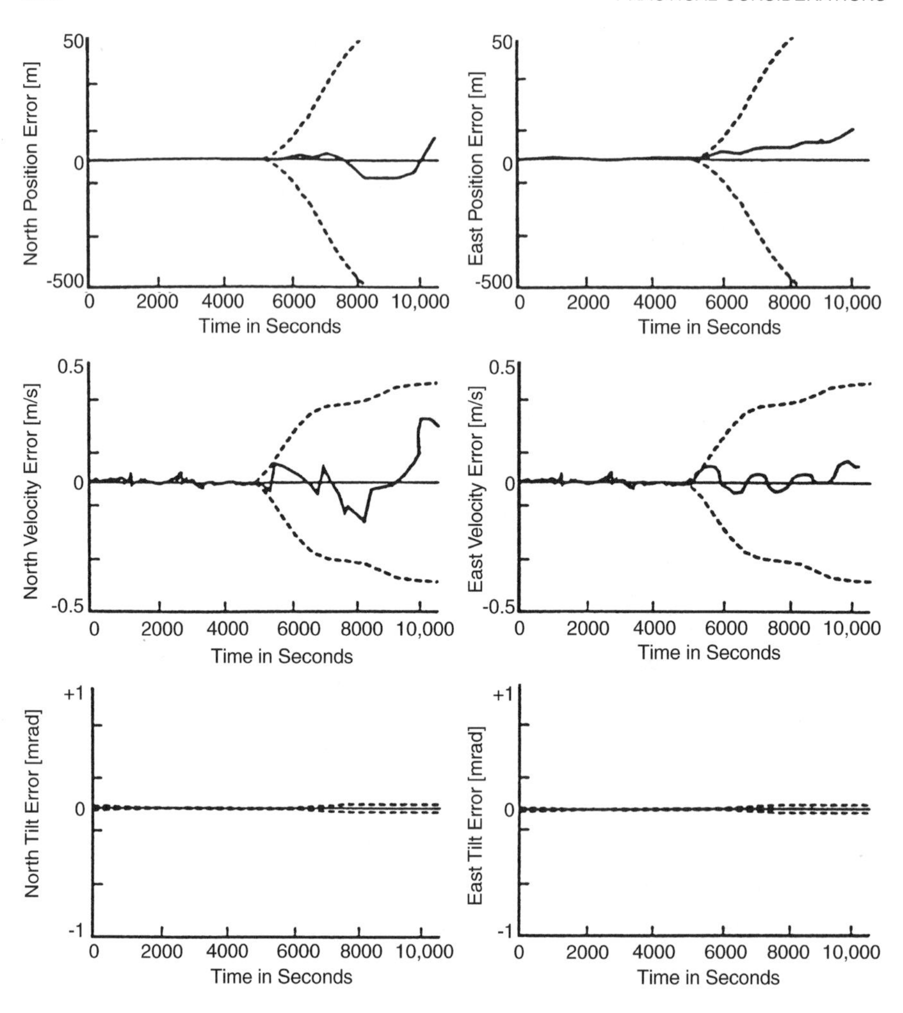

Fig. 7.32 *Integrated GPS/INS simulation results.*

7.2 Develop the equations for the dual-state error propagation by substituting Equations 7.90 and 7.91 into Equations 7.94 and 7.95 using Equation 7.97, explicitly.

7.3 Obtain the dual-state vector equation for the covariances of the system and error, where x_1 is a ramp plus random walk and x_2 is constant:

$$\dot{x}_1^S = x_2^S + w^S, \qquad \dot{x}_2^S = 0, \qquad z_k = x_k^1 + v_k,$$

using as the filter model a random walk

$$\dot{x}^F = w^F, \qquad z_K = x_k^F + v_k.$$

7.4 Derive the results of Example 7.4.

7.5 Prove that cov$[\tilde{x}_k^s]$ depends upon cov(x_k^s).

7.6 Prove the results shown for H_{PR} in Equation 7.185 and for H_{DR} in Equation 7.188.

7.7 Rework Problem 4.6 for the UDU^{T} formulation and compare your results with those of Problem 4.6.

7.8 Rework Problem 4.7 for the UDU^{T} formulation and compare your results with those of Problem 4.7.

7.9 Formulate the GPS plant model with three position errors, three velocity errors, and three acceleration errors and the corresponding measurement model with pseudorange and delta pseudorange as measurements.

7.10 Do Problem 4.6 with the Schmidt–Kalman filter (Section 7.6) and compare the results with Example 4.4.

Appendix A

MATLAB Software

The accompanying diskette contains MATLAB functions and scripts for implementating the Kalman filter and demonstrating its use. The ASCII file README.1ST in the root directory should be read before starting to use any of the software. It describes the current contents and directory structure of the files on the diskette.

A.1 NOTICE

This software is intended for demonstration and instructional purposes only. The authors and publishers make no warrranty of any kind, expressed or implied, that these routines meet any standards of mercantibility for commercial purposes. These routines should not be used as-is for any purpose or application that may result in loss or injury, and the publishers and authors shall not be liable in any event for incidental or consequential damages in connection with or arising out of the furnishing, performance, or use of these programs.

A.2 GENERAL SYSTEM REQUIREMENTS

The diskette and MATLAB scripts are designed for MATLAB environments on "WINtel" IBM-compatible systems (i.e., on an Intel-class processor operating under the Microsoft Windows operating system). Information on MATLAB can be obtained from

The MathWorks, Inc.

3 Apple Hill Drive
Natick, MA 01760 USA
Tel: 508-647-7000
Fax: 508-647-7101
E-mail: info@mathworks.com
Web: www.mathworks.com

You may also use the MATLAB editor to modify these scripts as needed. Comments in the listings contain additional information on the calling sequences and array dimensions.

A.3 DISKETTE DIRECTORY STRUCTURE

The diskette directories are organized by the chapters in which the supporting concepts are presented. The ASCII file WHATSUP.DOC in the root directory describes any changes made in the directory structure or software after printing. It should also be read before starting to use any of the software.

A.4 MATLAB SOFTWARE FOR CHAPTER 2

The directory CHAPTER2 contains expm1.m, referred to in Chapter 2, Section 2.6.3.

A.5 MATLAB SOFTWARE FOR CHAPTER 4

The directory CHAPTER4 contains software implementing the algorithms defined in Chapter 4. See the file WHATSUP.DOC in the root directory for descriptions of any changes.

The file demo1.m is a MATLAB script for demonstrating the effects of process noise and observations of the probability distribution of a single state variable as a function of time. The plots show the evolution of the probability density function while observations are made at discrete times.

The file exam43.m is a MATLAB script for demonstrating Example 4.3 in MATLAB.

The file exam44.m is a MATLAB script for demonstrating Example 4.4 in MATLAB.

The file obsup.m is a MATLAB script implementation of the Kalman filter observational update, including the state update and the covariance update (Riccati equation).

The file timeup.m is a MATLAB script implementation of the Kalman filter temporal update, including the state update and the covariance update (Riccati equation).

The file RTSvsKF.m is a MATLAB script for demonstrating the solutions to an estimation problem using

1. Kalman filtering, which uses only data up to the time of the estimate, and
2. Rauch–Tung–Striebel smoothing, which uses all the data.

A MATLAB implementation of a Rauch–Tung–Striebel smoother is included in the script, along with the corresponding Kalman filter implementation.

A.6 MATLAB SOFTWARE FOR CHAPTER 5

The file exam53.m is a MATLAB script implementation of Example 5.3.

A.7 MATLAB SOFTWARE FOR CHAPTER 6

The Matlab m-file shootout.m provides a demonstration of the relative fidelity of nine different ways to perform the covariance correction on Example 6.2.

To test how different solution methods perform as conditioning worsens, the observational update is performed for $10^{-9}\varepsilon^{2/3} \leq \delta \leq 10^{9}\varepsilon^{2/3}$ using nine different implementation methods:

1. the conventional Kalman filter, as published by R. E. Kalman;
2. Swerling inverse implementation, published by P. Swerling before the Kalman filter;
3. Joseph-stabilized implementation as given by P. D. Joseph;
4. Joseph-stabilized implementation as modified by G. J. Bierman;
5. Joseph-stabilized implementation as modified by T. W. DeVries;
6. the Potter algorithm (due to J. E. Potter);
7. the Carlson "triangular" algorithm (N. A. Carlson);
8. the Bierman "UD" algorithm (G. J. Bierman); and
9. the closed-form solution for this particular problem.

The first, second, and last methods are implemented within the m-file shootout.m. The others are implemented in m-files listed below.

The results are plotted as the RMS error in the computed value of P relative to the closed-form solution. In order that all results, including failed results, can be plotted, the value NaN (not a number) is interpreted as an underflow and set to zero, and the value Inf is interpreted as the result of a divide-by-zero and set to 10^4.

This demonstration should show that, for this particular problem, the accuracies of the Carlson and Bierman implementations degrade more gracefully than the others as $\delta \rightarrow \varepsilon$. This might encourage the use of the Carlson and Bierman methods for applications with suspected roundoff problems, although it does not necessarily demonstrate the superiority of these methods for all applications:

bierman.m performs the Bierman UD implementation of the Kalman filter measurement update and

carlson.m performs the Carlson "fast triangular" implementation of the Kalman filter measurement update.

There are several forms of this Riccati equation corrector implementation, which helps to preserve symmetry of P, among other things:

joseph.m performs the Joseph-stabilized implementation of the Kalman filter measurement update, as proposed by Peter Joseph [15];

josephb.m performs the Joseph-stabilized implementation of the Kalman filter measurement update, as modified by G. J. Bierman;

josephdv.m performs the Joseph-stabilized implementation of the Kalman filter measurement update, as modified by T. W. DeVries;

potter.m performs the Potter "square-root" implementation of the Kalman filter measurement update; and

utchol.m performs upper triangular Cholesky factorization for initializing the Carlson fast triangular implementation of the Kalman filter measurement update.

A.8 MATLAB SOFTWARE FOR CHAPTER 7

The file KFvsSKF.m includes MATLAB implementations of the Schmidt–Kalman filter and Kalman filter for a common problem, implements both, and plots the results for comparison.

The file thornton.m implements the Thornton temporal update compatible with the Bierman observational update using modified Cholesky factors of P.

The file schmidt.m performs the Schmidt temporal update compatible with the Carlson observational update using triangular Cholesky factors of P.

A.9 OTHER SOURCES OF SOFTWARE

Controls Toolbox. Available from The Mathworks, Controls Toolbox includes MATLAB routines for numerical solution of the algebraic Riccati equation for the Kalman filtering problem for linear time-invariant systems. These essentially provide the steady-state Kalman gain (Wiener gain).

Software for Kalman Filter Implementation. There are several sources of good up-to-date software specifically designed to address the numerical stability issues in Kalman filtering. Scientific software libraries and workstation environments for the design of control and signal processing systems typically use the more robust implementation methods available. In addition, as a noncommercial source of algorithms for Kalman filtering, the documentation and source codes of the collected algorithms from the *Transactions on Mathematical Software* (TOMS) of the Association for Computing Machinery are available at moderate cost on electronic media. The TOMS collection contains several routines designed to address the numerical stability issues related to Kalman filter implementation, and these are often revised to correct deficiencies discovered by users.

Utilities for Monte Carlo Simulation. The TOMS collection also contains several routines designed for pseudorandom number generation with good statistical properties. In addition, most reputable libraries contain good pseudorandom number generators, and many compilers include them as built-in functions. There are also several books (e.g., [93] or [90]) that come with the appropriate code on machine-readable media.

Appendix B

A Matrix Refresher

This overview of the notation and properties of matrices as data structures and algebras is for readers familiar with the general subject of linear algebra but whose recall may be a little rusty. A more thorough treatment can be found in most college-level textbooks on linear algebra and matrix theory.

B.1 MATRIX FORMS

B.1.1 Notation for Real Matrices

Scalars. For the purposes of this book, *scalars* are real numbers, although in computer implementations they must be approximated by floating-point numbers, which are but a finite subset of the rational numbers. We will use parentheses to denote open intervals (intervals not including the designated endpoints) on the real line, so that $(-\infty, +\infty)$ denotes the set of all real numbers. We will use square brackets to denote closed ends (ends including the designated endpoint) of intervals, so that $[0, +\infty)$ denotes the nonnegative real numbers.

Real Matrices. For positive integers m and n, an m-by-n real *matrix* A is a two-dimensional rectangular array of scalars, designated by the subscript notation a_{ij} and usually displayed in the following format:

$$A = \begin{bmatrix} a_{11} & a_{12} & a_{13} & \cdots & a_{1n} \\ a_{21} & a_{22} & a_{23} & \cdots & a_{2n} \\ a_{31} & a_{32} & a_{33} & \cdots & a_{3n} \\ \vdots & \vdots & \vdots & \ddots & \vdots \\ a_{m1} & a_{m2} & a_{m3} & \cdots & a_{mn} \end{bmatrix}.$$

The scalars a_{ij} are called the *elements* of A. We will use upper case letters to denote matrices and the corresponding lowercase letters to denote scalar elements of the associated matrices.

Indices and Subscripts. The first subscript (i) on the element a_{ij} refers to the *row* in which the element occurs, and the second subscript (j) refers to the *column* in which a_{ij} occurs in this format. The integers i and j in this notation are also called *indices* of the elements. The first index is called the *row index*, and the second index is called the *column index* of the element. The term (ij)th *position* in the matrix A refers to the position of a_{ij}, and a_{ij} is called the (ij)th *element* of A.

If juxtaposition of subscripts leads to confusion, they may be separated by commas. The element in the eleventh row and first column of the matrix A would then be denoted by $a_{11,1}$, not a_{111}.

Dimensions. The positive integers m and n are called the *dimensions* of A: m is called the *row dimension* of A and n is called the *column dimension* of A. The dimensions of A may also be represented as $m \times n$, which is to be read "m by n". The symbol "$\times$" in this notation does not indicate multiplication. (The number of elements in the matrix A equals the product mn, however, and this is important for determining memory requirements for data structures to hold A.)

B.1.2 Special Matrix Forms

Square Matrices and Diagonal Matrices. A matrix is called *square* if it has the same row and column dimensions. The *main diagonal* of a square matrix A is the set of elements a_{ij} for which $i = j$. The other elements are called *off-diagonal.* If all the off-diagonal elements of a square matrix A are zero, A is called a *diagonal* matrix. This and other special forms of square matrices are illustrated in Figure B.1.[1]

Sparse and Dense Matrices. A matrix with a "significant fraction" (typically, half or more) of zero elements is called *sparse.* Matrices that are decidedly not sparse are called *dense*, although both sparsity and density are matters of degree. Except for the Toeplitz and Hankel matrix,[2] the forms shown in Figure B.1 are sparse, although sparse matrices do not have to be square. Sparsity is an important characteristic for implementation of matrix methods because it can be exploited to reduce computer memory and computational requirements.

[1]The matrix forms in the third row of Figure B.1 belong to both forms in the column above. That is, diagonal matrices are both upper triangular and lower triangular, identity matrices are both unit upper triangular and unit lower triangular, and square zero matrices are both strictly upper triangular and strictly lower triangular.

[2]Although a Toeplitz matrix is fully dense, it can be represented by the $2n - 1$ distinct values of its elements.

UPPER TRIANGULAR — UNIT UPPER TRIANGULAR — STRICTLY UPPER TRIANGULAR

$$\begin{bmatrix} a_{11} & a_{12} & a_{13} & \cdots & a_{1n} \\ 0 & a_{21} & a_{22} & \cdots & a_{2n} \\ 0 & 0 & a_{33} & \cdots & a_{3n} \\ \vdots & \vdots & \vdots & \ddots & \vdots \\ 0 & 0 & 0 & \cdots & a_{nn} \end{bmatrix} \begin{bmatrix} 1 & a_{12} & a_{13} & \cdots & a_{1n} \\ 0 & 1 & a_{22} & \cdots & a_{2n} \\ 0 & 0 & 1 & \cdots & a_{3n} \\ \vdots & \vdots & \vdots & \ddots & \vdots \\ 0 & 0 & 0 & \cdots & 1 \end{bmatrix} \begin{bmatrix} 0 & a_{12} & a_{13} & \cdots & a_{1n} \\ 0 & 0 & a_{22} & \cdots & a_{2n} \\ 0 & 0 & 0 & \cdots & a_{3n} \\ \vdots & \vdots & \vdots & \ddots & \vdots \\ 0 & 0 & 0 & \cdots & 0 \end{bmatrix}$$

LOWER TRIANGULAR — UNIT LOWER TRIANGULAR — STRICTLY LOWER TRIANGULAR

$$\begin{bmatrix} a_{11} & 0 & 0 & \cdots & 0 \\ a_{21} & a_{22} & 0 & \cdots & 0 \\ a_{31} & a_{32} & a_{33} & \cdots & 0 \\ \vdots & \vdots & \vdots & \ddots & \vdots \\ a_{n1} & a_{n2} & a_{n3} & \cdots & a_{nn} \end{bmatrix} \begin{bmatrix} 1 & 0 & 0 & \cdots & 0 \\ a_{21} & 1 & 0 & \cdots & 0 \\ a_{31} & a_{32} & 1 & \cdots & 0 \\ \vdots & \vdots & \vdots & \ddots & \vdots \\ a_{n1} & a_{n2} & a_{n3} & \cdots & 1 \end{bmatrix} \begin{bmatrix} 0 & 0 & 0 & \cdots & 0 \\ a_{21} & 0 & 0 & \cdots & 0 \\ a_{31} & a_{32} & 0 & \cdots & 0 \\ \vdots & \vdots & \vdots & \ddots & \vdots \\ a_{n1} & a_{n2} & a_{n3} & \cdots & 0 \end{bmatrix}$$

DIAGONAL — IDENTITY — ZERO

$$\begin{bmatrix} d_1 & 0 & 0 & \cdots & 0 \\ 0 & d_2 & 0 & \cdots & 0 \\ 0 & 0 & d_3 & \cdots & 0 \\ \vdots & \vdots & \vdots & \ddots & \vdots \\ 0 & 0 & 0 & \cdots & d_n \end{bmatrix} \begin{bmatrix} 1 & 0 & 0 & \cdots & 0 \\ 0 & 1 & 0 & \cdots & 0 \\ 0 & 0 & 1 & \cdots & 0 \\ \vdots & \vdots & \vdots & \ddots & \vdots \\ 0 & 0 & 0 & \cdots & 1 \end{bmatrix} \begin{bmatrix} 0 & 0 & 0 & \cdots & 0 \\ 0 & 0 & 0 & \cdots & 0 \\ 0 & 0 & 0 & \cdots & 0 \\ \vdots & \vdots & \vdots & \ddots & \vdots \\ 0 & 0 & 0 & \cdots & 0 \end{bmatrix}$$

TOEPLITZ — HANKEL

$$\begin{bmatrix} d_0 & d_{-1} & d_{-2} & \cdots & d_{1-n} \\ d_1 & d_0 & d_{-1} & \cdots & d_{2-n} \\ d_2 & d_1 & d_0 & \cdots & d_{3-n} \\ \vdots & \vdots & \vdots & \ddots & \vdots \\ d_{n-1} & d_{n-2} & d_{n-3} & \cdots & d_0 \end{bmatrix} \begin{bmatrix} d_{1-n} & \cdots & d_{-2} & d_{-1} & d_0 \\ d_{2-n} & \cdots & d_{-1} & d_0 & d_1 \\ d_{3-n} & \cdots & d_0 & d_1 & d_2 \\ \vdots & & \vdots & \vdots & \vdots \\ d_0 & \cdots & d_{n-3} & d_{n-2} & d_{n-1} \end{bmatrix}$$

Fig. B.1 *Special forms of square matrices*

Zero Matrices. The ultimate sparse matrix is a matrix in which *all* elements are 0 (zero). It is called a *zero matrix,* and it is represented by the symbol "0" (zero). The equation $A = 0$ indicates that A is a zero matrix. Whenever it is necessary to specify the dimensions of a zero matrix, they may be indicated by subscripting: $0_{m\times n}$ will indicate an $m \times n$ zero matrix. If the matrix is square, only one subscript will be used: 0_n will mean an $n \times n$ zero matrix.

Identity Matrices. The identity matrix will be represented by the symbol I. If it is necessary to denote the dimension of I explicitly, it will be indicated by subscripting the symbol: I_n denotes the $n \times n$ identity matrix.

B.1.3 Vectors

Vectors and Matrices. A *vector* is essentially a matrix[3] in which one of the dimensions is 1. If the column dimension is 1, it is called a *column vector.* If the row dimension is 1, it is called called a *row vector.*[4] Once it is understood which of the dimensions is 1, the index for that dimension can be dropped from the notation. The other dimension may be prefixed to "-vector" to shorten the notation: the term *n-vector* refers to a matrix for which one dimension is 1 and the other is n.

Representational Differences. Although vectors may share many mathematical properties with matrices, they are used in this book to represent quite different concepts. To a considerable degree, vectors are used to represent physically measurable properties of dynamic systems, and matrices are used to represent transformations of those properties performed by sensors or by the passage of time. In order to make the distinction between vectors and matrices more apparent, we shall use lowercase letters to denote vectors and scalars and uppercase letters to denote matrices. However, as data structures, vectors are not fundamentally different from matrices or other arrays.

Row Vector Notation. Commas can be used for separating the elements of a row vector:

$$x = [x_1, x_2, x_3, \ldots, x_n],$$

where the notation "x_i" refers to the element in the ith column of x. (We will return to a related issue—the compositional efficiency of row vector notation—after the matrix transpose is defined.)

Column Vector Default. Whenever a vector x is not defined to be a row vector, it is implied that it is a column vector.

B.1.4 Conformable Matrix Dimensions

Syntax of Mathematical Expressions. *Syntax* is a set of rules governing the formation of patterns of symbols in a language. In mathematics, these symbols may stand for data structures such as scalars and matrices, operators such as addition and multiplication, or delimiters such as parentheses and brackets. Patterns of these symbols satisfying the rules of syntax are called *expressions*. Within the constraints imposed by syntax, an expression of a particular type (e.g., a scalar or a matrix expression) can be substituted for a symbol of that type, and vice versa.

[3]Defining a vector as a special case of a matrix is not the customary approach, but it obviates the need for separate definitions of "inner" and "outer" products.
[4]And if *both* dimensions are 1, it is called a *scalar*, not a "1-vector."

Syntax for Matrix Dimensions. For matrix expressions, there are additional rules of syntax related to the dimensions of the matrices. For example, whenever we write a matrix equation as $A = B$, we assume that the matrices (or matrix expressions) represented by the symbols A and B have the same dimensions.

Implied Conformability of Matrix Dimensions. Additional rules of syntax for matrix operations will be introduced with the operators. Matrices whose dimensions conform to these rules for a position in a particular expression are said to be *conformable* for that position in that expression. Whenever a symbol for a matrix appears in a matrix expression, it is implied that the matrix represented by that symbol is conformable for its position in that expression.

B.2 MATRIX OPERATIONS

B.2.1 Transposition

Transpose of a Matrix. All matrices are conformable for transposition. The *transpose* of A is the matrix A^{T} (with the superscript "T" denoting the transpose operation), obtained from A by interchanging rows and columns:

$$\begin{bmatrix} a_{11} & a_{12} & a_{13} & \cdots & a_{1n} \\ a_{21} & a_{22} & a_{23} & \cdots & a_{2n} \\ a_{31} & a_{32} & a_{33} & \cdots & a_{3n} \\ \vdots & \vdots & \vdots & \ddots & \vdots \\ a_{m1} & a_{m2} & a_{m3} & \cdots & a_{mn} \end{bmatrix}^{\mathrm{T}} = \begin{bmatrix} a_{11} & a_{21} & a_{31} & \cdots & a_{m1} \\ a_{12} & a_{22} & a_{32} & \cdots & a_{m2} \\ a_{13} & a_{23} & a_{33} & \cdots & a_{m3} \\ \vdots & \vdots & \vdots & \ddots & \vdots \\ a_{1n} & a_{2n} & a_{3n} & \cdots & a_{mn} \end{bmatrix}.$$

The transpose of an $m \times n$ matrix is an $n \times m$ matrix.

Transpose of a Vector. The transpose of a row vector is a column vector, and vice versa. It makes more efficient use of space on the page if we express a column vector v as the transpose of a row vector:

$$v = [v_1, v_2, v_3, \ldots, v_m]^{\mathrm{T}}.$$

Symmetric Matrices. A matrix A is called *symmetric* if $A^{\mathrm{T}} = A$, and *skew symmetric* (or *antisymmetric*) if $A^{\mathrm{T}} = -A$. Only square matrices can be symmetric or skew symmetric. Therefore, whenever a matrix is said to be symmetric or skew symmetric, it is implied that it is a square matrix.

B.2.2 Extracting Elements of Matrix Expressions

Subscripted Expressions. Subscripts represent an operation on a matrix that extracts the designated matrix element. Subscripts may also be applied to matrix expressions. The element in the (ij)th position of a matrix expression can be indicated by subscripting the expression, as in

$$\{A^{\mathrm{T}}\}_{ij} = a_{ji}.$$

Here, we have used braces { } to indicate the scope of the expression to which the subscripting applies. This is a handy device for defining matrix operations.

B.2.3 Multiplication by Scalars

All matrices are conformable for multiplication by scalars, either on the left or on the right. Multiplication is indicated by juxtaposition of symbols or by infix notation with the multiplication symbol ($\times$). Multiplication of a matrix A by a scalar s is equivalent to multiplying every element of A by s:

$$\{As\}_{ij} = \{sA\}_{ij} = sa_{ij}.$$

B.2.4 Addition and Multiplication of Conformable Matrices

Addition of Conformable Matrices Is Associative and Commutative. Matrices are conformable for addition if and only if they share the same dimensions. Whenever matrices appear as sums in an expression, it is implied that they are conformable. If A and B are conformable matrices, then addition is defined by adding corresponding elements:

$$\{A + B\}_{ij} = a_{ij} + b_{ij}.$$

Addition of matrices is *commutative* and *associative*. That is, $A + B = B + A$ and $A + (B + C) = (A + B) + C$.

Additive Inverse of a Matrix. The product of a matrix A by the scalar -1 yields its *additive inverse* $-A$:

$$(-1)A = -A, \quad A + (-A) = A - A = 0.$$

Here, we have followed the not uncommon practice of using the symbol "$-$" both as a unary (additive inverse) and binary (subtraction) operator. *Subtraction* of a matrix A from a matrix B is equivalent to adding the additive inverse of A to B:

$$B - A = B + (-A).$$

Multiplication of Conformable Matrices Is Associative But Not Commutative. Multiplication of an $m \times n$ matrix A by a matrix B on the right-hand side of A, as in the matrix product AB, is defined only if *the row dimension of B equals the column dimension of A*. That is, we can multiply an $m \times n$ matrix A by a $p \times q$ matrix B in this order only if $n = p$. In that case, the matrices A and B are said to be conformable for multiplication in that order, and the matrix product is defined element by element by

$$\{AB\}_{ij} \stackrel{\text{def}}{=} \sum_{k=1}^{n} a_{ik} b_{kj},$$

the result of which is an $m \times q$ matrix. Whenever matrices appear as a product in an expression, it is implied that they are conformable for multiplication.

Inner and Outer Products. There are special names given to products of vectors that are otherwise adequately defined as matrix products. The *inner product* or *dot product* of conformable column vectors x and y is the matrix product $x^{\mathrm{T}}y = y^{\mathrm{T}}x$. (For row vectors, the format is $xy^{\mathrm{T}} = yx^{\mathrm{T}}$.) The *outer products* have the transpose on the other vector, and the vectors need not have the same dimensions. If the vectors are treated as matrices, these products can be used without special treatment.

Products with Identity Matrices. Multiplication of any $m \times n$ matrix A by a conformable *identity matrix* yields the original matrix A as the product:

$$AI_n = A, \qquad I_m A = A.$$

B.2.5 Powers of Square Matrices

Square matrices are conformable with multiplication by themselves, and the resulting matrix products are again conformable for multiplication. Consequently, one can define the pth *power* of a square matrix A as

$$A^p = \underbrace{A \times A \times A \times \cdots \times A}_{p \text{ elements}}.$$

B.2.6 Matrix Inverses

Inverses of Nonsingular Square Matrices. If A and B are square matrices of the same dimension and such that their product

$$AB = I,$$

then B is the *matrix inverse* of A and A is the matrix inverse of B. (It turns out that $BA = AB = I$ in this case.) The inverse of a matrix A is unique, if it exists, and is denoted by A^{-1}. Not all matrices have inverses. *Matrix inversion* is the process of finding a matrix inverse, if it exists. If the inverse of a matrix A does not exist, A is called *singular*. Otherwise, it is called *nonsingular*.

Generalized Inverses. Even nonsquare or singular matrices can have *generalized inverses*. The *Moore–Penrose generalized inverse* of an $m \times n$ matrix A is the $n \times m$ matrix A^+ such that

$$\begin{aligned} AA^+A &= A, & A^+AA^+ &= A^+, \\ (AA^+)^{\mathrm{T}} &= AA^+, & (A^+A)^{\mathrm{T}} &= A^+A. \end{aligned}$$

B.2.7 Orthogonality

Orthogonal Vectors. For vectors, *orthogonality* is a pairwise property. Vectors x and y are called *orthogonal* or *normal* if their inner product is zero. If the inner product of a vector x with itself is 1, x is called a *unit vector*. Orthogonal unit vectors are called *orthonormal*.[5]

Orthogonal Matrices. A matrix A is called orthogonal if $A^{\mathrm{T}} = A^{-1}$. These matrices have several useful properties:

- Orthogonality of a matrix A implies that the row vectors of A are jointly orthonormal, and the column vectors of A are also jointly orthonormal.
- The dot products of vectors are invariant under multiplication by a conformable orthogonal matrix. That is, if A is orthogonal, then $x^{\mathrm{T}}y = (Ax)^{\mathrm{T}}(Ay)$ for all conformable x and y.
- Products and inverses of orthogonal matrices are orthogonal.

As a rule, multiplications by orthogonal matrices tend to be numerically well conditioned—compared to general matrix multiplications. (The inversion of orthogonal matrices is obviously extremely well conditioned.)

B.2.8 Square Matrix Subalgebras

Certain subclasses of $n \times n$ (square) matrices have the property that their products belong to the same class. Orthogonal matrices, for example, have the property that their products are also orthogonal matrices. These are said to form a multiplicative *subalgebra* of $n \times n$ matrices. Subalgebras have the property that their set intersections are also subalgebras. Upper triangular matrices and lower triangular matrices are two subalgebras of the square matrices that are used in implementing the Kalman filter. Their intersection is the set of diagonal matrices—another subalgebra. A lattice (partially ordered by set inclusion) of such multiplicative subalgebras is diagrammed in Figure B.2.

[5]The term "normal" has many meanings in mathematics. Its use here is synonymous with "of unit length". It is also used to mean "orthogonal".

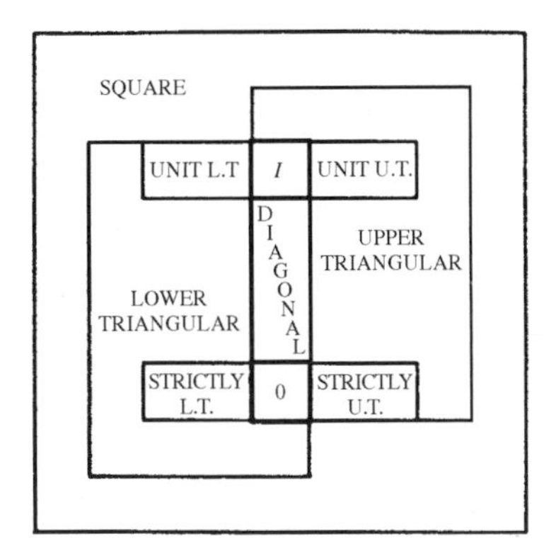

Fig. B.2 Subalgebras of square matrices.

B.3 BLOCK MATRIX FORMULAS

B.3.1 Submatrices, Partitioned Matrices, and Blocks

For any $m \times n$ matrix A and any subset $S_{\text{rows}} \subseteq \{1, 2, 3, \ldots, m\}$ of the row indices and subset $S_{\text{cols}} \subseteq \{1, 2, 3, \ldots, n\}$ of the column indices, the subset of elements

$$A' = \{a_{ij} | i \in S_{\text{rows}}, j \in S_{\text{cols}}\}$$

is called a *submatrix* of A.

A *partitioning* of an integer n is an exhaustive collection of contiguous subsets S_k of the form

$$\overbrace{1, 2, 3, \ldots, \ell_1}^{S_1}, \overbrace{(\ell_1 + 1), \ldots, \ell_2}^{S_2}, \ldots, \overbrace{(\ell_{p-1} + 1), \ldots, n}^{S_p}.$$

The collection of submatrices formed by partitionings of the row and column dimensions of a matrix is called a partitioning of the matrix, and the matrix is said to be partitioned by that partitioning. Each submatrix of a partitioned matrix A is called a *partitioned submatrix*, *partition*, *submatrix block*, *subblock*, or *block* of A. Each block of a partitioned matrix A can be represented by a conformable matrix expression, and A can be displayed as a *block matrix*:

$$A = \begin{bmatrix} B & C & D & \cdots & F \\ G & H & J & \cdots & L \\ M & N & P & \cdots & R \\ \vdots & \vdots & \vdots & \ddots & \vdots \\ V & W & X & \cdots & Z \end{bmatrix}$$

where $B, C, D, \ldots$ stand for matrix expressions. Whenever a matrix is displayed as a block matrix, it is implied that all block submatrices in the same row have the same row dimension and that all block submatrices in the same column have the same column dimension.

A block matrix of the form

$$\begin{bmatrix} A & 0 & 0 & \cdots & 0 \\ 0 & B & 0 & \cdots & 0 \\ 0 & 0 & C & \cdots & 0 \\ \vdots & \vdots & \vdots & \ddots & \vdots \\ 0 & 0 & 0 & \cdots & M \end{bmatrix},$$

in which the off-diagonal block submatrices are zero matrices, is called a *block diagonal matrix,* and a block matrix in which the block submatrices on one side of the diagonal are zero matrices is called a *block triangular matrix.*

Columns and Rows as Blocks. There are two special partitionings of matrices in which the block submatrices are vectors. The column vectors of an $m \times n$ matrix A are the block submatrices of the partitioning of A for which all column dimensions are 1 and all row dimensions are m. The row vectors of A are the block submatrices of the partitioning for which all row dimensions are 1 and all column dimensions are n. All column vectors of an $m \times n$ matrix are m-vectors, and all row vectors are n-vectors.

B.3.2 Rank and Linear Dependence

A *linear combination* of a finite set of n-vectors $\{v_i\}$ is a summation of the sort $\sum_i a_i v_i$ for some set of scalars $\{a_i\}$. If some linear combination $\sum a_i v_i = 0$ and at least one coefficient $a_i \neq 0$, the set of vectors $\{v_i\}$ is called *linearly dependent.* Conversely, if the the only linear combination for which $\sum a_i v_i = 0$ is the one for which all the $a_i = 0$, then the set of vectors $\{v_i\}$ is called *linearly independent.*

The *rank* of a $n \times m$ matrix A equals the size of the *largest* collection of its column vectors that is linearly independent. Note that any such linear combination can be expressed in the form Aa, where the nonzero elements of the column m-vector a are the associated scalars of the linear combination, and the number of nonzero components of a is the size of the collection of column vectors in the linear combination. The same value for the rank of a matrix is obtained if the test is applied to its row vectors, where any linear combination of row vectors can be expressed in the form $a^{\mathrm{T}}A$ for some column n-vector a.

An $n \times n$ matrix is nonsingular if and only if its rank equals its dimension n.

B.3.3 Conformable Block Operations

Block matrices with conformable partitionings may be transposed, added, subtracted, and multiplied in block format. For example,

$$\begin{bmatrix} A & B \\ C & D \end{bmatrix}^{\mathrm{T}} = \begin{bmatrix} A^{\mathrm{T}} & C^{\mathrm{T}} \\ B^{\mathrm{T}} & D^{\mathrm{T}} \end{bmatrix},$$

$$\begin{bmatrix} A & B \\ C & D \end{bmatrix} + \begin{bmatrix} E & F \\ G & H \end{bmatrix} = \begin{bmatrix} A+E & B+F \\ C+G & D+H \end{bmatrix},$$

$$\begin{bmatrix} A & B \\ C & D \end{bmatrix} \times \begin{bmatrix} E & F \\ G & H \end{bmatrix} = \begin{bmatrix} AE+BG & AF+BH \\ CE+DG & CF+DH \end{bmatrix}.$$

B.3.4 Frobenius–Schur Inversion Formula

The inverse of a partitioned matrix with square diagonal blocks may be represented in block form as[6]

$$\begin{bmatrix} A & B \\ C & D \end{bmatrix}^{-1} = \begin{bmatrix} E & F \\ G & H \end{bmatrix}.$$

where

$$E = A^{-1} + A^{-1}BHCA^{-1},$$
$$F = -A^{-1}BH,$$
$$G = -HCA^{-1},$$
$$H = [D - CA^{-1}B]^{-1}.$$

This formula can be proved by multiplying the original matrix times its alleged inverse and verifying that the result is the identity matrix.

B.3.5 Inversion Formulas for Matrix Expressions

Sherman–Morrison Formula.[7] A "rank 1" modification of a square matrix A is a sum of the form $A + bc^{\mathrm{T}}$, where b and c are conformable column vectors.[8] Its

[6]This formula has had many discoverers. Bodewig [55] cites nine such discoverers but gives credit to the German mathematicians Georg Ferdinand Frobenius (1849–1917) and Issai Shur (1875–1941) as the earliest discovers of record.

[7]The naming of this and the next formula follows the convention of Golub and Van Loan [89], which is at odds with the references of Henderson and Searle [76].

[8]A finite set of vectors $\{x_1, x_2, x_3, \ldots, x_n\}$ is said to be *linearly independent* if there is no linear combination $\sum_k a_k x_k = 0$ with no coefficient $a_k = 0$. The rank of a set of vectors is defined as the size of the maximum subset of them that is linearly independent. The rank of the row vectors of a matrix is called its *row rank*, and the rank of its column vectors is called its *column rank*. For a square matrix, the row rank equals the column rank. An $n \times n$ square matrix is nonsingluar if and only if its (row *or* column) rank is n. Obviously, the rank of a single vector is 1.

inverse is given by the formula

$$[A + bc^{\mathrm{T}}]^{-1} = A^{-1} - \frac{A^{-1}bc^{\mathrm{T}}A^{-1}}{1 + c^{\mathrm{T}}A^{-1}b}.$$

Sherman–Morrison–Woodbury Formula. This is the generalization of the above formula for conformable matrices in place of vectors:

$$[A + BC^{\mathrm{T}}]^{-1} = A^{-1} - A^{-1}B[I + C^{\mathrm{T}}A^{-1}B]^{-1}C^{\mathrm{T}}A^{-1}.$$

Hemes Inversion Formula. A further generalization of this formula (used in the derivation of the Kalman filter equations) includes an additional conformable square matrix factor in the modification[9]

$$[A + BC^{-1}D^{\mathrm{T}}]^{-1} = A^{-1} - A^{-1}B[C + D^{\mathrm{T}}A^{-1}B]^{-1}D^{\mathrm{T}}A^{-1}.$$

B.4 FUNCTIONS OF SQUARE MATRICES

B.4.1 Determinants and Characteristic Values

Elementary Permutation Matrices. An *elementary permutation matrix* is formed by interchanging rows or columns of an identity matrix I_n:

$$P_{[ij]} = \begin{array}{c} \\ \\ i \\ \\ j \\ \\ \\ \end{array} \begin{pmatrix} 1 & \cdots & 0 & \cdots & 0 & \cdots & 0 \\ \vdots & \ddots & \vdots & & \vdots & & \vdots \\ 0 & \cdots & 0 & \cdots & 1 & \cdots & 0 \\ \vdots & & \vdots & \ddots & \vdots & & \vdots \\ 0 & \cdots & 1 & \cdots & 0 & \cdots & 0 \\ \vdots & & \vdots & & \vdots & \ddots & \vdots \\ 0 & \cdots & 0 & \cdots & 0 & \cdots & 1 \end{pmatrix}.$$

(columns i and j marked above the third and fifth columns)

Multiplication of a vector x by $P_{[ij]}$ permutes the ith and jth elements of x. Note that $P_{[ij]}$ is an *orthogonal* matrix and that $P_{[ii]} = I_n$, the identity matrix.

[9]This is yet another formula with many discoverers. Fortmann [161] cites several of them. Bodewig [55, page 218] credits H. Hemes for its discovery, although Henderson and Searle [76] cite an earlier reference.

Determinants of Elementary Permutation Matrices. The *determinant* of an elementary permutation matrix $P_{[ij]}$ is defined to be -1, unless $i = j$ (i.e., $P_{[ij]} = I_n$):

$$\det(P_{[ij]}) \stackrel{\text{def}}{=} \begin{cases} -1, & i \neq j, \\ +1, & i = j. \end{cases}$$

Permutation Matrices. A *permutation matrix* is any product of elementary permutation matrices. These are also orthogonal matrices. Let $\mathscr{P}_n$ denote the set of all distinct $n \times n$ permutation matrices. There are $n! = 1 \times 2 \times 3 \times \cdots \times n$ of them, corresponding to the $n!$ permutations of n indices.

Determinants of Permutation Matrices. The determinant of a permutation matrix can be defined by the rule that the determinant of a product of matrices is the product of the determinants:

$$\det(AB) = \det(A)\det(B).$$

Therefore, the determinant of a permutation matrix will be either $+1$ or -1. A permutation matrix is called "even" if its determinant is $+1$ and "odd" if its determinant equals -1.

Determinants of Square Matrices. The determinant of any $n \times n$ matrix A can be defined as follows:

$$\det(A) \stackrel{\text{def}}{=} \sum_{P \in \mathscr{P}_n} \det(P) \prod_{i=1}^{n} \{AP\}_{ii}.$$

This formula has $\mathcal{O}(n \times n!)$ computational complexity (for a sum over $n!$ products of n elements each).

Characteristic Values of Square Matrices. For a free variable λ, the polynomial

$$\begin{aligned} p_A(\lambda) &\stackrel{\text{def}}{=} \det\,[A - \lambda I] \\ &= \sum_{i=0}^{n} a_i \lambda^i \end{aligned}$$

is called the *characteristic polynomial* of A. The roots of $p_A(\lambda)$ are called the *characteristic values* (or *eigenvalues*) of A. The determinant of A equals the product of its characteristic values, with each characteristic value occurring as many times in the product as the multiplicity of the associated root of the characteristic polynomial.

Definiteness of Symmetric Matrices. If A is symmetric, all its characteristic values are real numbers, which implies that they can be ordered. They are usually expressed in descending order:

$$\lambda_1(A) \geq \lambda_2(A) \geq \lambda_3(A) \geq \cdots \geq \lambda_n(A).$$

A real square symmetric matrix A is called

positive definite	if $\lambda_n(A) > 0$,
non–negative definite	if $\lambda_n(A) \geq 0$,
indefinite	if $\lambda_1(A) > 0$ and $\lambda_n(A) < 0$,
non–positive definite	if $\lambda_1(A) \leq 0$, and
negative definite	if $\lambda_1(A) < 0$.

Non–negative–definite matrices are also called *positive semidefinite*, and non–positive–definite matrices are also called *negative semidefinite*.

Characteristic Vectors. For each real characteristic value $\lambda_i(A)$ of a real symmetric A, there is a corresponding *characteristic vector* (or *eigenvector*) $e_i(A)$ such that $e_i(A) \neq 0$ and $Ae_i(A) = \lambda_i(A)e_i(A)$. The characteristic vectors corresponding to distinct characteristic values are mutually orthogonal.

B.4.2 Matrix Trace

The *trace* of a square matrix is the sum of its diagonal elements. It also equals the sum of the characteristic values and has the property that the trace of the product of conformable matrices is independent of the order of multiplication—a very useful attribute:

$$\text{trace}\,(AB) = \sum_i \{AB\}_{ii} \tag{B.1}$$

$$= \sum_i \sum_j A_{ij} B_{ji} \tag{B.2}$$

$$= \sum_j \sum_i B_{ji} A_{ij} \tag{B.3}$$

$$= \text{trace}(BA). \tag{B.4}$$

Note the product AB is conformable for the trace function only if it is a square matrix, which requires that A and B^{T} have the same dimensions. If they are $m \times n$ (or $n \times m$), then the computation of the trace of their product requires mn multiplications, whereas the product itself would require $m^2 n$ (or mn^2) multiplications.

B.4.3 Algebraic Functions of Matrices

An algebraic function may be defined by an expression in which the independent variable (a matrix) is a free variable, such as the truncated power series

$$f(A) = \sum_{k=-n}^{n} B_k A^k,$$

where the negative power $A^{-p} = \{A^{-1}\}^p = \{A^p\}^{-1}$. In this representation, the matrix A is the independent (free) variable and the other matrix parameters (B_k) are assumed to be known and fixed.

B.4.4 Analytic Functions of Matrices

An analytic function is defined in terms of a convergent power series. It is necessary that the power series converge to a limit, and the matrix norms defined in Section B.5 must be used to define and prove convergence of a power series. This level of rigor is beyond the scope of this book, but we do need to use one particular analytic function—the exponential function.

Exponential Function. The power series

$$e^A = \sum_{k=0}^{\infty} \frac{1}{1 \cdot 2 \cdot 3 \cdots k} A^k$$

does converge[10] for all square matrices A. It defines the exponential function of the matrix A. This definition is sufficient to prove some elementary properties of the exponential function for matrices, such as:

- $e^{0_n} = I_n$ for 0_n the $n \times n$ zero matrix.
- $e^{I_n} = eI_n$ for I_n the $n \times n$ identity matrix.
- $e^{A^{\mathrm{T}}} = \{e^A\}^{\mathrm{T}}$.
- $(d/dt)e^{At} = Ae^{At} = e^{At}A$.
- The exponential of a skew-symmetric matrix is an orthogonal matrix.
- The characteristic vectors of A are also the characteristic vectors of e^A.
- If λ is a characteristic value of A, then e^{λ} is a characteristic value of e^A.

[10]However, convergence is not fast enough to make this a reasonable general-purpose formula for approximating the exponential of A. More reliable and efficient methods can be found in the book by Golub and Van Loan [89].

B.4.5 Similarity Transformations and Analytic Functions

For any $n \times n$ nonsingular matrix A, the transform $X \to A^{-1}XA$ is called a *similarity transformation* of the $n \times n$ matrix X. It is a useful transformation for analytic functions of matrices

$$f(X) = \sum_{k=0}^{\infty} a_k X^k,$$

because

$$\begin{aligned} f(A^{-1}XA) &= \sum_{k=0}^{\infty} a_k (A^{-1}XA)^k \\ &= A^{-1}\left(\sum_{k=0}^{\infty} a_k X^k\right)A \\ &= A^{-1}f(X)A. \end{aligned}$$

If the characteristic values of X are distinct, then the similarity transform performed with the characteristic vectors of X as the column vectors of A will diagonalize X with its characteristic values along the main diagonal:

$$\begin{aligned} A^{-1}XA &= \operatorname*{diag}_{\ell}\{\lambda_\ell\}, \\ f(A^{-1}XA) &= \operatorname*{diag}_{\ell}\{F(\lambda_\ell)\}, \\ f(X) &= A \operatorname*{diag}_{\ell}\{F(\lambda_\ell)\}A^{-1}. \end{aligned}$$

(Although this is a useful analytical approach for demonstrating functional dependencies, it is not considered a robust numerical method.)

B.5 NORMS

B.5.1 Normed Linear Spaces

Vectors and matrices can be considered as elements of *linear spaces,* in that they can be added and multiplied by scalars. A *norm* is *any* nonnegative real-valued function $\|\cdot\|$ defined on a linear space such that, for any scalar s and elements x and y of the linear space (vectors or matrices),

$$\begin{aligned} \|x\| = 0 \quad &\text{iff} \quad x = 0, \\ \|x\| > 0 \quad &\text{if} \quad x \neq 0, \\ \|sx\| &= |s|\,\|x\|, \\ \|x + y\| &\leq \|x\| + \|y\|, \end{aligned}$$

where iff stands for "if and only if". These constraints are rather loose, and many possible norms can be defined for a particular linear space. A linear space with a specified norm is called a *normed linear space.* The norm induces a *topology* on the linear space, which is used to define continuity and convergence. Norms are also used in numerical analysis for establishing error bounds, and in sensitivity analysis for bounding sensitivities. The multiplicity of norms is useful in these applications, because the user is free to pick the one that works best for her or his particular problem.

We define here many of the more popular norms, some of which are known by more than one name.

B.5.2 Hölder Norms

The inner product of a column n-vector x with itself is

$$\begin{aligned} x^{\mathrm{T}}x &= \operatorname{trace} xx^{\mathrm{T}} \\ &= \sum_{i=1}^{n} x_i^2 \\ &\stackrel{\text{def}}{=} \|x\|_E^2, \end{aligned}$$

the square of the *Euclidean norm* of x. This is but one of a class of norms called *Hölder norms*,[11] ℓ_p *norms*,[12] or simply *p-norms*:

$$\|x\|_p \stackrel{\text{def}}{=} \left[\sum_{i=1}^{n} |x_i|^p\right]^{1/p},$$

and in the limit (as $p \to \infty$) as the *sup*[13] norm, or ∞-norm:

$$\|x\|_\infty \stackrel{\text{def}}{=} \max_i |x_i|.$$

These norms satisfy the *Hölder inequality*:

$$|x^{\mathrm{T}}y| \le \|x\|_p \|y\|_q \quad \text{for} \quad \frac{1}{p} + \frac{1}{q} = 1.$$

They are also related by inequalities such as

$$\|x\|_\infty \le \|x\|_E \le \|x\|_1 \le n\|x\|_\infty.$$

The Euclidean norm (2-norm) is the default norm for vectors. When no other norm is identified, the implied norm is the Euclidean norm.

B.5.3 Matrix Norms

Many norms have been defined for matrices. Two general types are presented here. Both are derived from vector norms, but by different means.

[11]Named for the German mathematician Otto Ludwig Hölder (1859–1937).

[12]This "little ℓ" notation is used for infinite-dimensional normed vector spaces (sequences), which include finite-dimensional normed vector spaces as a subclass.

[13]"sup" (sounds like "soup") stands for *supremum*, a mathematical term for the *least upper bound* of a set of real numbers. The maximum (max) is the supremum over a finite set.

Generalized Vector Norms. Vector norms can be generalized to matrices by treating the matrix like a doubly subscripted vector. For example, the Hölder norms for vectors can be generalized to matrices as

$$\|A\|_{(p)} = \left\{ \sum_{i=1}^{m} \sum_{j=1}^{n} |a_{i,j}|^p \right\}^{1/p}.$$

The matrix (2)-norm defined in this way is also called the *Euclidean norm*, *Schur norm*, or *Frobenius norm*. We will use the notation $\|\cdot\|_F$ in place of $\|\cdot\|_{(2)}$ for the Frobenius norm.

The reason for putting the parentheses around the subscript p in the above definition is that there is another way that the vector p-norms are used to define matrix norms, and it is this alternative definition that is usually allowed to wear an unadorned p subscript. These alternative norms also have the following desirable properties.

Desirable Multiplicative Properties of Matrix Norms. Because matrices can be multiplied, one could also apply the additional constraint that

$$\|AB\|_M \leq \|A\|_M \, \|B\|_M$$

for conformable matrices A and B and a matrix norm $\|\cdot\|_M$. This is a good property to have for some applications. One might also insist on a similar property with respect to multiplication by vector x, for which a norm $\|\cdot\|_{V_1}$ may already be defined:

$$\|Ax\|_{V_2} \leq \|A\|_M \|x\|_{V_1}.$$

This property is called *compatibility* between the matrix norm $\|\cdot\|_M$ and the vector norms $\|\cdot\|_{V_1}$ and $\|\cdot\|_{V_2}$. (Note that there can be two distinct vector norms associated with a matrix norm: one in the normed linear space containing x, and one in the space containing Ax.)

Matrix Norms Subordinate to Vector Hölder Norms. There is a family of alternative matrix "p-norms" [but not (p)-norms] defined by the formula

$$\|A\|_p \stackrel{\text{def}}{=} \sup_{\|x\| \neq 0} \frac{\|Ax\|_p}{\|x\|_p},$$

where the norms on the right-hand side are the vector Hölder norms and the induced matrix norms on the left are called *subordinate* to the corresponding Hölder norms. The 2-norm defined in this way is also called the *spectral norm* of A. It has the properties

$$\| \operatorname{diag}_i \{\lambda_i\} \|_2 = \max_i |\lambda_i| \quad \text{and} \quad \|Ax\|_2 \leq \|A\|_2 \, \|x\|_2.$$

The first of these properties implies that $\|I\|_2 = 1$. The second property is compatibility between the spectral norm and the vector Euclidean norm. (Subordinate matrix norms are guaranteed to be compatible with the vector norms used to define them.) All matrix norms subordinate to vector norms also have the property that $\|I\| = 1$.

Computation of Matrix Hölder Norms. The following formulas may be used in computing 1-norms and ∞-norms of $m \times n$ matrices A:

$$\|A\|_1 = \max_{1 \le j \le n} \left\{ \sum_{i=1}^{m} |a_{ij}| \right\},$$

$$\|A\|_\infty = \max_{1 \le i \le m} \left\{ \sum_{j=1}^{n} |a_{ij}| \right\}.$$

The norm $\|A\|_2$ can be computed as the square root of the largest characteristic value of $A^{\mathrm{T}}A$, which takes considerably more effort.

Default Matrix Norm. When the type of norm applied to a matrix is not specified (by an appropriate subscript), the default will be the spectral norm (Hölder matrix 2-norm). It satisfies the following bounds with respect to the Frobenius norm and the other matrix Hölder norms for $m \times n$ matrices A:

$$\|A\|_2 \le \|A\|_F \le \sqrt{n}\,\|A\|_2,$$

$$\frac{1}{\sqrt{m}}\|A\|_1 \le \|A\|_2 \le \sqrt{n}\|A\|_1,$$

$$\frac{1}{\sqrt{n}}\|A\|_\infty \le \|A\|_2 \le \sqrt{m}\|A\|_\infty,$$

$$\max_{\substack{1 \le i \le m \\ 1 \le j \le n}} |a_{ij}| \le \|A\|_2 \le \sqrt{mn} \max_{\substack{1 \le i \le m \\ 1 \le j \le n}} |a_{ij}|.$$

B.6. CHOLESKY DECOMPOSITION

This decomposition is named after André Louis Cholesky, a French[14] geodesist and artillery officer and a casualty of World War I. He discovered a method for solving linear least-squares problems that uses a method for factoring a symmetric, positive-definite matrix P as a product of triangular factors. He was perhaps not the first discoverer[15] of the factoring technique, but his use of the method for solving least-squares problems was unique. His results were published posthumously by a fellow officer, Commendant Benoit [139], and credited to Cholesky.

[14]Because Cholesky was French, his name should perhaps be pronounced "Show-less-KEY" with the accent on the last syllable.

[15]Zurmühl [80] cites an earlier discovery by M. H. Doolittle, published in a U.S. Coast and Geodetic Report in 1878.

Choleksy decomposition is used in several ways for implementing Kalman filters.

B.6.1 Matrix Square Roots and Cholesky Factors

A square root of a matrix M is a matrix S such that $M = S^2 = SS$. The matrix square root is sometimes confused with a Cholesky factor, which is not the same thing.

A Cholesky factor of a symmetric positive-definite matrix P is a matrix C such that

$$CC^{\mathrm{T}} = P. \tag{B.5}$$

Note that it does not matter whether we write this equation in the alternative form $F^{\mathrm{T}}F = P$, because the two solutions are related by $F = C^{\mathrm{T}}$.

Cholesky Factors Are Not Unique. If C is a Cholesky factor of P, then for any conformable orthogonal matrix M, the matrix

$$A \stackrel{\mathrm{def}}{=} CM$$

satisfies the equation

$$\begin{aligned} AA^{\mathrm{T}} &= CM(CM)^{\mathrm{T}} \\ &= CMM^{\mathrm{T}}C^{\mathrm{T}} \\ &= CC^{\mathrm{T}} \\ &= P. \end{aligned}$$

That is, A is also a legitimate Cholesky factor.

The ability to transform one Cholesky factor into another using orthogonal matrices turns out to be very important in square-root filtering (in Section 6.5).

B.6.2 Cholesky Factoring Algorithms

There are two possible forms of Cholesky factorization, corresponding to two possible forms of the defining equation:

$$\begin{aligned} P &= L_1 L_1^{\mathrm{T}} = U_1^{\mathrm{T}} U_1 && \text{(B.7)} \\ &= U_2 U_2^{\mathrm{T}} = L_2^{\mathrm{T}} L_2, && \text{(B.8)} \end{aligned}$$

where the Cholesky factors U_1, U_2 are upper triangular and their respective transposes L_1, L_2 are lower triangular.

The first of these is implemented by the built-in MATLAB function chol(P), with argument P a symmetric positive-definite matrix. The call chol(P) returns an upper triangular matrix U_1 satisfying Equation B.7. The MATLAB m-file utchol.m on the accompanying diskette implements the solution to Equation B.8. The call utchol(P) returns an upper triangular matrix U_2 satisfying Equation B.8.

There are also two possible forms for each of the two factoring algorithms, depending on whether the second level of the indexing loop is by rows or columns, but this detail has no significant effect on the result.

B.6.3 Modified Cholesky Factorization

The algorithm for Cholesky factorization of a matrix requires taking square roots, which can be avoided by using a *modified Cholesky factorization* in the form

$$P = UDU^{\mathrm{T}}, \tag{B.9}$$

where D is a diagonal matrix with positive diagonal elements and U is a *unit triangular matrix* (i.e., U has 1s along its main diagonal). This algorithm is implemented in the file modchol.m on the accompanying diskette.

B.6.4 Rank 1 Modifications of Cholesky Factors

A "rank 1 modification" of a Cholesky factor C_0 such that $C_0 C_0^{\mathrm{T}} = A$ is a Cholesky factor of $A \pm vv^{\mathrm{T}}$, where v is a column vector and vv^{T} is a rank 1 matrix. The built-in MATLAB function cholupdate.m performs that function, given C_0 and v, for factorization in the form $C_{\text{output}}^{\mathrm{T}} C_{\text{output}} = C_{\text{input}}^{\mathrm{T}} C_{\text{input}} + vv^{\mathrm{T}}$ (i.e., transposing the first factor, rather than the second factor). The MATLAB rank 1 Cholesky factor modification functions potter.m, carlson.m, and bierman.m on the accompanying diskette are specialized for application to Kalman filtering, as described in Chapter 6.

B.7 ORTHOGONAL DECOMPOSITIONS OF MATRICES

Decompositions are also called *factorizations* of matrices. They are formulas for representing a matrix as a product of matrix factors with useful properties. The two factorizations described here have either triangular or diagonal factors in addition to orthogonal factors.

Decomposition methods are algorithms for computing the factors, given the matrix to be "decomposed".

B.7.1 *QR* Decomposition (Triangularization)

The *QR decomposition* of a matrix A is a representation in the form

$$A = QR,$$

where Q is an orthogonal matrix and R is a triangular matrix. Methods for the QR decomposition of a matrix are described in Chapter 6.

B.7.2 Singular-Value Decomposition

The *singular-value decomposition* of an $m \times n$ matrix A is a representation in the form $A = T_m D T_n$, where T_m and T_n are orthogonal matrices (with square dimensions as specified by their subscripts) and D is an $m \times n$ matrix filled with zeros everywhere except along the main diagonal of its maximal upper-left square submatrix. This decomposition will have either of the three forms:

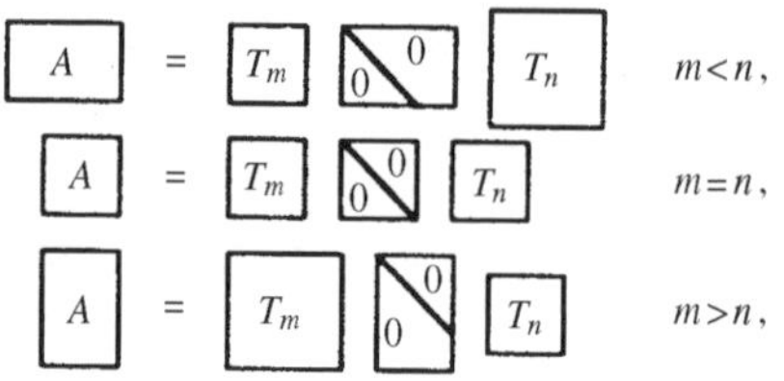

$A = T_m \; D \; T_n \qquad m<n,$

$A = T_m \; D \; T_n \qquad m=n,$

$A = T_m \; D \; T_n \qquad m>n,$

depending on the relative values of m and n. The middle matrix D has the block form

$$D = \begin{cases} [\text{diag}_i\{\sigma_i\} | 0_{m\times(n-m)}] & \text{if } m < n, \\ \text{diag}_i\{\sigma_i\} & \text{if } m = n, \\ \begin{bmatrix} \text{diag}_i\{\sigma_i\} \\ 0_{(m-n)\times n} \end{bmatrix} & \text{if } m > n, \end{cases}$$

$$\sigma_1 \geq \sigma_2 \geq \sigma_3 \geq \cdots \geq \sigma_p \geq 0,$$
$$p = \min(m, n).$$

That is, the diagonal nonzero elements of D are in *descending order* and nonnegative. These are called the *singular values* of A. For a proof that this decomposition exists, and an algorithm for computing it, see the book by Golub and Van Loan [89].

The singular values of a matrix characterize many useful matrix properties, such as:

$\|A\|_2 = \sigma_1(A)$.

rank $(A) = r$ such that $\sigma_r > 0$ and either $\sigma_{r+1} = 0$ or $r = p$. (The rank of a matrix is defined in Section B.3.2.)

The condition number of A equals σ_1/σ_p.

The condition number of the matrix A in the linear equation $Ax = b$ bounds the sensitivity of the solution x to variations in b and the sensitivity of the solution to roundoff errors in determining it. The singular-value decomposition may also be used to define the "pseudorank" of A as the smallest singular value σ_i such that $\sigma_i > \varepsilon\sigma_1$, where ε is a processor- and precision-dependent constant such that $0 < \varepsilon \ll 1$.

These relationships are useful for the analysis of state transition matrices Φ of Kalman filters, which can be singular or close enough to being singular that numerical roundoff can cause the product $\Phi P \Phi^{\mathrm{T}}$ to be essentially singular.

B.7.3 Eigenvalue–Eigenvector Decompositions of Symmetric Matrices

Symmetric QR Decomposition. The so-called symmetric QR decomposition of an $n \times n$ symmetric real matrix A has the special form $A = TDT^{\mathrm{T}}$, where the right orthogonal matrix is the transposed left orthogonal matrix and the diagonal matrix

$$D = \mathrm{diag}_i\{\lambda_i\}.$$

That is, the diagonal elements are the characteristic values of the symmetric matrix. Furthermore, the column vectors of the orthogonal matrix T are the associated characteristic vectors e_i of A:

$$\begin{aligned} A &= TDT^{\mathrm{T}} \\ &= \sum_{i=1}^{n} \lambda_i e_i e_i^{\mathrm{T}}, \\ T &= [e_1 \quad e_2 \quad e_3 \quad \cdots \quad e_n]. \end{aligned}$$

These relationships are useful for the analysis of covariance matrices, which are constrained to have nonnegative characteristic values, although their numerical values may stray enough in practice (due to computer roundoff errors) to develop negative characteristic values.

B.8 QUADRATIC FORMS

Bilinear and Quadratic Forms. For a matrix A and all conformable column vectors x and y, the functional mapping $(x, y) \rightarrow x^{\mathrm{T}}Ay$ is called a *bilinear form.* As a function of x and y, it is linear in both x and y and hence *bilinear.* In the case that $x = y$, the functional mapping $x \rightarrow x^{\mathrm{T}}Ax$ is called a *quadratic form*. The matrix A of a quadratic form is always a square matrix.

B.8.1 Symmetric Decomposition of Quadratic Forms

Any square matrix A can be represented uniquely as the sum of a symmetric matrix and a skew-symmetric matrix:

$$A = \tfrac{1}{2}(A + A^{\mathrm{T}}) + \tfrac{1}{2}(A - A^{\mathrm{T}}),$$

where $\frac{1}{2}(A + A^{\mathrm{T}})$ is called the symmetric part of A and $\frac{1}{2}(A - A^{\mathrm{T}})$ is called the skew-symmetric part of A. The quadratic form $x^{\mathrm{T}}Ax$ depends only on the symmetric part of A:

$$x^{\mathrm{T}}Ax = x^{\mathrm{T}}\left\{\tfrac{1}{2}(A + A^{\mathrm{T}})\right\}x.$$

Therefore, one can always assume that the matrix of a quadratic form is symmetric, and one can express the quadratic form in summation form as

$$\begin{aligned}
x^{\mathrm{T}}Ax &= \sum_{i=1}^{n}\sum_{j=1}^{n} a_{ij}x_ix_j \\
&= \sum_{i=j} a_{ij}x_ix_j + \sum_{i\neq j} a_{ij}x_ix_j \\
&= \sum_{i=1}^{n} a_{ii}x_i^2 + 2\sum_{i<j} a_{ij}x_ix_j
\end{aligned}$$

for symmetric A.

Ranges of Quadratic Forms. The domain of a quadratic form for an $n \times n$ matrix is n-dimensional Euclidean space, and the range is in $(-\infty, +\infty)$, the real line. Assume that $x \neq 0$. Then:

If A is positive definite, the range of $x \to x^{\mathrm{T}}Ax$ is $(0, +\infty)$,
If A is non–negative definite, the range of $x \to x^{\mathrm{T}}Ax$ is $[0, +\infty)$,
If A is indefinite, the range of $x \to x^{\mathrm{T}}Ax$ is $(-\infty, +\infty)$,
If A is non–positive definite, the range of $x \to x^{\mathrm{T}}Ax$ is $(-\infty, 0]$,
If A is negative definite, the range of $x \to x^{\mathrm{T}}Ax$ is $(-\infty, 0)$.

If $x^{\mathrm{T}}x = 1$, then $\lambda_n(A) \leq x^{\mathrm{T}}Ax \leq \lambda_1(A)$. That is, the quadratic form maps the unit n-sphere onto the closed interval $[\lambda_n(A), \lambda_1(A)]$.

B.9 DERIVATIVES OF MATRICES

B.9.1 Derivatives of Matrix-Valued Functions

The derivative of a matrix with respect to a scalar is the matrix of derivatives of its elements:

$$F(t) = \begin{bmatrix} f_{11}(t) & f_{12}(t) & f_{13}(t) & \cdots & f_{1n}(t) \\ f_{21}(t) & f_{22}(t) & f_{23}(t) & \cdots & f_{2n}(t) \\ f_{31}(t) & f_{32}(t) & f_{33}(t) & \cdots & f_{3n}(t) \\ \vdots & \vdots & \vdots & \ddots & \vdots \\ f_{m1}(t) & f_{m2}(t) & f_{m3}(t) & \cdots & f_{mn}(t) \end{bmatrix},$$

$$\frac{d}{dt}F(t) = \begin{bmatrix} \frac{d}{dt}f_{11}(t) & \frac{d}{dt}f_{12}(t) & \frac{d}{dt}f_{13}(t) & \cdots & \frac{d}{dt}f_{1n}(t) \\ \frac{d}{dt}f_{21}(t) & \frac{d}{dt}f_{22}(t) & \frac{d}{dt}f_{23}(t) & \cdots & \frac{d}{dt}f_{2n}(t) \\ \frac{d}{dt}f_{31}(t) & \frac{d}{dt}f_{32}(t) & \frac{d}{dt}f_{33}(t) & \cdots & \frac{d}{dt}f_{3n}(t) \\ \vdots & \vdots & \vdots & \ddots & \vdots \\ \frac{d}{dt}f_{m1}(t) & \frac{d}{dt}f_{m2}(t) & \frac{d}{dt}f_{m3}(t) & \cdots & \frac{d}{dt}f_{mn}(t) \end{bmatrix}.$$

The rule for the derivative of a product applies also to matrix products:

$$\frac{d}{dt}[A(t)B(t)] = \left[\frac{d}{dt}A(t)\right]B(t) + A(t)\left[\frac{d}{dt}B(t)\right],$$

provided that the order of the factors is preserved. If $F(t)$ is square and nonsingular, then $F(t)F^{-1}(t) = I$, a constant. As a consequence, its derivative will be zero. This fact can be used to derive the formula for the derivative of a matrix inverse:

$$\begin{aligned} 0 &= \frac{d}{dt}I \\ &= \frac{d}{dt}[F(t)F^{-1}(t)] \\ &= \left[\frac{d}{dt}F(t)\right]F^{-1}(t) + F(t)\left[\frac{d}{dt}F^{-1}(t)\right], \end{aligned}$$

$$\frac{d}{dt}F^{-1}(t) = -F^{-1}\left[\frac{d}{dt}F(t)\right]F^{-1}. \tag{B.10}$$

B.9.2 Gradients of Quadratic Forms

If $f(x)$ is a differentiable scalar-valued function of an n-vector x, then the vector

$$\frac{\partial f}{\partial x} = \left[\frac{\partial f}{\partial x_1}, \frac{\partial f}{\partial x_2}, \frac{\partial f}{\partial x_3}, \ldots, \frac{\partial f}{\partial x_n} \right]^{\mathrm{T}}$$

is called the *gradient* of f with respect to x. In the case that f is a quadratic form with symmetric matrix A, then the ith component of its gradient will be

$$\begin{aligned}
\left[\frac{\partial}{\partial x}(x^{\mathrm{T}}Ax) \right]_i &= \frac{\partial}{\partial x_i}\left(\sum_j a_{jj}x_j^2 + 2\sum_{j<k} a_{jk}x_j x_k \right) \\
&= \left(2a_{ii}x_i + 2\sum_{i<k} a_{ik}x_k + 2\sum_{j<i} a_{ji}x_j \right) \\
&= \left(2a_{ii}x_i + 2\sum_{i \neq k} a_{ik}x_k \right) \\
&= 2\sum_{k=1}^{n} a_{ik}x_k \\
&= (2Ax)_i,
\end{aligned}$$

That is, the gradient vector can be expressed as

$$\frac{\partial}{\partial x}(x^{\mathrm{T}}Ax) = 2Ax.$$

References

Books

Books on Kalman Filtering and Its Applications

1. B. D. O. Anderson and J. B. Moore, *Optimal Filtering*, Prentice-Hall, Englewood Cliffs, NJ, 1979.
2. R. N. Ansher, Ed., *Physics and Beyond*, Harper & Row, New York, 1971.
3. A. C. Antoulas, Ed., *Mathematical System Theory, The Influence of R. E. Kalman*, Springer-Verlag, Berlin, 1991.
4. A. V. Balakrishnan, *Kalman Filtering Theory*, Optimization Software, New York, 1987.
5. R. H. Battin, *Astronautical Guidance*, McGraw-Hill, New York, 1964.
6. R. H. Battin, *An Introduction to the Mathematics and Methods of Astrodynamics*, American Institute of Aeronautics and Astronautics, New York, 1987.
7. G. J. Bierman, *Factorization Methods for Discrete Sequential Estimation*, Academic, New York, 1977.
8. J. H. Blakelock, *Automatic Control of Aircraft and Missiles*, Wiley, New York, 1965.
9. S. M. Bozic, *Digital and Kalman Filtering: An Introduction to Discrete-Time Filtering and Optimal Linear Estimation*, Wiley, New York, 1979.
10. K. Brammer and G. Siffling, *Kalman-Bucy Filters*, Artech House, Norwood, MA, 1989.
11. R. G. Brown, *Introduction to Random Signal Analysis and Kalman Filtering*, Wiley, New York, 1983.
12. R. G. Brown and P. Y. C. Hwang, *Introduction to Random Signals and Applied Kalman Filtering*, 2nd ed., Wiley, New York, 1992.

13. R. G. Brown and P. Y. C. Hwhang, *Introduction to Random Signals and Applied Kalman Filtering: With MATLAB Exercises and Solutions*, 3rd ed., Wiley, New York, 1997.
14. A. E. Bryson, Jr. and Y.-C. Ho, *Applied Optimal Control*, Blaisdell, Waltham, MA, 1969.
15. R. S. Bucy and P. D. Joseph, *Filtering for Stochastic Processes, with Applications to Guidance*, Wiley, New York, 1968.
16. D. E. Catlin, *Estimation, Control, and the Discrete Kalman Filter*, Springer-Verlag, New York, 1989.
17. H. F. Chen, *Recursive Estimation and Control for Stochastic Systems*, Wiley, New York, 1985.
18. C. K. Chui and G. Chen, *Kalman Filtering with Real-Time Applications*, Springer-Verlag, New York, 1987.
19. K. L. Chung, *A Course in Probability Theory*, Harcourt Brace, New York, 1968.
20. M. H. A. Davis, *Linear Estimation and Stochastic Control*, Halsted, New York, 1977.
21. A. Gelb, J. F. Kasper, Jr., R. A. Nash, Jr., C. F. Price, and A. A. Sutherland, Jr., *Applied Optimal Estimation*, MIT Press, Cambridge, MA, 1974.
22. M. S. Grewal, L. R. Weill, and A. P. Andrews, *Global Positioning Systems, Inertial Navigation and Integration*, Wiley, New York, 2000.
23. A. H. Jazwinski, *Stochastic Processes and Filtering Theory*, Academic, New York, 1970.
24. T. Kailath, *Lectures on Weiner and Kalman Filtering*, Springer-Verlag, New York, 1981.
25. T. Kailath, *Linear Least Squares Estimation*, Dowden, Hutchinson and Ross, Stroudsburg, PA, 1977.
26. T. Kailath, "State-space modelling: Square root algorithms," in *Systems and Control Encyclopedia* (M. G. Singh, Ed.), Pergamon, Elmsford, NY, 1984.
27. H. Kwaknernaak and R. Sivan, *Linear Optimal Control Systems*, Wiley, New York, 1972.
28. C. T. Leondes, Ed., *Control and Dynamic Systems*, Vols. 19–21: *Nonlinear and Kalman Filtering Techniques*, Parts 1–3, Academic, New York, 1983–1984.
29. F. H. Lewis, *Optimal Estimation with an Introduction to Stochastic Control Theory*, Wiley, New York, 1986.
30. P. S. Maybeck, *Stochastic Models, Estimation, and Control*, Vol. 1, Academic, New York, 1979.
31. P. S. Maybeck, *Stochastic Models, Estimation, and Control*, Vol. 2, Academic, New York, 1982.
32. J. L. Melsa and D. L. Cohen, *Decision and Estimation Theory*, McGraw-Hill, New York, 1978.
33. J. M. Mendel, *Discrete Techniques of Parameter Estimation: The Equation Error Formulation*, Marcel Dekker, New York, 1973.
34. J. M. Mendel, *Lessons in Digital Estimation Techniques*, Prentice-Hall, Englewood Cliffs, New Jersey, 1987.
35. J. M. Mendel, "Kalman filtering and other digital estimation techniques." in *IEEE Individual Learning Package*, New York, 1987.
36. N. E. Nahi, *Estimation Theory and Applications*, Wiley, New York, 1969; reprinted by Krieger, 1975.
37. M. B. Nevelson and R. Z. Hazminskii, *Stochastic Approximation and Recursive Estimation*, American Mathematical Society., Transl. Math. Mono., No. 47, Chicago, 1971.

38. K. Ogata, *State Space Analysis of Control Systems*, Prentice-Hall, Englewood Cliffs, NJ, 1967.
39. A. Papoulis, *Probability, Random Variables, and Stochastic Processes*, McGraw-Hill, New York, 1988.
40. E. Parzen, *Stochastic Processes*, Holden-Day, San Francisco, 1962.
41. J. C. Pinson, "Inertial guidance for cruise vehicles," in *Guidance and Control of Aerospace Vehicles* (C. T. Leondes, Ed.), McGraw-Hill, New York, 1963.
42. P. A. Ruymgaart and T. T. Soong, *Mathematics of Kalman-Bucy Filtering*, Springer-Verlag, Berlin, 1988.
43. G. T. Schmidt, "Linear and nonlinear filtering techniques," in *Control and Dynamic Systems* (C. T. Leondes, Ed.), Vol. 12, Academic, New York, 1976.
44. D. G. Schultz and J. L. Melsa, *State Functions and Linear Control Systems*, McGraw-Hill, New York, 1967.
45. S. F. Schmidt, "Applications of state-space methods to navigation problems," in *Advances in Control Systems* (C. T. Leondes, Ed.), Vol. 3, Academic, New York, 1966.
46. H. W. Sorenson, "Kalman filtering techniques," in *Advances in Control Systems*, Vol. 3, Academic, New York, 1966, pp. 219–292.
47. H. W. Sorenson, Ed., *Kalman Filtering: Theory and Application*, IEEE Press, New York, 1985.
48. R. F. Stengel, *Stochastic Optimal Control: Theory and Application*, Wiley, New York, 1986.
49. A. S. Willsky, *Digital Signal Processing and Control and Estimation Theory: Points of Tangency, Areas of Intersection, and Parallel Directions*, MIT Press, Cambridge, MA, 1979.
50. P. C. Young, *Recursive Estimation and Time Series*, Springer-Verlag, New York, 1984.

Books on Mathematical Foundations for the Kalman Filter

51. L. Arnold, *Stochastic Differential Equations: Theory and Applications*, Wiley, New York, 1974.
52. J. Baras and V. Mirelli, *Recent Advances in Stochastic Calculus*, Springer-Verlag, New York, 1990.
53. P. Billingsley, *Probability and Measure*, Wiley, New York, 1986.
54. S. Bittanti, A. J. Laub, and J. C. Willems, Eds., *The Riccati Equation*, Springer-Verlag, New York, 1991.
55. E. Bodewig, *Matrix Calculus*, North-Holland, Amsterdam, 1959.
56. R. W. Brockett, *Finite Dimensional Linear Systems*, Wiley, New York, 1970.
57. E. A. Coddington and N. Levinson, *Theory of Ordinary Differential Equations*, McGraw-Hill, New York, 1955.
58. W. D. Davenport and W. L. Root, *Random Signals and Noise*, McGraw-Hill, New York, 1958.
59. P. M. DeRusso, R. J. Roy, and C. M. Close, *State Variables for Engineers*, Wiley, New York, 1965.
60. R. C. Dubes, *The Theory of Applied Probability*, Prentice-Hall, Englewood Cliffs, NJ, 1968.

61. R. V. Gamkrelidze, Ser. Ed., *Encyclopedia of Mathematical Sciences*, Vol. 3: *Dynamical Systems III* (V. I. Arnold, Ed.), Springer-Verlag, Berlin, 1988.
62. F. R. Gantmacher, *The Theory of Matrices*, Vol. 1, Chelsea, New York, 1990.
63. B. Harris, *Theory of Probability*, Addison-Wesley, Reading, MA., 1966.
64. K. Itô and H. P. McKean, Jr., *Diffusion Processes and Their Sample Paths*, Academic, New York, 1965.
65. T. Kailath, *Linear Systems*, Prentice-Hall, Englewood Cliffs, NJ, 1980.
66. I. Karatzas and S. Shreve, *Brownian Motion and Stochastic Calculus*, Springer-Verlag, New York, 1991.
67. J. H. Laning, Jr., and R. H. Battin, *Random Processes in Automatic Control*, McGraw-Hill, New York, 1956.
68. M. Loève, *Probability Theory*, 3rd ed., Van Nostrand Reinhold, New York, 1963.
69. Y. L. Luke, *The Special Functions and Their Approximations*, Vol. 2, Academic, New York, 1969.
70. P. Masani, Ed., *Norbert Weiner: Collected Works*, Vols. 1–4, MIT Press, Cambridge, MA, 1976, 1979, 1981, 1985.
71. P. Masani, "The life and work of Norbert Wiener," in *Norbert Wiener: Collected Works*, Vol. 4, MIT Press, Cambridge, MA, 1985.
72. K. S. Miller, *Some Eclectic Matrix Theory*, Krieger, Malabar, FL, 1987.
73. M. M. Rao, "Probability," in *Encyclopedia of Physical Science and Technology*, Academic, New York, 1987.
74. W. T. Reid, *Riccati Differential Equations*, Academic, New York, 1972.
75. F. C. Schweppe, *Uncertain Dynamic Systems*, Prentice-Hall, Englewood Cliffs, NJ, 1973.
76. S. R. Searle, *Linear Models*, Wiley, New York, 1971.
77. K. Sobczyk, *Stochastic Differential Equations with Applications to Physics and Engineering*, Kluwer Academic, Dordrecht, 1991.
78. R. L. Stratonovich, *Topics in the Theory of Random Noise*, (R. A. Silverman, Ed.), Gordon & Breach, New York, 1963.
79. L. A. Zadeh and C. A. Desoer, *Linear System Theory*, McGraw-Hill, New York, 1963.
80. R. Zurmühl, *Matrizen*, Academic Press, Orlando, FL., 1961.

Books on Numerical Methods

81. F. S. Acton, *Numerical Methods That Usually Work*, Mathematical Association of America, Washington DC, 1991 (revision of 1970 edition published by Harper & Rowe, New York).
82. E. Anderson, Z. Bai, C. Bischof, J. Demmel, J. Dongarra, J. Du Croz, A. Greenbvaum, S. Hammerling, A. McKenney, S. Ostrouchov, and D. Sorensen, *LAPACK Users' Guide*, Society for Industrial and Applied Mathematics, Philadelphia, 1992.
83. G. Dahlquist and Å. Björck, *Numerical Methods*, (N. Anderson, Trans.), Prentice-Hall, Englewood Cliffs, NJ, 1974.
84. J. J. Dongarra, C. B. Moler, J. R. Bunch, and G. W. Stewart, *LINPACK Users' Guide*, Society for Industrial and Applied Mathematics, Philadelphia, 1979.
85. G. E. Forsythe, M. A. Malcolm, and C. B. Moler, *Computer Methods for Mathematical Computations*, Prentice-Hall, Englewood Cliffs, NJ, 1977.

86. C. F. Gauss, *Theory of Motion of the Heavenly Bodies* (English translation), Dover Publications, New York, 1963.

87. C. F. Gauss, *Abhandlungen zur Methode der kleinsten Quadrate* (German translation by A. Borsch and P. Simon), P. Stankiewicz, Berlin, 1887.

88. W. C. Gear, *Numerical Initial Value Problems in Ordinary Differential Equations*, Prentice-Hall, Englewood Cliffs, NJ, 1971.

89. G. H. Golub and C. F. Van Loan, *Matrix Computations*, 2nd ed., Johns Hopkins University Press, Baltimore, MD, 1989.

90. D. Kahaner, C. Moler, and S. Nash, *Numerical Methods and Software*, Prentice-Hall, Englewood Cliffs, NJ, 1989.

91. C. L. Lawson and R. J. Hanson, *Solving Least Squares Problems*, Prentice-Hall, Englewood Cliffs, NJ, 1974.

92. J. C. Nash, *Compact Numerical Methods for Computers: Linear Algebra and Function Minimization*, Adam Hilger, Bristol, 1990.

93. W. H. Press, B. P. Flannery, S. A. Teukolsky, and W. T. Vettering, *Numerical Recipes in FORTRAN*, Cambridge University Press, Cambridge, UK, 1987.

94. A. Ralston and P. Rabinowitz, *A First Course in Numerical Analysis*, McGraw-Hill, New York, 1978.

95. J. Stoer and R. Bulirsh, *Introduction to Numerical Analysis*, Springer-Verlag, New York, 1980.

96. S. Van Huffel and J. Vanderwalle, *The Total Least Squares Problem: Computational Aspects and Analysis*, Society for Industrial and Applied Mathematics, Philadelphia, PA, 1991.

Miscellaneous Books

97. R. M. L. Baker and M. W. Makemson, *An Introduction to Astrodynamics*, Academic, New York, 1960.

98. G. M. Jenkins and D. G. Watts, *Spectral Analysis and Its Applications*, Holden-Day, San Francisco, 1968.

99. T. Kailath, "Equations of Wiener-Hopf type in filtering theory and related applications," in *Norbert Weiner: Collected Works* (P. Masani, Ed.), Vol. 3, MIT Press, Cambridge, MA, 1981.

100. M. Schwartz and L. Shaw, *Signal Processing, Discrete Spectral Analysis, Detection, and Estimation*, McGraw-Hill, New York, 1975.

101. L. Strachey, *Eminent Victorians*, Penguin Books, London, 1988.

102. D. J. Struik, *A Concise History of Mathematics*, Dover, New York, 1987.

103. N. Wiener, *Time Series*, MIT Press, Cambridge, MA, 1964 (originally published in 1949 as *Extrapolation, Interpolation, and Smoothing of Stationary Time Series with Engineering Applications*).

104. N. Wiener, *Ex-Pridigy: My Childhood and Youth*, MIT Press, Cambridge, MA, 1964.

105. N. Wiener, *I Am a Mathematician*, MIT Press, Cambridge, MA, 1964.

Theses and Reports

106. W. S. Agee and R. H. Turner, *Triangular Decomposition of a Positive Definite Matrix Plus a Symmetric Dyad, with Applications to Kalman Filtering*, White Sands Missile Range Tech. Rep. No. 38, Oct. 1972.
107. ANSI/IEEE Std. 754-1985, *IEEE Standard for Binary Floating-Point Arithmetic*, The Institute of Electrical and Electronics Engineers, New York, 1985.
108. J. L. Center, J. A. D'Appolito, and S. I. Marcus, *Reduced-Order Estimators and Their Application to Aircraft Navigation*, Tech. Rep., TASC TR-316-4-2, The Analytic Sciences Corporation, Reading, MA, Aug. 1974.
109. E. Eskow and R. B. Schnabel, "Algorithm 695: Software for a new modified Cholesky factorization," *Collected Algorithms of the ACM*, Association for Computing Machinery, New York, 1991.
110. R. C. DiPietro and F. A. Farrar, *Comparative Evaluation of Numerical Methods for Solving the Algebraic Matrix Riccati Equation*, Rep. No. R76-140268-1, United Technologies Research Center, East Hartford, CT, 1976.
111. R. M. DuPlessis, *Poor Man's Explanation of Kalman Filtering, or How I Stopped Worrying and Learned to Love Matrix Inversion*, Rep. No. QN014239, North American Rockwell, Autonetics Division, Anaheim, CA, 1967.
112. M. S. Grewal and A. P. Andrews, *Application of Kalman Filtering to GPS, INS, & Navigation (Notes)*, Kalman Filtering Consulting Associates, Anaheim, CA, 2000.
113. K. Itô, *Lectures on Stochastic Processes*, Tata Institute of Fundamental Research, Bombay, India, 1961.
114. R. E. Kalman, *Phase-Plane Analysis of Nonlinear Sampled-Data Servomechanisms*, S.M. thesis, Dept. of Electrical Engineering, Massachusetts Institute of Technology, Cambridge, MA, 1954.
115. P. G. Kaminski, *Square Root Filtering and Smoothing for Discrete Processes*, Ph.D. thesis, Stanford University, 1971.
116. C. T. Leondes, Ed., *Theory and Applications of Kalman Filtering*, AGARDograph No. 139, NATO Advisory Group for Aerospace Research and Development, London, Feb. 1970.
117. C. T. Leondes, Ed., *Advances in the Techniques and Technology of the Application of Nonlinear Filters and Kalman Filters*, AGARDograph No. 256, NATO Advisory Group for Aerospace Research and Development, Paris, 1982.
118. L. A. McGee and S. F. Schmidt, *Discovery of the Kalman Filter as a Practical Tool for Aerospace and Industry*, National Aeronautics and Space Administration, Technical Memorandum 86847, Nov. 1985.
119. J. E. Potter, *A Matrix Equation Arising in Statistical Estimation Theory*, Report No. CR-270, 1965, National Aeronautics and Space Administration, 1965.
120. J. M. Rankin, *Kalman Filtering Approach to Market Price Forecasting*, Ph.D. Thesis, Iowa State University, Ames, IA, 1986.
121. J. M. Richardson and K. A. Marsh, "Nonlinear filtering theory and its applications," *Rockwell International First Annual Signal Processing Conference Proceedings*, 1988, pp. 266–279.
122. G. T. Schmidt, Ed., *Practical Aspects of Kalman Filtering Implementation*, AGARD–LS–82, NATO Advisory Group for Aerospace Research and Development, London, May 1976.

123. S. F. Schmidt, "Computational techniques in Kalman filtering," in *Theory and Applications of Kalman Filtering*, AGARDograph 139, NATO Advisory Group for Aerospace Research and Development, London, Feb. 1970.
124. C. L. Thornton, *Triangular Covariance Factorizations for Kalman Filtering*, Ph.D. Thesis, University of California at Los Angeles, School of Engineering, 1976.
125. C. L. Thornton and G. J. Bierman, *A Numerical Comparison of Discrete Kalman Filtering Algorithms: An Orbit Determination Case Study*, JPL Technical Memorandum 33-771, Pasadena, 1976.
126. W. S. Widnall and P. A. Grundy, *Inertial Navigation System Error Models*, Tech. Rep. TR-03-73, Intermetrics, Cambridge, MA, May 1973.
127. M. A. Woodbury, *Inverting Modified Matrices*, Memorandum Report 42, Statistical Research Group, Princeton University, Princeton, NJ, 1950.

Journal Articles and Conference Papers

128. D. W. Allan, "Statistics of atomic frequency standards," *IEEE Proceedings*, Vol. 54, pp. 221–230, 1966.
129. B. D. O. Anderson, "Second-order convergent algorithms for the steady-state Riccati equation," *International Journal of Control*, Vol. 28, pp. 295–306, 1978.
130. A. Andrews, "A square root formulation of the Kalman covariance equations," *AIAA Journal*, Vol. 6, pp. 1165–1166, 1968.
131. A. Andrews, "Marginal optimization of observation schedules," *AIAA Journal of Guidance and Control*, Vol. 5, pp. 95–96, 1982.
132. R. B. Asher, P. S. Maybeck, and R. A. K. Mitchell, "Filtering for precision pointing and tracking with application for aircraft to satellite tracking," in *Proceedings of the IEEE Conference Decision and Control*, Houston, TX, 1975, pp. 439–446.
133. M. Athans et al., Guest Eds., "Special issue on linear-quadratic-Gaussian problem," *IEEE Transactions on Automatic Control*, Vol. AC-16, 1971.
134. R. W. Bass, V. D. Norum, and L. Schwartz, "Optimal multichannel nonlinear filtering," *Journal of Mathematical Analysis and Applications*, 1966.
135. R. H. Battin, "Space guidance evolution—a personal narrative," *AIAA Journal of Guidance and Control*, Vol. 5, pp. 97–110, 1982.
136. J. F. Bellantoni and K. W. Dodge, "A square root formulation of the Kalman-Schmidt filter," *AIAA Journal*, Vol. 5, pp. 1309–1314, 1967.
137. T. R. Benedict and G. W Bordner, "Synthesis of an optimal set of radar track-while scan smoothing equation," *IEEE Transactions on Automatic Control*, Vol. AC-7, pp. 27–32, 1962.
138. J. M. Bennet, "Triangular factors of modified matrices," *Numerische Mathematik*, Vol. 7, pp. 217–221, 1963.
139. Commandant Benoit, "Sur une méthode de résolution des équations normales provenant de l'application de la méthode des moindes carrés a un système d'équations linéaires en numbre inférieur a celui des inconnues—application de la méthode a la resolution d'un système defini d'équations linéaires (Procédé du Commandant Cholesky)," *Bulletin Géodêsique et Géophysique Internationale*, Vol. 2, Toulouse, pp. 67–77, 1924.
140. G. J. Bierman, "A new computationally efficient fixed-interval discrete time smoother," *Automatica*, Vol. 19, pp. 503–561, 1983.

141. Å. Björck, "Solving least squares problems by orthogonalization," *BIT*, Vol. 7, pp. 1–21, 1967.

142. F. R. Bletzacker et al., "Kalman filter design for integration of Phase III GPS with an inertial navigation system," *Computing Applications Software Technology Technical Papers*, Los Alamitos, CA, 1988.

143. H. W. Bode and C. E. Shannon, "A simplified derivation of linear least-squares smoothing and prediction," *IRE Proceedings*, Vol. 48, pp. 417–425, 1950.

144. J.-L. Botto and G. V. Moustakides, "Stabilizing the fast Kalman filter algorithms," *IEEE Transactions on Acoustics, Speech and Signal Processing*, Vol. ASSP-37, pp. 1342–1348, 1989.

145. K. Brodie, D. Eller and G. Seibert, "Performance analysis of integrated navigation systems," in *Proceedings of the 1985 Winter Simulation Conference* (D. Gantz, G. Glais, and S. Solomon, Eds.), pp. 605–609, San Francisco, CA., 1985.

146. A. E. Bryson, Jr., and D. E. Johansen, "Linear filtering for time-varying systems using measurements containing colored noise," *IEEE Transactions on Automatic Control*, Vol. AC-10, pp. 4–10, 1965.

147. R. S. Bucy, "Nonlinear filtering theory," *IEEE Transactions on Automatic Control*, Vol. AC-10, pp. 198–206, 1965.

148. R. S. Bucy, "Optimal filtering for correlated noise." *Journal of Mathematical Analysis Applications*, Vol. 20, pp. 1–8, 1967.

149. N. A. Carlson, "Fast triangular formulation of the square root filter," *AIAA Journal*, Vol. 11, No. 9, pp. 1259–1265, 1973.

150. G. Chen and C. K. Chui, "A modified adaptive Kalman filter for real-time applications," *IEEE Transactions on Aerospace and Electronic Systems*, Vol. 27, pp. 149–154, 1991.

151. C. Y. Choe and B. D. Tapley, "A new method for propagating the square root covariance in triangular form," *AIAA Journal*, Vol. 13, pp. 681–683, 1975.

152. C. H. Choi and A. J. Laub, "Efficient matrix-valued algorithms for solving stiff differential Riccati equations," *IEEE Transactions on Automatic Control*, Vol. AC-35, pp. 770–776, 1990.

153. H. Cox, "On the estimation of state variables and parameters for noisy dynamic systems," *IEEE Transactions on Automatic Control*, Vol. AC-9, pp. 5–12, 1964.

154. E. J. Davison and M. C. Maki, "The numerical solution of the matrix Riccati differential equation," *IEEE Transactions on Automatic Control*, Vol. AC-19, pp. 71–73, 1973.

155. L. Dieci, "Numerical integration of the differential Riccati equation and some related issues," *SIAM Journal of Numerical Analysis*, Vol. 29, pp. 781–815, 1992.

156. P. Dyer and S. McReynolds, "Extension of square-root filtering to include process noise," *Journal of Optimization Theory and Applications*, Vol. 3, pp. 444–458, 1969.

157. A. Einstein, "Über die von molekularkinetischen Theorie der Wärme geforderte Bewegung von in ruhenden Flüssigkeiten suspendierten Teilchen," *Annelen der Physik*, Vol. 17, pp. 549–560, 1905.

158. A. F. Fath, "Computational aspects of the linear optimal regulator problem," *IEEE Transactions on Automatic Control*, Vol. AC-14, pp. 547–550, 1969.

159. R. J. Fitzgerald, "Divergence of the Kalman filter," *IEEE Transactions on Automatic Control*, Vol. AC-16, pp. 736–743, 1971.

160. A. D. Fokker, "Die mittlerer Energie rotierender elektrischer Dipole im Strahlungsfeld," *Annelen der Physik*, Vol. 43, pp. 810–820, 1914.

161. T. E. Fortmann, "A matrix inversion identity," *IEEE Transactions on Automatic Control*, Vol. AC-15, p. 599, 1970.

162. F. M. Gaston and G. W. Irwin, "Systolic Kalman filtering: An overview," *IEE Proceedings*, Vol. 137, pp. 235–244, 1990.

163. W. M. Gentleman, "Least squares computations by Givens transformations without square roots," *Journal of the Institute for Mathematical Applications*, Vol. 12, pp. 329–336, 1973.

164. W. Givens, "Computation of plane unitary rotations transforming a general matrix to triangular form," *Journal of the Society for Industrial and Applied Mathematics*, Vol. 6, pp. 26–50, 1958.

165. G. H. Golub, "Numerical methods for solving linear least squares problems," *Numerische Mathematik*, Vol. 7, pp. 206–216, 1965.

166. M. S. Grewal and H. J. Payne, "Identification of parameters in a freeway traffic model," *IEEE Transactions on Systems, Man, and Cybernetics*, Vol. SMC-6, pp. 176–185, 1976.

167. M. S. Grewal, "Application of Kalman filtering to the calibration and alignment of inertial navigation systems," in *Proceedings of PLANS '86—Position Location and Navigation Symposium*, Las Vegas, NV, Nov. 4–7, 1986, IEEE, New York, 1986.

168. M. S. Grewal and R. S. Miyasako, "Gyro compliance estimation in the calibration and alignment of inertial measurement units," in *Proceedings of Eighteenth Joint Services Conference on Data Exchange for Inertial Systems*, San Diego, CA, Oct. 28–30, 1986, IEEE Joint Services Data Exchange, San Diego, CA, 1986.

169. M. S. Grewal, R. S. Miyasako and J. M. Smith, "Application of fixed point smoothing to the calibration, alignment, and navigation data of inertial navigation systems," in *Proceedings of IEEE PLANS '88—Position Location and Navigation Symposium*, Orlando, FL, Nov. 29–Dec. 2, 1988, pp. 476 479, IEEE, New York, 1988.

170. M. S. Grewal, V. D. Henderson, and R. S. Miyasako, "Application of Kalman filtering to the calibration and alignment of inertial navigation systems," *IEEE Transactions on Automatic Control*, Vol. AC-38, pp. 4–13, 1991.

171. H. Heffes, "The effects of erroneous models on the Kalman filter response," *IEEE Transactions on Automatic Control*, Vol. AC-11, pp. 541–543, 1966.

172. A. S. Householder, "Unitary triangularization of a nonsymmetric matrix," *Journal of the Association for Computing Machinery*, Vol. 5, pp. 339–342, 1958.

173. J. R. Huddle and D. A. Wismer, "Degradation of linear filter performance due to modeling error," *IEEE Transactions on Automatic Control*, Vol. AC-13, pp. 421–423, 1968.

174. T. L. Jordan, "Experiments on error growth associated with some linear least-squares procedures," *Mathematics of Computation*, Vol. 22, pp. 579–588, 1968.

175. J. M. Jover and T. Kailath, "A parallel architecture for Kalman filter measurement update and parameter update," *Automatica*, Vol. 22, pp. 783–786, 1986.

176. T. Kailath, "A view of three decades of linear filtering theory," *IEEE Transactions on Information Theory*, Vol. IT-20, No. 2, pp. 146–181, 1974.

177. T. Kailath, "An innovations approach to least squares estimation, Part I: Linear filtering in additive white noise," *IEEE Transactions on Automatic Control*, Vol. AC-13, pp. 646–655, 1968.

178. T. Kailath and R. A. Geesey, "An innovations approach to least squares estimation—Part

IV: Recursive estimation given lumped covariance functions," *IEEE Transactions on Automatic Control*, Vol. AC-16, pp. 720–726, 1971.

179. R. E. Kalman, "A new approach to linear filtering and prediction problems," *ASME Journal of Basic Engineering*, Vol. 82, pp. 34–45, 1960.
180. R. E. Kalman, "New methods and results in linear prediction and filtering theory," in *Proceedings of the Symposium on Engineering Applications of Random Function Theory and Probability*, Wiley, New York, 1961.
181. R. E. Kalman and Richard S. Bucy, "New results in linear filtering and prediction theory," *ASME Journal of Basic Engineering*, Series D, Vol. 83, pp. 95–108, 1961.
182. R. E. Kalman, "New methods in Wiener filtering," in *Proceeding of the First Symposium on Engineering Applications of Random Function Theory and Probability*, Wiley, New York, 1963.
183. P. G. Kaminski, A. E. Bryson, Jr., and S. F. Schmidt, "Discrete square root filtering: A survey of current techniques," *IEEE Transactions on Automatic Control*, Vol. AC-16, pp. 727–736, 1971.
184. M. H. Kao and D. H. Eller, "Multiconfiguration Kalman filter design for high-performance GPS navigation," *IEEE Transactions on Automatic Control*, Vol. AC-28, 1983.
185. C. S. Kenney and R. B. Liepnik, "Numerical integration of the differential matrix Riccati equation," *IEEE Transactions on Automatic Control*, Vol. AC-30, pp. 962–970 1985.
186. D. W. Klein, "Navigation software design for the user segment of the NAVSTAR GPS," paper presented at the AIAA Guidance and Control Conference, San Diego, CA, Aug. 1976.
187. A. A. Kolmogorov, Über die analytichen Methoden in der Wahrscheinlichkeitsrechnung," *Mathematische Annelen*, Vol. 104, pp. 415–458, 1931.
188. M. M. Konstantinov and G. B. Pelova, "Sensitivity of the solutions to differential matrix Riccati equations," *IEEE Transactions on Automatic Control*, Vol. AC-36, pp. 213–215, 1991.
189. E. Kreindler and P. E. Sarachik, "On the concepts of controllability and observability of linear systems," *IEEE Transactions on Automatic Control*, Vol. AC-9, pp. 129–136, 1964.
190. L. R. Kruczynski, "Aircraft navigation with the limited operational phase of the Navstar Global Positioning System," *Journal of the Institute of Navigation*, Vol. 1, Global Positioning System: Papers Published in *Navigation*, Institute of Navigation, Alexandria, VA, 1980.
191. H. J. Kushner, "On the differential equations satisfied by conditional probability densities of Markov processes," *SIAM Journal on Control*, Ser. A, Vol. 2, pp. 106–119, 1964.
192. D. G. Lainiotis, "Estimation: A brief survey," *Information Sciences*, Vol. 7, pp. 191–202, 1974.
193. A. J. Laub, "A Schur method for solving algebraic Riccati equations," *IEEE Transactions on Automatic Control*, Vol. AC-19, pp. 913–921, 1979.
194. A. M. Legendre, "Methode de moindres quarres, pour trouver le milieu de plus probable entre les resultats des differentes obvservations," *Memoires Institute de France*, pp. 149–154, 1810.
195. N. Levinson, "Wiener's Life," *Bulletin of the American Mathematical Society*, Vol. 72, No. 1, Pt. II, pp. 1–32, 1966.

196. L. Ljung, "Asymptotic behavior of the extended Kalman filter as a parameter estimator for linear systems," *IEEE Transactions on Automatic Control*, Vol. AC-24, pp. 36–50, 1979.
197. A. G. J. MacFarlane, "An eigenvector solution of the optimal linear regulator," *Journal of Electronic Control*, Vol. 14, pp. 643–654, 1963.
198. G. Matchett, "GPS-aided shuttle navigation," paper presented at the IEEE National Aerospace and Electronics Conference, 1978.
199. S. R. McReynolds, "Fixed interval smoothing: Revisited," *AIAA Journal of Guidance, Control, and Dynamics*, Vol. 13, pp. 913–921, 1990.
200. P. S. Maybeck, J. G. Reid, and R. N. Lutter, "Application of an extended Kalman filter to an advanced fire control system," in *Proceedings of the IEEE Conference on Decision and Control*, New Orleans, LA, 1977, pp. 1192–1195.
201. J. S. Meditch, "A survey of data smoothing for linear and nonlinear systems," *Automatica*, Vol. 9, pp. 151–162, 1973.
202. R. K. Mehra, "A comparison of several nonlinear filters for reentry vehicle tracking," *IEEE Transactions on Automatic Control*, Vol. AC-16, pp. 307–319, 1971.
203. J. M. Mendel and D. L. Geiseking, "Bibliography on the linear-quadratic-Gaussian problem," *IEEE Transactions on Automatic Control*, Vol. AC-16, pp. 847–869, 1971.
204. M. Morf and T. Kailath, "Square root algorithms for least squares estimation," *IEEE Transactions on Automatic Control*, Vol. AC-20, pp. 487–497, 1975.
205. C. Moler and C. Van Loan, "Nineteen dubious ways to compute the exponential of a matrix," *SIAM Review*, Vol. 20, pp. 801–836, 1978.
206. H. Padé, "Sur la réprésentation approchée d'une fonction par des fractions rationelle," Annals d'écoles Vol. 9s Suppl., 1892.
207. M. Planck, Über einen Satz der statistischen Dynamik und seine Erweiterung in der Quantentheorie," *Sitzungsberichte d. König. Preussischen Akademie der Wissenschaft*, pp. 324–341, 1917.
208. J. E. Potter and R. G. Stern, "Statistical filtering of space navigation measurements," in *Proceedings of 1963 AIAA Guidance and Control Conference*, AIAA, New York, 1963.
209. J. E. Potter, "Matrix quadratic solutions," *SIAM Journal of Applied Mathematics*, Vol. 14, pp. 496–501, 1966.
210. M.-A. Poubelle, I. R. Petersen, M. R. Gevers, and R. R. Bitmead, "A miscellany of results on an equation of Count J. F. Riccati," *IEEE Transactions on Automatic Control*, Vol. AC-31, pp. 651–654, 1986.
211. C. F. Price and R. S. Warren, "An analysis of the divergence problem in the Kalman filter," *IEEE Transactions on Automatic Control*, Vol. AC-13, pp. 699–702, 1968.
212. H. E. Rauch, F. Tung, and C. T. Striebel, "Maximum likelihood estimates of linear dynamic systems," *AIAA Journal*, Vol. 3, pp. 1445–1450, 1965.
213. J. F. Riccati, "Animadversationnes in aequationes differentiales secundi gradus," *Acta Eruditorum Quae Lipside Publicantur Supplementa*, Vol. 8, pp. 66–73, 1724.
214. J. M. Richardson and K. A. Marsh, "Point process theory and the surveillance of many objects," in *Proceedings of the 1991 Symposium on Maximum Entropy and Bayesian Methods*, Seattle University, 1991.
215. F. H. Schlee, C. J. Standish, and N. F. Toda, "Divergence in the Kalman filter," *AIAA Journal*, Vol. 5, pp. 1114–1120, 1967.

216. S. F. Schmidt, "The Kalman filter: Its recognition and development for aerospace applications," *AIAA Journal of Guidance and Control*, Vol. 4, pp. 4–7, 1981.
217. F. C. Schweppe, "Evaluation of likelihood functions for Gaussian signals," *IEEE Transactions on Information Theory*, Vol. IT-11, pp. 61–70, 1965.
218. J. Sherman and W. J. Morrison, "Adjustment of an inverse matrix corresponding to a change in one element of a given matrix," *Annals of Mathematical Statistics*, Vol. 21, pp. 124–127, 1950.
219. R. A. Singer, "Estimating optimal tracking filter performance for manned maneuvering targets," *IEEE Transactions on Aerospace and Electronic Systems*, Vol. AES-6, pp. 473–483, 1970.
220. R. A. Singer and K. W. Behnke, "Real-time tracking filter evaluation and selection for tactical applications," *IEEE Transactions on Aerospace and Electronic Systems*, Vol. AES-7, pp. 100–110, 1971.
221. R. A. Singer, "Estimating optimal tracking filter performance for manned maneuvering targets," *IEEE Transactions on Aerospace Electronic Systems*, Vol. AES-6, 1970.
222. D. Slepian, "Estimation of signal parameters in the presence of noise," *IRE Transactions on Information Theory*, Vol. IT-3, pp. 68–69, 1954.
223. H. W. Sorenson, "On the error behavior in linear minimum variance estimation problems," *IEEE Transactions on Automatic Control*, Vol. AC-12, pp. 557–562, 1967.
224. H. W. Sorenson, "Least-squares estimation: From Gauss to Kalman,"*IEEE Spectrum*, pp. 63–68, 1970.
225. H. W. Sorenson, "On the development of practical nonlinear filters," *Information Sciences*, Vol. 7, pp. 253–270, 1974.
226. R. L. Stratovovich, "Application of the theory of Markoff processes in optimal signal discrimination," *Radio Engineering and Electronic Physics*, Vol. 1, pp. 1–19, 1960.
227. P. Swerling, "First order error propagation in a stagewise differential smoothing procedure for satellite observations," *Journal of Astronautical Sciences*, Vol. 6, pp. 46–52, 1959.
228. V. Strassen, "Gaussian elimination is not optimal," *Numerische Matematik*, Vol. 13, p. 354, 1969.
229. S. I. Sudharsanan and M. K. Sundhareshan, "Neural network computational algorithms for least squares estimation problems," in *Proceedings of a Joint Conference on Neural Networks*, Washington DC, IEEE, New York, June 1989.
230. T. M. Upadhyay and J. G. Damoulakis, "Sequential piecewise recursive filter for GPS low dynamics navigation," *IEEE Transactions on Aerospace Electronics Systems*, Vol. AES-16, pp. 481–491, 1980.
231. M. Vanbegin, P. Van Dooren, and M. Verhaegen, "Algorithm 675: FORTRAN subroutines for computing the square root covariance filter and square root information filter in dense or Hessenberg forms," *ACM Transactions on Mathematical Software*, Vol. 15, pp. 243–256, 1988.
232. M. Verhaegen and P. Van Dooren, "Numerical aspects of different Kalman filter implementations," *IEEE Transactions on Automatic Control*, Vol. AC-31, pp. 907–917, 1986.
233. R. C. Ward, "Numerical computation of the matrix exponential with accuracy estimate," *SIAM Journal of Numerical Analysis*, Vol. 14, pp. 600–611, 1977.

234. K. Watanabe and S. G. Tzafestas, "New computationally efficient formula for backward-pass fixed interval smoother and its UD factorization algorithm, "*IEE Proceedings*, Vol. 136D, pp. 73–78, 1989.

235. D. M. Wiberg and L. A. Campbell, "A discrete-time convergent approximation of the optimal recursive parameter estimator," in *Proceedings of the IFAC Identification and System Parameter Identification Symposium*, Vol. 1, pp. 140–144, 1991.

236. W. S. Widnall and P. K. Sinha, "Optimizing the gains of the baro-inertial vertical channel," *AIAA Journal of Guidance and Control*, Vol. 3, pp. 172–178, 1980.

237. L. A. Zadeh and J. R. Ragazzini, "An extension of Wiener's theory of prediction," *Journal of Applied Physics*, Vol. 21, pp. 645–655, 1950.

Index

Accelerometer, 86
  error, 86
Adrian, R., 7, 62
Agee, W. S., 18, 239
Antisymmetric matrix, 359
Antonelli, G., 202
A posteriori, 21
A priori, 21
Autocorrelation, 77–78, 83

Bacon, R., 1
Bad data, 282–283
Baker, R., 5
Balakrishnan, A. V., 14
Bass, R. W., xviii, 14, 19
Battin, r. H., 5, 14–15
Bayes, T., 6, 11, 64, 66
Bayes' rule, 64
Benefit function, 329
Bennet, J. M., 18
Bernoulli, J., 6, 11
Bernoulli trials, 11
Bierman, G. J., 5–6, 187, 214, 245, 231, 246, 257, 259, 353–353
Bilinear form, 377
Birkhoff, G. D., 71
Björck, J., 246
Bletzacker, F. R., 333
Block matrix, 218, 363
Bode, J., 6
Bode's Law, 6
Bohr, N. H. D., 107
Borel, F., 62
Brahmagupta, 11
Brown, R., 57
Brownian motion, 57
Bryson, A., E., Jr., 257
Bucy, R. S., 15, 18–19, 126

Cardano, G., 7, 11
Carlson, N. A., 18, 206, 239–240, 252
Carroll, Lewis, 270
Cauchy, A. L., 271
Cholesky, A.-L., 6, 373
Cholesky decomposition, 73, 218, 373
  algorithm, 220–222
  applications, 219
  MATLAB chol, 374
  modchol, 375
  modified, 222, 375
  utchol, 353, 374
Cholesky factor, 18, 218, 373–374
  modified, 220, 222, 375
  non uniqueness, 218, 374
  rank-one modification, 217
  triangular, 219
Cholesky matrix inversion method, 219, 223–225
Cholupdate.m, 375

Coefficient matrix, 29, 31
Cohen, E. R., xii
Companion matrix, 32
Complete observability, 43
Computational complexity, 225–228, 241–244, 247, 251, 253–254, 259, 318–319, 325–327
Computer roundoff, 204
Controllability, 14, 46
Convolution integral, 77
Correlated noise, 129
Correlation time, 79–80
Covariance analysis, 125
Covariance filter, 262
Covariance matrix, 63, 69
Covariance propagation, 88, 120, 126–127
  continuous, 89
  discrete, 90, 93, 120
Cross spectral density, 76

D'Alembert, J. R., 116
Dang, D., xiii
Data rejection filter, 298
De Moivre, A., 11
De Vries, T. W., xviii, 258, 327, 352–353
Decomposition, matrix, 216
Decomposition
  additive, 216
  Cholesky, 73, 218, 373
  eigenvalue–eigenvector, 377
  fraction, 133, 148
  symmetric, 377
  symmetric–antisymmetric, 216
Decorrelation, 217, 221–223
Derivative of matrix, 379
Determinant, 367
Diagonal of matrix, 356, 375
Difference equation, 34
Dimensions of matrix, 356
Dirac, P. A. M., 70
Dirac δ function, 70, 114, 170
Discrete time, 30
Divergence of the Kalman filter, 209, 271–272, 277
Divide by marix, 208
Dongarra, J. J., 216
Dot derivatives, 20
Doubling method, 329
Dual state analysis, 335
Dyer, P., 18, 243, 264
Dynamic coefficient matrix, 29, 31
Dynamic system, 4, 26
  homogeneous, 34

Einstein, A., 12, 19, 57
Elementary matrix, 217, 227
Eps (MATLAB), 205
Ergodic process, 71
Error
  overflow, 352
  roundoff, 204–205
  underflow, 352
Error budget, 332
Estimator
  linear, 117
  optimal, 116
  unbiased, 116
Exception handler, 283
Expected value, 66
Exponential of matrix, 48, 369
Extended Kalman filter, 15, 19, 170, 175–176, 180
  nonconvergence, 272

Factorization, 18, 216
Feedback loop, 211
Fermat, P. de, 6, 11
Filter
  Chandrasekhar, 214
  convergence, 271
  data rejection, 298
  divergence, 272
  extended Kalman, 15, 19, 170, 175–176, 180
  information, 262
  Kalman, 116, 121
    divergence, 209–212, 217
    extended, 15, 19, 170, 175–176, 180
    nonlinear, 169
    stabilized, 214
  Kalman–Bucy, 126–128
  Kalman–Schmidt, 19, 178
  Kolmorogov, 23
  Schmidt–Kalman, 309
  shaping, 58, 75, 84–85, 104, 129
  square root, 17–19, 23, 238
  suboptimal, 299
  unbiased, 116
  Wiener, 23, 300
Fisher, R. A., 9, 262
Fisher information matrix, 262
Fokker–Planck equation, 19
Fourier, J. B., 23
Fourier transform, 12
Freeway model, 123
Frobenius, G. F., 365
Fuller, R. B., 25

Gaffney, Rev. J., xviii
Galileo, 5–6
Gauss, C. F., 5–8, 23, 62
Gaussian distribution, 62, 104
  multivariate, 63, 104
Gaussian processes, 72
Gelb, A., 22, 130
Generating function, 63
Gentleman, W. M., 19, 217, 230
Gibbs, J. W., 71
Givens, W., 19, 217, 230
Givens Rotation, 230
Golub, G., 49, 238, 264
Gram, J. P., 9, 246
Gram–Schmidt orthogonalization, 217, 246
  modified, 246
Gramian, 9–10, 23, 43
Guillemin, E. A., 13
Gunckel, T. L. II, xviii
Gyroscope drift, 87

Hamilton, W. R., 135
Hanson, R. J., 264
Harmonic resonator, 27–29, 37–41, 91, 94, 97, 142, 194
Heckman, D. W., xviii
Hegel, G., 6
Herschel, F., 6
Hölder, O. L., 371
Hölder norm, 371
Householder, A.S., 19, 227, 234
Householder decomposition, 236–238
Householder transformation, 230, 234
Hubbs, R. A., xviii
Huygens, C., 6, 11

Identity matrix, 357
Ill-conditioning, 207–208
Indefinite matrix, 359, 378
Information filter, 262
  advantages, 263
  disadvantages, 263
Information matrix, 45, 225, 262
  Fisher, 262
Information states, 263
Input coupling matrix, 31, 34
INS, 342
Instability of the Kalman filter, 209, 271–272, 298
Inverarity, G., xviii
Inverse of matrix, 208, 361
  derivative, 379
  Moore–Penrose, 362

Itô, K., 19, 57
Itô calculus, 19, 58, 77

Jacobian matrix, 344
Jazwinski, A. H., 122
Joseph, P. D., 17–18, 120, 257, 352, 353
Joseph stabilized filter, 119–120, 122

Kailath, T., xviii, 5, 18, 122, 259
Kalman, R. E., xviii, 5–6, 13, 352
Kalman filter
  computer requirements, 326
  convergence, 271
  correlated noise, 129
  divergence, 272
  essential equations, 121
  extended, 15, 19, 170, 175–176
  feedback, 211
  ill-conditioned, 205
  memeory requirements, 316, 319
  mis-modeling, 284
  nonconvergence, 275
  notation, 21–22
  origins, 2
  precision requirements, 317
  scaling problems, 319
  serial processing, 124
  stability, 298
  throughput, 325
  versus Wiener, 16, 130
  wordlength, 317
Kalman–Bucy filter, 126–128
Kalman gain
  off-line, 327
  scheduling, 328
Kalman–Schmidt filter, 19, 178, 309–316
Kaminski, P. G., 18, 226
Khinchin, H. Ya., 12
Knuth, D. E., xviii
Kolmogorov, A. N., 3, 6, 12–13
Kronecker, L., 70
Kronecker Δ function, 70
Kutta, W. M., 41

Lainiotis, D. G., 5
Lambert, J. H., 7
Lamport, L., xviii
Langevin, P., 57
Lanning, J. H., 14
Laplace, P. S., Marquis de, 6, 11, 62
Laub, A. J., xviii
Lawson, C. H., 264

Least squares, 5, 7, 9, 23
Lee, R. C. K., 17
Legendre, A.-M., 6–7, 11
Linear dynamic system, 51
Linear stochastic process, 104
Lippman, G., 62
Loss function, 131

Makemson, M. W., 5
Mantissa, 204
Marginal benefit, 340
Margianl optimization, 337–338, 341
Marginal risk., 338
Markov, A. A., 6, 72
Mathworks, The, ix
MATLAB, ix
Matrix, 356
  antisymmetric, 359
  block, 218, 363
  characteristic
    polynomial, 367
    value, 367
    vector, 368
  Cholesky factor, 73, 218, 373
  coefficient, 29, 31
  companion, 32
  condition number, 207
  covariance, 63, 69
    indefinite, 294, 359
    propagation, 88, 121
  decomposition, 133, 216
    additive, 216
    Cholesky, 73, 218, 373
    eigenvalue–eigenvector, 377
    fraction, 133, 148
    symmetric, 377
    symmetric–antisymmetric, 216
  derivative, 379
  determinant, 367
  diagonal, 356, 375
  dimensions, 356
  divide, 208
  dynamic coefficient, 29, 31
  elementary, 217, 221
  exponential, 48, 369
  factorization, 18
  fraction, 133, 148
  Gramian, 23, 43
  Hölder norm, 371
  identity, 357
  indefinite, 359, 378
  information, 45, 225, 262
  input coupling, 31, 34
  inverse, 208, 361
    derivative, 379
    Moore–Penrose, 362
  Jacobian, 344
    measurement sensitivity, 33, 53, 117
  Moler, 266
  Moore–Penrose inverse, 362
  negative definite, 359, 378
  non-negative definite, 359, 378
  non-positive definite, 359, 378
  nonsingular, 361
  norm, 371
    compatible, 372
    Euclidean, 372
    Frobenius, 372
    Hölder, 372–373
    Shur, 372
    spectral, 372
    subordinate, 372
  notation, 355
  observability, 10, 23, 42–44, 53
  orthogonal, 362
  partitioned, 363
  permutation, 366–367
  positive definite, 359, 378
  pseudorank, 376
  QR decomposition, 229, 375
  rank, 364–365, 376
  rank-one modificiation, 217, 255, 365
  semidefinite, 359
  similarity transform, 370
  singular, 361
  singular value, 376
    decomposition, 376
  skew symmetric, 359
  sparse, 356
  square, 356
  square root, 18, 218, 374
  state transition, 34, 36
  symmetric, 359
  symmetric product, 318
  Toeplitz, 357
  trace, 359
  transpose, 8, 359
  triangular, 18, 219, 357
    unit, 357, 375
  triangularization, 216–217
  unit triangular, 220, 357, 375
  zero, 357
Maxwell, J. C., 6, 11–12, 71
Maybeck, P. S., 298, 346
McReynolds, S. R., 18, 203, 243, 264

Mean, 67, 69
Measurement
  noise, 78, 85, 104
  sensitivity, 33, 53, 117
  vector, 33
Measurement sensitivity matrix, 33, 53, 117
Meditch, J. S., 203
Memory requirements, 319–323
Mendel, J., 5
Modified Cholesky factor, 220, 222, 375
Moler, C. B., 48–49
Moler matrix, 266
Monte Carlo analysis, 219, 273, 333, 336, 354
Morf, M., 18, 259
Murphy's Law, 281–282
MWGS algorithm, 246–252

Nease, R. F., xviii
Negative definite matrix, 359, 378
Newton, I., 6, 25–26
Newton, I.
  dot notation, 20, 28
Newton, I.
  Second Law, 27
Newton, I.
  Third Law, 27
Newton–Raphson solution, 143
Noise
  correlated, 129
  decorrelated, 129
  Gaussian, 72
  measurement, 78, 85, 104
  plant, 104
  process, 104
  white, 70–71
Nonlinear estimation, 169
Nonlinear filtering, 169
Non-negative definite matrix, 359, 378
Non-positive definite matrix, 359, 378
Nonsingular matrix, 361
Norm of matrix, 371
  compatible, 372
  Euclidean, 372
  Frobenius, 372
  Hölder, 372–373
  Shur, 372
  spectral, 372
  subordinate, 372
Normal equation, 9
Numerical stability, 205

Observability, 10, 42–44
  complete, 43
  matrix, 10, 23, 42–44, 53
Ogata, K., 44
Orthogonal matrix, 362
Orthogonality principle, 97, 100–102, 116
Overflow, 352

Padé, H., 48
Padé approximation, 48
Partitioned matrix, 363
Pascal, B., 6, 11
Performance analysis, 125
Permutation matrix, 366–367
Piazzi, G., 5, 29
Pinson, J. C., xviii
Planck, M. K. E. L., 19
Plant noise, 104
Point process, 20
Positive definite matrix, 359, 378
Potter, J. E., 15, 18, 141, 146, 211, 226, 239, 352
Power spectral density, 58, 74, 77
Predictor, 115, 128
Prentice-Hall, ix
Probability
  conditional, 63–64
  density, 61
  distribution, 61
    Gaussian, 62, 104
  joint, 63–64
Process noise, 104
PSD, 58, 74, 77
Pseudorank of matrix, 376

Q factor, 92
QR decomposition, 19, 229
Quadratic form, 377
  gradient, 380
Quadratic loss function, 131

Ragazzini, J. R., 13–14
Random process, 68, 103
  autocorrelation, 77–78, 80 81, 83
  autoregressive, 82
  Bernoulli, 103
  correlated, 69
  correlation, 69
  covariance, 69
  cross spectral density, 76
  discrete, 69

Random process (*Continued*)
  ergodic, 71, 103
  Gaussian, 72, 103
  i.i.d., 103
  linear models, 80, 83–84
  linear predictive model, 82
  Markov, 72, 103
  mean, 69
  orthogonal, 70, 103
  PSD, 58, 74, 77
  shaping filter for, 75, 80, 82, 84
  simulating, 73
  stationary, 71, 84, 103
  stochastic differential equations, 77, 80
  white, 70–71
  WSS, 71–72, 84
Random sequence, 79, 83
Random variable, 58
  expected value, 66
  mean, 66
Rank of matrix, 364–365, 376
Rank one modification, 217, 255, 365
Rauch, H. E., 162
Resonator, 27–29, 37–41, 91, 94, 97, 142, 194
RIAS, 14
Riccati, J. F., 116
Riccati equation, 15, 88–89, 116, 127, 133
  algebraic, 139, 142
  convergence, 141
  differential, 127, 133, 142
  discrete, 148
  doubling method, 329
  linearizing, 133
  Newton-Raphson solution, 143
  numerical stability, 17
  scalar, 135, 141, 149–151
Richardson, J. M., xviii, 20
Risk, 131, 337
  marginal, 338
  trace formulation, 339
Rissanen, J., xviii
Robustness, 207
Roundoff error, 204–214
  unit, 205, 214
    MATLAB, 205
RP (random process), 68
RS (random sequence), 79
Runge, K. D. T., 41
Runyon, G. E., xviii
RV (random variable), 66
Sampling frequency, 97
Schmidt, E., 246
Schmidt, S. F., 5, 15, 17, 19, 178, 309
Schmidt–Kalman filter, 309–317
  complexity, 316
  derivation, 310
  equations, 317
  history, 309
Schmidt–Kalman gain, 313
Semidefinite matrix, 359
Shaping filter, 58, 75, 84–85, 104, 129
Singular matrix, 361
Singular value, 376
  decomposition, 376
Skew symmetric matrix, 359
Smith, J., xviii
Smoother, 116, 160
  fixed-interval, 160, 162
  fixed-lag, 162, 164
  fixed-point, 160, 163
  Rauch–Tung–Striebel, 162
Sorenson, H. W., 5, 122, 169
Sparse matrix, 356
Square matrix, 356
Square root filter, 17–19, 23, 238
  Bierman-Thornton, 245
  Carlson–Schmidt, 238
  Morf–Kailath, 259
  Potter, 256
Square root matrix, 18, 218, 228, 374
State space, 23, 25
State transition matrix, 34, 36
State variables, 25, 28
State vector, 29
STM (state transition matrix), 34, 36
Stochastic calculus, 19, 77
Strachey, L., 202
Strassen, V., 253
Stratonovich, R. L., 14, 19
Striebel, C. T., 162
Sup, 371
Supremum, 371
Swerling, P., 14, 252, 269, 352
Symmetric matrix, 359
Symmetric product, 218

Thornton, C., 18, 214, 246, 259
Throughput, 325
Tietz, J., 6
Time-invariant systems, 30, 44, 52
Time-varying systems, 30, 41, 52
Toeplitz matrix, 357
Trace of matrix, 359
Transpose of matrix, 8, 359

Triangular matrix, 18, 219, 357
  unit, 357, 375
Triangularization, 19, 216–217, 229–238
  Givens, 230–234
  Householder, 234–238
Tung, K., 162
Turner, R. H., 18, 239
Type 1 servo, 285
Type 2 servo, 288

UD factorization, 220
Underflow, 352
Unit roundoff error, 205, 214
Unit triangular matrix, 219–220, 357, 375
Van Dooren, P., 17, 209, 212–214
Van Loan, C., 48–49, 228
Vector, 358
  dot product, 361
  inner product, 361
  outer product, 361
  norm, 370
    Euclidean, 371
    Hölder, 371
  orthogonal, 362
  orthonormal, 362
  state, 29
  unit, 362
Verhaegen, M., 17, 209, 212–214
Von Neumann, J., 71

Ward, R., C., 48, 51
Watanabe, K., 203
Well conditioned problem, 207
White noise, 70–71
Wiberg, D. F., xviii–19
Wiener filter, 12, 130
Winer, N., 3, 6, 12, 57, 71
Wordlength, 320
WSS, 71–72

Zadeh, L. A., 13
Zero matrix, 357
Zurmüh, R., 373